Co- and Post-Translational Modification of Proteins

Chemical Principles and Biological Effects

Co- and Post-Translational Modification of Proteins
Chemical Principles and Biological Effects

Donald J. Graves
Iowa State University

Bruce L. Martin
Iowa State University

Jerry H. Wang
University of Calgary

New York   Oxford
Oxford University Press
1994

Oxford University Press

Oxford   New York   Toronto
Delhi   Bombay   Calcutta   Madras   Karachi
Kuala Lumpur   Singapore   Hong Kong   Tokyo
Nairobi   Dar es Salaam   Cape Town
Melbourne   Auckland   Madrid

and associated companies in
Berlin   Ibadan

Copyright © 1994 by Oxford University Press, Inc.

Published by Oxford University Press, Inc.,
200 Madison Avenue, New York, New York 10016

Oxford is a registered trademark of Oxford University Press

Library of Congress Cataloging-in-Publication Data
Graves, Donald J.
Co- and post-translational modification of proteins:
chemical principles and biological effects/
by Donald J. Graves, Bruce L. Martin, and Jerry H. Wang.
p. cm. Includes bibliographical references and index.
ISBN 0-19-505549-7
1. Post-translational modification. I. Martin, Bruce L., 1959-
II. Wang, Jerry H.. III. Title.
QH450.6.G73 1993 574.19'245–dc20 92-25711

9 8 7 6 5 4 3 2 1
Printed in the United States of America
on acid-free paper

We dedicate this book
To our dear wives,
Marge, Donna, and Teresa
for their love and support

Acknowledgment

The authors wish to acknowledge their appreciation to all those research investigators who sent us current literature relevant to this text. We appreciate the conscientious work done by Sally Van De Pol in preparing the manuscript and to Carol Greiner, the staff of the College of Agriculture Information Service, Iowa State University, for careful editing, and Susan Zeigler for her help putting the manuscript into camera-ready form. The authors acknowledge analysis of portions of the manuscript by students in the Graves' lab and in Protein Chemistry BB542 at Iowa State University and to L. Firsov, P. Graves, and L. Graves. D.J.G. acknowledges the support of Dr. Edwin G. Krebs and the Howard Hughes Medical Institute at the University of Washington and to the Medical Research Council of Canada for a MRC Visiting Scientist Award.

Preface

The purpose of this book is to provide basic information about the subject of post- and co-translational modification reactions. Post- or co-translational refers to modifications made to proteins after or during, respectively, the process of protein biosynthesis. It is not intended to be a comprehensive treatment describing all the different types of modification. This is not feasible in this framework. Our intent is to show general biological principles, methods of study, chemical and physical concepts that are important for understanding this subject and to guide readers in their own study be it in the classroom or research laboratory. Certain modification reactions were chosen to illustrate specific points. No single section will be found on glycosylation, phosphorylation, prenylation, etc., but examples of these and other types of modification will be found throughout the chapters. Each chapter contains its own list of references to aid further study.

The book can be used in a class of general molecular biology as a companion text or in an advanced course on signal transduction, or in a class of enzymology and protein chemistry. A major use could be in the research laboratory studying protein modification. In this instance, the book may be used to guide the investigator to determine what modification has taken place, where it occurs, and the consequences of the modification on the protein structure and biological function.

Ames, Iowa   D.J.G.
Ames, Iowa   B.L.M.
Calgary, Alberta   J.H.W.

Contents

CHAPTER 7: BIOLGICAL EFFECTS OF COVALENT MODIFICATION 263

Co- and Post-Translational Modification of Proteins

Chemical Principles and Biological Effects

1

Introduction: General Features of Protein Modification

The enormous diversity of protein structures found in living systems is related to the fact that these macromolecules are composed of a number of different monomeric units, 19 α-amino acids, and 1 α-imino acid. The variation in amino acid composition, sequence, and size of proteins is specified by the genomic sequences of DNA. Moreover, these sequences can be cut, rearranged, and spliced to code for new structures. The living cell has an enormous potential for making proteins with different sequences, sizes, and shapes.

Yet living organisms use other processes to introduce additional versatility into protein structures that can change their properties and functions. Reactions are known that can introduce new organic and inorganic functional groups into proteins. *And these added groups, such as phosphates, sulfates, carbohydrates, lipids, and so on, bring a new chemistry to the modified protein which influences how these proteins act.*[1,2] But modifications may also involve the cleavage of peptide bonds, coupling of the polypeptide, for example, ubiquitin, or addition of amino acid residues to the side chains or at the amino or the carboxyl terminus of the protein. These modifications, too, influence structure and function.

The enzyme catalyzed reactions may occur during synthesis of the protein or after completion of the protein chain and are referred to as co- and post-translational modifications, respectively. But how do these processes occur? How are they controlled? What are the consequences of the reactions to both structure and function of the proteins, and how do these modification reactions influence cellular events?

IRREVERSIBLE REACTIONS

Enzyme catalyzed modification reactions can be reversible or irreversible, and these two processes can serve different purposes in the cell (Chapter 7). Limited proteolysis, an example of an irreversible process, has been studied in detail, and the studies provide us with an excellent understanding of the processes *in vivo*, and of important structural-function relations in protein chemistry, as well as a foundation for the study of other modification reactions.[3,4]

In 1899, Schepowalnecokow in Pavlov's laboratory described the presence in duodenal juice of a factor named enterokinase, which would activate proteolysis in pancreatic juice. Enterokinase was shown later to be an enzyme, and Kunitz in 1939 described the activation of trypsinogen to trypsin, a process now known to be caused by the cleavage of a lys-ile bond and the release of a hexapeptide, val-$(asp)_4$-lys. from the amino terminus. Trypsin, *in vitro*, can catalyze the same reaction as can enterokinase, so it might be asked how enterokinase, and not trypsin, was established as the agent causing physiological activation. First, the effectiveness of the two processes should be considered. Because of its strong recognition of the neighboring aspartyl sequence, which inhibits the action of trypsin, the enterokinase reaction is approximately 2,000 times more effective than is the trypsin reaction. Thus, the sequence close to the scissile bond may determine the effectiveness of the post-translational modification reaction. Similarly, neighboring groups in a linear sequence can have a profound effect on side chain modification reactions of other post-translational modification reactions. Second, location and concentration of enterokinase *in vivo* favors the reaction. The binding of the enzyme to the brush border membrane of the intestine could position the enzyme for effective encounter in the lumen with trypsinogen released from the acinar cells of the pancreas (Figure 1.1).[5] This mechanism is not found for trypsin.

Trypsin then acts by limited proteolysis on other zymogens[6,3] released from the pancreas including chymotrypsinogen, proelastase, procarboxypeptidase, and prophospholipase, to facilitate the digestive process. A major point then in the study of modification reactions is to establish the physiological significance of the reaction. That a reaction occurs *in vitro* does not mean that it also occurs *in vivo*.

The study of linked processes such as those involved in zymogen

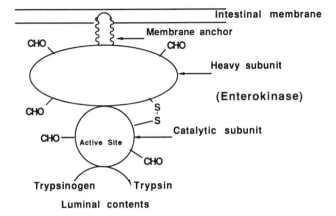

Figure 1.1. Schematic representation of the structure of enterokinase. The position of the heavy subunit ensures that the catalytic subunit lies on the surface of the brush border membrane and that the active site is directed towards the luminal contents. Reprinted with permission from *Trends in Biochemical Science* 14:110 (1989). Copyright 1989, Elsevier Trends Journals.

activation in digestion or those involved in blood coagulation, fibrinolysis, and the complement activation in the antigen–antibody response illustrates how signals can be amplified by covalent modification reactions and mechanisms that can be used to regulate the reactions. An initial event may trigger the conversion of an inactive zymogen to an active enzyme in the first step of a cascade. Active Xa produced in this reaction (Figure 1.2) may catalyze the conversion of many molecules of Y to Ya, which in turn acts in a following step, and so on, to form a protein with a specific biological activity. The multilevel path provides for an enormous signal amplification. A few molecules of an activator at the beginning of a cascade could cause the formation of millions of molecules of a final product. But the multilevel path allows for control of the overall process at the various steps, in short, acting as a fine tuning mechanism. A scheme (Figure 1.3) presented for a coagulation cascade illustrates these features.[7] Two pathways, intrinsic and extrinsic (dark arrows), exist and converge with the formation of Factor Xa, a protease that acts with other factors to convert prothrombin to thrombin. These two paths serve different purposes providing a means for amplification of an initial signal for clotting but also that blood clotting occurs in a controlled way and only when needed.

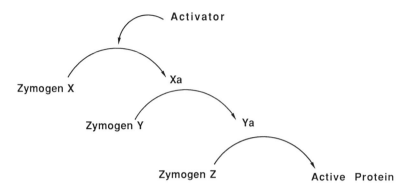

Figure 1.2. Cascade of zymogen activation.

The extrinsic path activated in vascular injury by the contact of tissue factor with blood is thought to be involved in initiating coagulation. Tissue factor bound to zymogen Factor VII aids its conversion to its active form VIIa. Amplification occurs. Factor VII, which is present in plasma at a concentration of less than 1 µg/ml, influences the activation of Factor X (8 µg/ml). Activated Factor X in turn affects prothrombin (150 µg/ml), which then acts on fibrinogen (up to 4,000 µg/ml) to produce fibrin. Clotting can occur effectively through this pathway in 12 seconds, but clotting through the intrinsic pathway can take several minutes.[8] The second pathway, the intrinsic pathway, may not be involved initially but functions later in the growth and maintenance of the fibrin clot. Both paths are carefully controlled and the absence of critical factors leads to bleeding disorders. The important principles established in the studies of zymogen activation apply to other types of reactions discussed later in the book.

A second feature illustrated by the coagulation cascade is the multitude of modification reactions that occur and how these reactions can influence protein structure, ligand binding, and enzyme activity. For example, thrombin acts to remove sulfated fibrinopeptide B from fibrinogen, which allows for interactions in fibrin, essential for polymer formation, not possible in fibrinogen. Thrombin also acts on other factors including Factor XIII, which after the limited proteolysis, functions as a transglutaminase to cross-link the fibrin chains. And there are the vitamin K dependent carboxylation reactions in prothrombin and other proteins that promote calcium binding. Hydroxylation of aspartic acid and asparagine, glycosylation, and disulfide bond formation occur—the system serves as a model itself for the study of co- and post translational modification reactions.

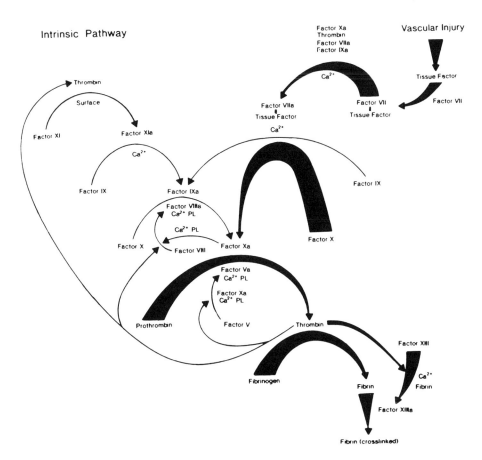

Figure 1.3. Coagulation cascade and fibrin formation by the intrinsic and extrinsic pathways. The initiation of the coagulation cascade occurs following vascular injury and the exposure of tissue factor to he blood. This triggers the extrinsic pathway (right side), shown in heavy arrows. The intrinsic pathway (left side) can be triggered when thrombin is generated, leading to the activation of factor XI. The two pathways converge by the formation of factor Xa. The activated clotting factors (except thrombin) are designated by lowercase a, that is, IXa, Xa, XIa, and so on. PL refers to phospholipid. The phospholipid bound to tissue factor apoprotein is not shown. Reprinted with permission from *Biochemistry* 30:10363 (1991). Copyright 1991, American Chemical Society.

IDENTIFICATION OF CO- AND POST-TRANSLATIONAL MODIFICATION PRODUCTS

Several general questions that apply to both modification reactions are: (1) what chemical groups are being attached? (2) how are these molecules linked to the protein, and in some cases, to each other? and (3) where multiple modifications occur of one or different types, how does one modification influence another?

Early studies of the structure of glycoproteins illustrate how the first two questions were addressed and suggest steps that could be taken to analyze new modification products of proteins. Colorimetric analyses showed that carbohydrate was found in crystalline egg albumin.[9] A first concern is: is the modifier, carbohydrate, just bound tightly to the protein by strong noncovalent interactive forces, or is it linked by covalent bonds? Denaturation of the protein by acid or heat did not release carbohydrate supports the latter possibility. But, digestion of egg albumin with a mixture of proteases, and isolation of a peptide containing carbohydrate and amino acids, a glycopeptide, provided proof that the linkage was covalent.[10,11,12] Acid hydrolysis of the glycopeptide produced mannose, glucosamine, acetic acid, ammonia, serine, threonine, leucine, and aspartic acid. To what amino acid is the carbohydrate attached? Is one of these two sugars or a modified sugar, destroyed by acid hydrolysis, involved in a linkage to the peptide chain? Good solid chemistry championed in various laboratories around the world led to the answers.[13,14,15,16]

Extensive treatment with carboxypeptidase yielded carbohydrate fractions containing aspartic acid, but with only traces of serine and threonine. Thus, neither of the hydroxylic side chains of these two residues in this instance serve as acceptors for the carbohydrate. Aspartic acid or asparagine, which breaks down upon acid hydrolysis to aspartic acid, could be linked to the carbohydrate. Reaction with 2,4–dinitrofluorobenzene and acid hydrolysis showed DNP-aspartic acid eliminating the involvement of an attachment of carbohydrate through the α-amino group.[17] As free aspartic acid could not be released by treating the peptide with carboxypeptidase, this result suggested that the carboxyl group of aspartic acid is in a peptide linkage, and that the attachment of carbohydrate was through the side chain.

Partial acid hydrolysis of the glycopeptide showed that mannose could be released but that carbohydrate was still attached to the peptide, suggesting glucosoamine or a related structure was involved in the linkage. Several structures were suggested,[18] two of which showed an attachment through a nitrogen atom, one at C-1 and the other at C-2, as illustrated in Figure 1.4.

Chemical synthesis of 2-acetamido-1-(L-β-aspartamido)-1,2-dideoxy-

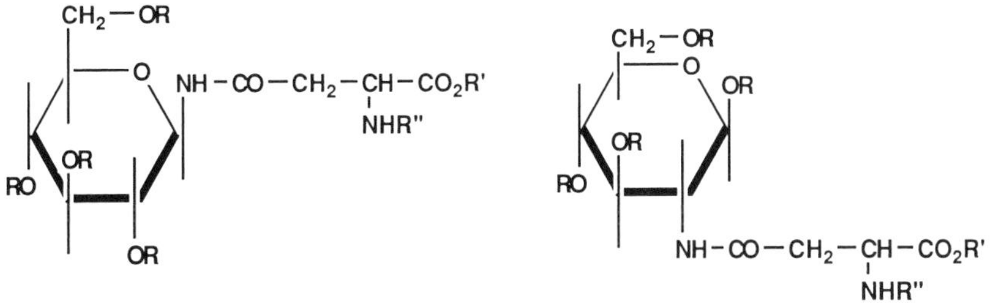

Figure 1.4. Proposed N-acylglycosylamine structures.

β-D-glucose and comparison of its properties with the isolated material from egg albumin established firmly that the linkage was through C-1.[19] The biosynthetic origin, however, was not known, and the product could have been derived from the condensation of a 1-amino sugar with the β-carboxyl group of aspartate or from the reaction of asparagine with a sugar devoid of nitrogen at C-1. Later, the fact that oligosaccharides containing n-acetylglucosamine, carried by dolichol pyrophosphate, can be transferred to asparagine in a polypeptide chain showed unequivocally that the amide functional group of asparagine was the acceptor.

Glycosylation of the side chain of proteins is not one simple addition and much enzymatic processing goes on before the oligosaccharide is complete. Because not all chains in different proteins are processed similarly, microheterogeneity associated with this type of modification occurs. Although some groups are added and removed from the carbohydrate chain, the overall process is not reversible (Chapter 7).

REVERSIBLE REACTIONS

Phosphorylation, nucleotidylation, and specific types of ADP-ribosylation, methylation, and acetylation are examples of reversible modification reactions. Phosphorylation–dephosphorylation reactions continue to be studied extensively because of their involvement in numerous biological processes.

Certain proteins found in milk (casein) and eggs (vitinellic acid,[20] ovalbumin, and phosvitin) contain phosphorous. But how is the phosphorous attached? Posternak and Posternak[21] suggested that a peptide obtained from vitinellic acid contained serine and phosphate. But the proof of the structure, the phosphoseryl ester, was provided by Professor Lipmann in Levene's laboratory when he obtained the barium salt of phosphoserine from a partial acid hydrolysis of vitinellic acid.[22,23] He wrote in his memoirs, "When 40 years ago (about 1932) I chose to probe into the binding of the phosphate in these, unwittingly, it now turns out, I struck a gold mine." Indeed he did!

Phosphorylation and dephosphorylation reactions are essential to many life processes. The first example showing the role of phosphorylation-dephosphorylation in a regulatory process was in glycogen metabolism.[24] The Cori laboratory demonstrated that in skeletal muscle glycogen phosphorylase existed in two forms, phosphorylase b and a.[25,26] Because phosphorylase was activated by AMP, it was believed that the active form, phosphorylase a, contained bound AMP. An enzyme reported to cause conversion of phosphorylase a to phosphorylase b was proposed to release AMP and hence was named the PR (Prosthetic Removing) enzyme. But no evidence could be obtained to substantiate this proposed idea. In the 1940s and 1950s evidence was accumulating from studies of Sutherland and the Cori's that the interconversion of phosphorylase was regulated by glucagon and epinephrine in liver slices and by epinephrine in muscle diaphragm,[27,28] but the exact nature of the shuttle between inactive and active phosphorylase was not known. In the early 1950s two professors (Krebs and Fischer) at the University of Washington embarked on a joint project to investigate the activation of muscle phosphorylase. Little did they know that this beginning would lead to a lifetime of productive, collaborative research, that their studies would have a great impact on those of others, and finally the Nobel Prize in Physiology or Medicine in 1992. As shown in Figure 1.5, activation was due to protein phosphorylation by phosphorylase kinase and inactivation caused by the PR enzyme was due to dephosphorylation.[29]

Similar results were obtained from Sutherland's laboratory with enzymes from liver tissue. From these studies came the remarkable discovery of cyclic AMP. Later, it was learned that the hormonal effects of glucagon and epinephrine could be explained by the action of this nucleotide on a protein kinase, which activated phosphorylase kinase. Other discoveries of phosphorylation-dephosphorylation systems and the role of these reactions in regulatory phenomena followed, and the list continues to grow with no end in sight.

Reversible reactions such as phosphorylation-dephosphorylation, ADP-ribosylation-deADP-ribosylation, and adenylation-deadenylation are usually catalyzed by separate enzymes or active sites. Thus, precise control of the cycle by effectors is allowed. For example, adenylation of glutamine synthetase is activated by a-ketoglutarate but its deadenylation is inhibited by the same effector. Mechanisms for amplification of a signal exist with such systems.

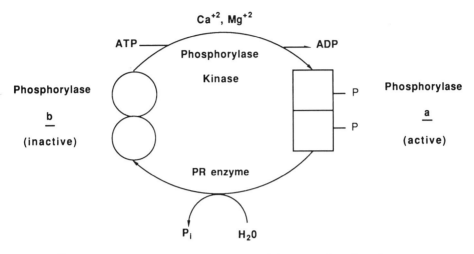

Figure 1.5. Interconversion reactions of glycogen phosphorylase.

Even though different enzyme activities may be involved in the reversal of the reactions *in vivo*, a study of a "nonphysiological" reaction, for example, the reverse reaction catalyzed by specific enzymes *in vitro*, is invaluable. Again, Lipmann's research provided an important lead. By using phosphorylated phosvitin and yeast phosphokinase, a transfer of phosphate to ADP with the formation of ATP could be demonstrated.[30] From these studies and those of others of the reactions in both directions,[31] it was learned that the seryl phosphate bond in proteins[32] had a higher free energy of hydrolysis than did related phosphate esters of low molecular weight. Figure 1.6 shows how the value was obtained by determining the equilibrium constant of the reaction. The free energy of hydrolysis of the seryl phosphate ester is about -6.5 kcal./mol. Similarly, tyrosyl phosphate in proteins was found to be "energy-rich".[33] These results suggest that the phosphate ester in the protein experiences a different environment than in the free amino acid. This idea is confirmed in the x-ray analysis of the seryl phosphate site in glycogen phosphorylase. Knowledge of the types of interactions are important to understand what phosphate or other modifying groups do in the protein. Although there is no direct relationship between reactivity and free energy change, knowledge of the environment about the residue to be modified may provide a clue why certain enzymes may catalyze a reaction effectively with a protein substrate in comparison to a reaction with small molecules.

$\Delta\ G^0$ obs.

P-protein + ADP $\rightleftharpoons$ protein + ADP   +1.9 kcal mol^{-1}

ATP + H$_2$O $\rightleftharpoons$ ADP + P$_\text{l}$ -8.4 kcal mol^{-1}

_____

Sum: P-protein + H$_2$O $\rightleftharpoons$ protein + P$_\text{l}$ -6.5 kcal mol^{-1}

Figure 1.6. Free energy of hydrolysis of the phosphoseryl ester in glycogen phosphorylase.

ENZYMATIC AND NONENZYMATIC REACTIONS

Nonenzymatic covalent modifications, like enzyme catalyzed reactions, do occur with protein substrates, and these reactions can be physiologically significant. Some of the chemical modifications may be reversed by an enzymatic process. Others may lead to other reactions and/or signal the proteolytic degradation of the modified protein.

The cross-linking that occurs in proteins such as collagen and elastin illustrates the interplay of enzymatic and nonenzymatic reactions and the influence of these linked reactions on the properties of modified proteins. First, the lysine side chains must be modified by enzymatic hydroxylation and/or oxidation to prepare the residues for chemistry that will cross-link polypeptide chains. Second, the nature of the linkage and its stability depends upon the enzymatic processing. Chemical cross-linking begins by forming an imine bond from the reaction of the derived aldehyde with the ε-amino group (Figure 1.7). The Schiff's base may rearrange and form a keto derivative by an Amadori rearrangement.[34] If, however, the aldehyde is not derived from hydroxylysine but from lysine, the stabilized structure due to the Amadori rearrangement is not achieved. This is because the imine cross-link is acid labile, and mild acid could promote interchain cleavage. Interestingly, a linking region (telopeptide) in skin collagen contains less hydroxylated residues than in similar parts of bone and cartilage collagen. The differences in the type of cross-link formed by these regions are assumed to account for differences in stability as assessed by the higher acid solubility of collagen from skin than from cartilage, and bone.[35]

A second example of an interplay between nonenzymatic and enzymatic reactions is found in the response of proteins in cells undergoing oxidative stress. Enzymatic reactions and nonenzymatic reactions may generate superoxide ($\cdot O_2^-$) and hydroxyl ($\cdot OH$) radical, or other activated species, which in turn can promote side chain modification of amino acids in proteins such as methionine, cysteine, histidine, proline, arginine, and lysine.[36,37] Selective modification can be mediated by the binding of metal ions like Fe^{+2} at specific metal binding sites, causing oxidative side chain modification, and/or, in some instances, peptide bond cleavage.[38,39] Oxidation of a methionyl residue to the sulfoxide is reversible by a reductant such as dithiothreitol, but the next level of oxidation, the sulfone, is not reversible. An enzyme has been found wide spread in nature that can reverse the oxidation.[40] This enzyme acts on the sulfoxide, can use dithiothreitol or a more natural reducing system, and could serve to rescue inactive proteins oxidized at methionine, particularly those with low turnovers, such as lens protein.

α-I proteinase inhibitor, an inhibitor of elastase, can be inactivated *in vitro* by oxidation of methionine to its sulfoxide in the active site.[41] Similar inactivation occurs *in vivo*, and enzymatic reduction reverses this process. The oxidation of the inhibitor is believed to contribute to the uncontrolled proteolysis associated with the disease pulmonary emphysema. Considerable attention has been paid to the relation of smoking to emphysema. It has been observed that the inhibitory capacity of α-PI is reduced in smokers. The level of protein is not changed, but oxidation has occurred, as shown by the appearance of methionine sulfoxide.[42,43]

The formation of mixed disulfides in proteins is a nonenzymatic reaction occurring both *in vitro* and *in vivo*. Proteins have different capacities for modification, but carbonic anhydrase is particularly sen-

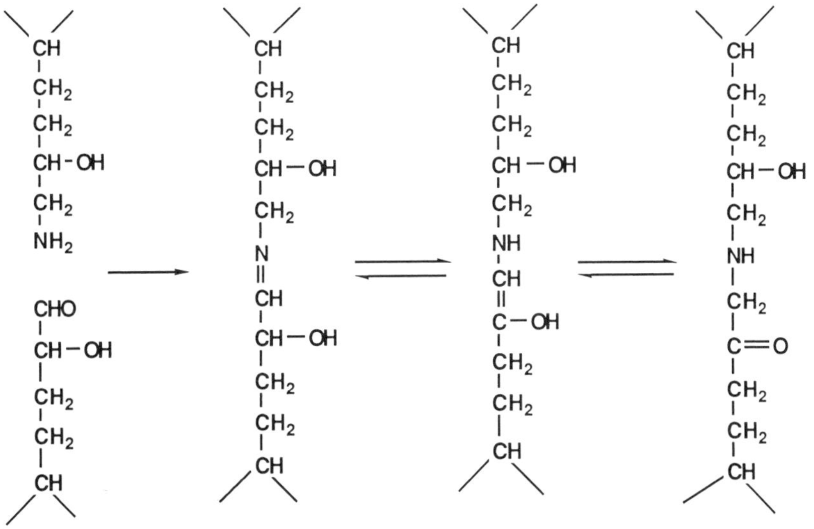

Figure 1.7. Cross-link formation between a hydroxyallylysine and a hydroxylysine.

sitive and is modified rapidly by S-thiolation in hepatocytes.[44,45] Enzymatic activities (dethiolases) that can reverse this reaction have been described in various sources.[46] Thus, it has been suggested that the formation of mixed disulfides in certain instances may serve to protect the sulfyhdryl group of proteins from irreversible changes by oxyradicals formed during oxidative stress.[47] But much remains to be learned about oxidative changes and the reversal of these processes under different physiological conditions.

The reactions of a protein with a modifier can have consequences on subsequent modification reactions as summarized by the following scheme (Figure 1.8). Enzyme catalyzed and nonenzyme catalyzed reactions producing modified protein may not be reversible. In one instance, the modification may prepare the protein for further modification reactions to yield a completed functional-mature protein. In another case, the modification may target the protein for selective degradation.[48,36] For example, oxidative damage including formation of methionine sulfoxide and mixed disulfides in proteins may not be reversed. Ribonuclease containing methionine sulfoxide is particularly sensitive to processing by ubiquitination and subsequent proteolysis.[49] Fructose-1,6-bisphosphate aldolase is particularly sensitive to proteolysis after its formation of a mixed disulfide with glutathione disulfide.[50] What are the chemical factors in the protein that lead to selective modification? What mechanisms exist that influence whether the reaction can be reversed or determine whether the protein is processed further or degraded? Approaches to these and related questions are addressed in this book.

RECOMBINANT PROTEINS AND ENZYME MODIFICATION REACTIONS

Expression of genes encoding for natural proteins and mutant forms can provide valuable information about processes involved in modification reactions and give materials needed to evaluate what modifications do to protein structure and function.

But problems do exist with expression systems and enzymatic modification reactions in host cells. For example, the enzymatic machinery needed for modification may not be present in cells used to express heterologus proteins; for example, prokaryotes lack enzymes needed for reactions such as myristoylation, glycosylation, and ADP-ribosylation of expressed proteins derived from eukaryotic genes. And in systems where modification can occur: (1) the co- and post-translational modification reactions may not be complete, (2) groups could be added at unnatural sites, and (3) new groups may be added.

Expression of a murine cDNA clone for the catalytic subunit of cyclic AMP-dependent protein kinase occurs effectively in E. coli but the expressed protein is missing the natural cotranslational modification product, myristate, at the amino terminus.[51] To achieve myristoylation in E. coli, a dual plasmid has been constructed to contain genes for both protein kinase and the enzyme, myristoyl-CoA:protein N-myristoyl transferase. Coexpression of the yeast gene for the transferase and the cDNA for the kinase yielded protein which was virtually all modified with myristate. The bacterial system has advantages over eukaryotic systems, because it can be used to prepare unmodified kinase, the myristoylated derivative, and analogs for studies of structure-function relations.[52]

Lipoate acyl transferase is an essential component of 2-oxy dehydrogenase multienzyme complexes. Reed[53] described the bound lipoate to act as a swinging arm bringing reactants to the active sites of three enzymes. Expression of a sub-gene for the lipoyl domain of the pyruvate dehydrogenase complex of Bacillus stearothermophilus was undertaken to provide sufficient materials for physical studies. Analysis of the expressed protein showed that 80 percent was not lipoylated, but most of the remaining protein was correctly lipoylated on Lys-42. A small portion of a new derivative presumed to be octonoate was also found.[54] Octanoylation of the lipoyl domains was proven to occur in lipoyl-deficient strains of E. coli. A new modification is found. It may yield information about the biosynthesis and processes of lipoylation.[55] These results point out the need for investigators to chemically characterize what the modified protein is. Without this information, wrong conclusions could be drawn about the role a modification may have on protein structure and function.

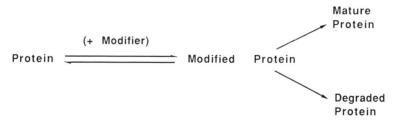

Figure 1.8. Consequences of modification on subsequent reactions.

REFERENCES

1. Wold, F., *Annu. Rev. Biochem.* 50, 783–814 (1980).
2. Freedman, R., and Hawkins, H.C., Eds., *The Enzymology of Post-translational Modification of Proteins*, Vols. 1 and 2, New York: Academic Press (1985).
3. Neurath, H., *Trends in Biochemical Science* 14, 268–271 (1989).
4. Neurath, H., and Walsh, K.A., *Proc. Natl. Acad. Sci. USA* 73, 3825–3832 (1976).
5. Light, A., and Janska, H., *Trends in Biochemical Science* 14, 110–112 (1989).
6. Neurath, H., *Adv. in Protein Chemistry* 12, 319–386 (1957).
7. Davie, E.W., Fujikawa, K., and Kisiel, W., *Biochemistry* 30, 10363–10370 (1991).
8. Voet, D., and Voet, J.G., in *Biochemistry*, p. 194, New York: John Wiley and Sons (1990).
9. Sorensen, M., *C.R. Lab. Carlsberg* 20, 1–19 (1934/35).
10. Neuberger, A., *Biochem. J.* 32, 1435–1451 (1938).
11. Neuberger, A., *Biochem. J.* 32, 1452–1456 (1938).
12. Neuberger, A., *Trends in Biochemical Science* 13, 398–399 (1988).
13. Bogdanov, V.P., Kaverzneva, E.D., and Andreyeva, A.P., *Biochim. Biophys. Acta* 83, 69–73 (1964).
14. Yamashina, I., and Makino, M., *J. Biochem. (Tokyo)* 51, 359–364 (1962).
15. Nuenke, R.A., and Cunningham, L.W., *J. Biol. Chem.* 236, 2452–2460 (1961).
16. Marks, G.S., Marshall, R.D., and Neuberger, A., *Biochem. J.* 87, 274–281 (1963).
17. Johansen, P.G., Marshall, R.D., and Neuberger, A., *Biochem. J.* 78, 518–527 (1961).
18. Marks, G.S., and Neuberger, A., *J. Chem. Soc.* 4872–4879 (1961).
19. Marshall, R.D., and Neuberger, A., *Biochemistry* 3, 1596–1600 (1964).
20. Levene, P., and Alsberg, *J. Biol. Chem.* 2, 127–138 (1906/07).
21. Posternak, S., and Posternak, T., *Compt. Rend. Acad.* 187, 313 (1928).
22. Lipmann, F., and Levene, P., *J Biol. Chem.* 98, 109–114 (1932).
23. Lipmann, F., *Wanderings of a Biochemist*, New York: John Wiley and Sons (1971).
24. Fischer, E.H., and Krebs, E.G., *Biochimca Biophysica Acta* 1000, 297–301 (1989).
25. Cori, G.T., and Green, A.A., *J .Biol. Chem.* 151, 31–38 (1943).
26. Cori, G.T., and Cori, C.F., *J. Biol. Chem.* 158, 341–345 (1945).
27. Sutherland, E.W., and Cori, C.F., *J. Biol. Chem.* 188, 531–543 (1951).
28. Sutherland, E.W., in *Phosphorous Metabolism* 1, 53–61 (1951).
29. Krebs, E.G., and Fischer, E.F., *Adv. in Enzymol.* 24, 263–290 (1962).
30. Rabinowitz, M., and Lipmann, F., *J. Biol. Chem.* 235, 1043–1050 (1960).
31. El-Maghrabi, M.R., Hatson, W.S., Flockhart, D.A., Claus, T.H., and Pilkus, S.J., *J. Biol. Chem.* 255, 668–675 (1980).
32. Shizuta, Y., Beavo, J.A., Bechtel, P.J., Hofmann, F., and Krebs, E.G., *J. Biol. Chem.* 250, 6891–6896 (1975).

33. Fukami, Y., and Lipmann, F., *Proc. Natl. Acad. Sci., USA* 80, 1872–1876 (1983).
34. Kuhn, K., in *Structure and Function of Collagen Types* (Mayne, R., and Burgeson, R.E., Eds.), New York: Academic Press (1987), p.1–42.
35. Robins, S.P., and Bailey, A.J., *Biochem J.* 163, 339–346 (1977).
36. Wolff, S.P., Garner, A., and Dean, R.T., *Trends in Biochemical Sci.* 11, 27–31 (1986).
37. Stadtman, E.R., *Biochemistry* 29, 6323–6331 (1990).
38. Amici, A., Levine, R.L., Tsai, L., and Stadtman, E.R., *J. Biol. Chem.* 264, 3341–3346 (1989).
39. Kim, K., Rhee, S.G., and Stadtman, E.R., *J. Biol. Chem.* 260, 15394–15397 (1985).
40. Brot, N., Fliss, H., Coleman, T., and Weissbach, H., *Methods in Enzymol.* 107, 352–360 (1984).
41. Matheson, N.R., Wong, P.S., and Travis, J., *Biochem. Biophys. Res. Commun.* 88, 402–409 (1979).
42. Brot, N., and Weissbach, H., *Arch. Biochem Biophys.* 223, 271–281 (1983).
43. Carp, H., Miller, F., Hoidal, J.R., and Janoff, A., *Proc. Natl. Acad. Sci.* 79, 2041–2045 (1982).
44. Rokutan, K., Thomas, J., and Sies, H., *Eur. J. Biochem.* 179, 233–239 (1989).
45. Chai, Y-C., Jung, C-H., Lii, C-K., Ashraf, S.S., Hendrich, S., Wolf, B., Sies, H., and Thomas, J.A., *Arch. Biochem. Biophys.* 284, 270–278 (1991).
46. Miller, R.M., Park, E-M., and Thomas, J.A., *Arch. Biochem. Biophys.* 287, 112–120 (1991).
47. Park, E-M., and Thomas, J.A., *Biochim. Biophys. Acta* 964, 151–160 (1988).
48. Rivett, A.J., *Curr. Top. Cell. Regul.* 28, 291–337 (1986).
49. Hershko, A., Heller, H., Eytan, E., and Reiss, Y., *J. Biol. Chem.* 261, 11992–11999 (1986).
50. Offerman, M.K., McKay, M.J., Marsh, M.W., and Bond, J.S., *J. Biol. Chem.* 259, 8886–8891 (1984).
51. Slice, L.W., and Taylor, S.S., *J. Biol. Chem.* 264, 8445–8446 (1989)
52. Duruonio, R.J., Machelski, J.E., Heuckeroth, R.O., Olins, P.O., Devine, C.S., Yonemoto, W., Slice, L.W., Taylor, S.S., and Gordon, J.I., *Proc. Natl. Acad. Sci.* 87, 1506–1510 (1990)
53. Reed, L., *Acc. Chem. Res.* 7, 40–46 (1974).
54. Dardel, F., Packman, L.C., and Perham, R.N., *FEBS Lett.* 264, 206–210 (1990).
55. Ali, S.T., Moir, A.J.G., Ashton, P.R., Engel, P.C., and Guest, J.R., *Mol. Microbiology* 4, 943–950 (1990).

2

Enzymology of Modification

What are the basic mechanisms used by enzymes that catalyze the chemical modification of proteins? Are there unique features for these enzyme systems? How can we learn about them? First, a large array of catalytic mechanisms exist for protein modification. Table 2.1 presents a list of the types of reactions catalyzed, and a few examples to illustrate how complex and diverse these reactions are. All types of enzymatic reactions seen with small molecules are found in protein modification. However, it is fair to conclude that the majority of the reactions may be classified as transferases. To understand how these and other reactions occur and are regulated is a challenge for the enzymologist, protein chemist, and molecular biologist.

Enzymatic reactions catalyze reactions representing every type of chemical reaction. Moreover, enzymes have developed unique systems for catalyzing these reactions. In enzymes containing post-translational modifications, the modification may function to temper properties of the enzyme. Enzymes may also have different amino acids that define specific properties; these modified amino acids have been incorporated translationally. A specific example is the substitution of selenocysteine for cysteine. Replacement of sulfur with selenium has dramatic effects on the properties of the amino acid; selenol (pK_a = 5.2) is much more acidic than cysteine (pK_a = 8.3).[1] Along with oxygen and sulfur, selenium is a member of Group VIA of the periodic table. Selenium and sulfur are more similar in properties than are oxygen and sulfur.[2] A significant difference

Table 2.1. Classification of enzymatic reactions

Types	Enzyme	Protein Product
Oxidoreductases	lysyl oxidase	allysine
	prolyl hydroxylase	hydroxy proline
	methionine sulfoxide reductase	methioninine
		thiolase cysteine
Transferases	glutamyl methyltransferase	glutamyl methyl ester
	tyrosine kinase	phosphotyrosine
	histone acetyltransferase	ε-amino acetyl lysine
	arginyl ADP-ribosyl-transferase	ADP-ribosyl arginine
Hydrolases	phosphoseryl phosphatase	serine/threonine
	arginyl deiminase	citrulline
	peptide-N-asparagine amidase	aspartic acid
	signal peptidase	new amino terminus
Lyases	ADP-ribosyl glutamate lyase	glutamate
	Peptidylamidoglycolate lyase	amidated glycine
Isomerases	protein disulfide isomerase	new disulfide bonds
	prolylpeptide isomerase	cis-trans.isomers of prolyl linkages
Ligases	tubulin-tyrosine ligase	carboxy terminal tyrosine
	ubiquitin-protein ligase	ubiquitnylated lysyl residue

is oxidation potential, a property of biological relevance. Selenium hydrides tend to remain reduced, whereas sulfur compounds tend to be oxidized. Because it is deprotonated more readily, the nucleophilic strength of selenocysteine is likely to be less than cysteine, but may be more reactive than cysteine at physiological values of pH. Similarity in size of the two centers suggests that substitution of selenocysteine will cause little steric problems for the protein.

Of the few enzymes that incorporate selenocysteine, one of the two enzymes that catalyzes iodothyronine deiodination is included.[3,4] Two different enzymes, designated types I and II, catalyze this reaction. Type I iodothyronine deiodinase contains selenocysteine; type II has a cysteine residue. The change in amino acid is responsible for certain biochemical differences between the two isozymes. Type I deiodinase is inhibited by 6-propylthiouracil, whereas the type II is not inhibited. The type II isozymes is also less sensitive to gold thioglucose. Mutants of the type I enzyme in which selenocysteine has been changed to cysteine become less sensitive to these inhibitors.[5] The sensitivity to these inhibitors is probably defined by the reactivity of selenocysteine presumably because of its protonation state. More importantly, a SeCys-Cys mutation in deiodinase I is accompanied by changes in its catalytic properties.[5] In particular, V_{max} is increased 6-fold in the mutant. The increase in rate is not unexpected based on the chemical properties.

KINETICS OF POST-TRANSLATIONAL MODIFICATION REACTIONS

Kinetic studies provide a means to obtain important information about post-translational modification reactions. In fact, the use of kinetics is necessary to understand how these processes occur and are regulated. K_m and V_{max} values, turnover numbers, inhibition and/or activation constants, pathways of substrate addition and product release, intermediates, rate constants, rate-limiting steps, and so on may be delineated. And particular novel aspects of post-translational modification reactions, such as autocatalysis, substrate control, synergism, and cascades, may be revealed and defined.

The Kinetic Mechanism

Uni-Bi reactions

A substrate is converted to two products in this category. Dephosphorylation, deadenylation, deADP-ribosylation, and peptidyl bond cleavage are examples of post-translational modification reactions fitting this kinetic mechanism.

Two general mechanisms are possible: (1) Random Uni Bi, which occurs when the two products are released in a random order from the enzyme. A preferred pathway could exist for release of products, but if either product can dissociate from a central complex containing the two products, the mechanism is considered random; and (2) Ordered Uni Bi, with which the name specifies, a compulsory path of product release exists.

A kinetic study showed that dephosphorylation catalyzed by a tyrosine phosphatase was noncompetitively inhibited by a nonphosphorylated polymer, poly Glu-Tyr (4:1).[6] That the inhibition was noncompetitive was interpreted as evidence that the inhibitor was binding away from the catalytic site. Yet, if the kinetic mechanism were known, the results might be interpreted differently. Consider the pattern of product inhibition for two cases in which P and Q represent inorganic phosphate and alcohol, respectively. In the rapid equilibrium mechanism, both inhibitors are competitive with respect to the substrate, but in the steady state ordered mechanism, in which Q dissociates before P, only P is competitive. Q is a noncompetitive inhibitor. Thus, noncompetitive inhibition by the nonphosphorylated polymer would not necessarily mean that binding occurs outside the active site region. Further kinetic studies could be conducted to eliminate one of these scenarios. Still, it should be noted that this kinetic mechanism predicts what results ought to be obtained from kinetic studies, but the kinetics do not prove that this mechanism is correct. Other schemes, however, could fit the results.

Kinetic studies may suggest further types of experimentation. An ordered reaction scheme may describe a mechanism by which the enzyme forms a phosphoryl intermediate reacting with water to generate free phosphate after the product alcohol has been released. A specific example of this situation will be discussed later this chapter. A covalent intermediate was isolated and characterized using radiolabeled substrate. If the intermediate is unstable, an approach using ^{32}P-labeled substrate in an attempt to isolate the labeled protein may not be successful. Pre-steady state kinetics and isotopic exchange reactions may be used because they have proved useful in delineating mechanisms.

Bi Bi reactions

Three general types of mechanisms exist: (1) Random Bi Bi, (2) Ordered Bi Bi, and (3) ping-pong. These mechanisms may be readily evaluated by initial velocity studies. Consider the rapid equilibrium random Bi Bi and the steady state ordered Bi Bi mechanisms. The two cases have the same general rate equation and cannot be distinguished readily by initial velocity measurements unless kinetic studies are done with products present, with competitive inhibitors, or with alternative substrates. The ping-pong mechanism, however, is distinctly different and predicts that double reciprocal plots will yield a family of parallel lines in contrast to the converging line patterns of the first two mechanisms. The ping-pong mechanism occurs when a covalent intermediate is formed and the second substrate binds only after the first product has dissociated. For example, if the enzyme is adenylated in the catalytic reaction and if the product dissociates before the acceptor substrate (perhaps a protein with a modifiable tyrosine) binds, parallel lines will be observed in the reciprocal plots. Protein crosslinking catalyzed by transglutaminases reportedly utilizes a ping-pong kinetic mechanism.[7] Evidence for acyl enzyme has been collected. The primary kinetic plots do not, however, match the parallel line pattern typical for a ping-pong reaction. Validation of the mechanism was obtained by isotope-exchange experiments. The actual mechanism is a modified double displacement reaction in which the acyl enzyme intermediate can be hydrolyzed by H_2O. The net result of this side reaction is the deamidation of a glutamine residue. A complicating feature is the requirement that the initial product, NH_3, must remain bound to the enzyme until after the second substrate binds. Hence, the kinetic studies can provide valuable insight into how the reaction proceeds.

Significance of the Kinetic Parameters, K_m and V_{max}

The K_m value is an operational constant and defines the concentration of the substrate needed for one-half the maximal velocity of the enzymatic reaction. It can be equal to, greater than, or smaller than the dissociation constant. Yet K_m values are often thought of in terms of affinities. A comparison of K_m values of different substrates may suggest that substrates with lower K_m values bind more tightly to the enzyme than those with higher values. But the interpretation we make of these constants depends upon our knowledge of the kinetic mechanism. For example, in ordered Bi Bi, one of the K_m constants can never be a simple dissociation constant because no interaction of the second substrate occurs with the free enzyme.

In the random mechanisms, Uni Bi and Bi Bi, if the conversion of the binary or ternary complexes, respectively, is rate limiting, then dissociation constants for the substrates may be obtained by kinetic studies. Consider the following equation for the two substrate systems.

$$\frac{V_M}{V_I} = \frac{K_A}{A} + \frac{K_B}{B} + \frac{K_{ia}K_B}{(A)(B)} + 1$$

Of these constants, only K_{ia} is a true dissociation constant. K_a and K_b are Michaelis constants and are measures of the substrates A or B interacting with their respective binary complexes EB and EA. In a kinetic study an apparent K_a or K_b may be obtained. Hence, care needs to be taken in interpreting these values, particularly if a comparison is made with other rate data obtained with different substrates. Only if the second substrate is saturating can the constants K_a or K_b be determined at one fixed value of the second substrate. The dissociation constants and the Michaelis constants may be obtained by secondary plots of the slopes and intercepts obtained from the primary plots (Figure 2.1). K_{ib} may be obtained from the relation, $K_a.K_{ib} = K_b.K_{ia}$.

The V_{max} value may give information about rate constants of the catalytic reaction. Division of V_{max} by the enzyme concentration will yield the turnover number of the enzyme, but only if the enzyme is pure. If the rate limiting step is the interconversion of the enzyme substrate complex to enzyme product complex (rapid equilibrium type), then we may obtain the rate constant for this reaction. If the reaction is limited by other events—for example, product dissociation, or isomerization of enzyme complexes (steady state type)—then the formulation of V_{max} values in terms of kinetic constants is much more complicated. Thus, care needs to be taken in interpreting the V_{max} value unless the kinetic mechanism and the enzyme purity is known.

A comparison of V_{max}/K_m values for different substrates can give some idea of the relative efficiency of the enzyme for different substrates. The ratio may approach the second order constant for association of the substrate with the enzyme. But because the kinetic relation of K_m and V_{max} are complex for different mechanisms, simplification to a second order constant is not obvious. Yet the comparisons of ratios can be meaningful and can eliminate complexities associated with formation of nonproductive complexes potentially obfuscating the interpretation of both K_m and V_{max} values.

STRUCTURAL PROBES OF MODIFYING ENZYMES

Kinetic investigations provide valuable insight about the reaction progress, but less information about the nature of the chemistry involved. There are approaches that can provide insight into the reaction course. Spectroscopic techniques, such as nuclear magnetic resonance and FT-infrared spectroscopy, can be applied to explore structural aspects of these reactions, including the nature of intermediates. Defined chemical analogs, including stereo and geometrical isomers and

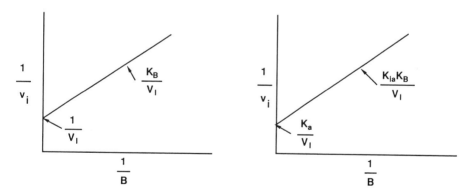

Figure 2.1. Kinetic plot for a Bi-Bi kinetic mechanism. On the left is a secondary plot of the y-intercept from a Lineweaver-Burk primary plot. On the right is shown a secondary plot of the slope from the primary plot.

affinity reagents, are useful in defining the chemical nature of the reaction and the active site residues involved in catalysis. Mutational analysis (genetic studies and site directed mutagenesis) and immuno-precipitation experiments can define limits to the biological roles of the enzymes as well as the interaction with other regulatory proteins.

Spectroscopic Studies

Extensive structural information can be gleaned by physical techniques, particularly if multiple approaches are combined. For example, eukaryotic carboxyl methylation occurs specifically at aspartyl residues chemically altered in a defined way. Chemical identification of *in vivo* methylated products indicates that labeling is found at D-aspartyl residues.[8,9] *In vitro* experiments suggest that the substrate for methylation is an isoaspartyl residue produced from the deamidation of an asparagine residue.[10,11] For both situations, a cyclic imide intermediate was implicated, but no direct structural proof was provided. Evidence for a cyclic imide as an intermediate during deamidation was obtained using FT-IR photoacoustic spectroscopy (FTIR-PAS) and fast atom bombardment mass spectrometry (FAB-MS) analysis of model peptides containing an Asn-Gly linkage, which were subjected to conditions shown to cause deamidation.[12] FAB-MS analysis revealed a pattern with a peak consistent with the loss of a single water molecule (MW 18); the peak was not consistent with an anhydride intermediate inasmuch as the pattern did not suggest the elimination of ammonia (MW 17). Proof of the cyclic imide structure in peptides was provided directly by FTIR-PAS. Analysis of putative imide structures relative to succinimide as a standard established its identity as a cyclic imide. Similar results were obtained with deamidated calmodulin although the assignment was not unequivocal.[13]

Spectroscopy also was used to establish the structure of the deamidated product. NMR spectroscopic investigations of calbindin, a 75 amino acid calcium binding protein, demonstrated the presence of deamidated forms of the protein.[14] The deamidation was localized to a specific asparagine residue, Asn_{56}, and confirmed the development of an isoaspartate linkage in the protein. The distinct nature of the NMR signal facilitated the identification.

Defined Chemical Probes

In addition to physical analysis, chemical probes with defined properties can be exploited to define the nature of the reaction. Of particular relevance to enzymatic reactions is the application of stereospecific substrates and analogs. Replacement of a specific acceptor residue, normally the L-isomer, with the D-stereoisomer provides a first approximation of the geometry of the active site. Eukaryotic protein carboxyl methylation again provides an example. Although *in vivo*, methylation seems to occur most readily on D-aspartyl groups, synthetic peptides with a D-aspartyl residue did not serve as substrates *in vitro* for a purified carboxyl methyltransferase.[15,16] Peptides with L-aspartate also were not substrates.[15,16] Instead, only L-isoaspartyl containing peptides were good substrates.[15,16] The D-aspartyl and L-isoaspartyl peptides have some similar conformational structures at the aspartyl sites[15] (Figure 2.2). The generation of isoaspartyl residues is described in Chapter 3.

Glutamyl carboxylation also has been investigated with the D-isomer of the acceptor residue. Specifically, alternate substrates with a diastereometric mixture of cyclic glutamyl analogs were utilized[17] (Figure 2.3). Surprisingly, the D,L mixed peptide substrates were effectively used with little difference in the kinetic parameters relative to the L-isomer peptide. Similar results were obtained regardless if the *cis* or *trans* structural isomer was used.[17] Each of the analogs associated with the carboxylase with lower apparent affinity, but there was little effect on turnover. These studies are consistent with a model involving a relatively open (loose) catalytic site in which the bulkiness of the different analogs does not completely preclude substrate association. Glutamyl residues stereospecifically labeled with tritium function in the carboxylase catalyzed proton abstraction in the absence of carbon dioxide.[18] With this substrate, the half reaction was shown to proceed

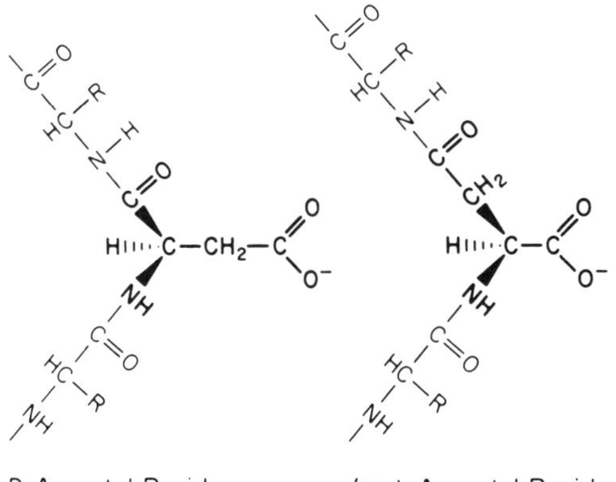

D-Aspartyl Residue   *Iso*-L-Aspartyl Residue

Figure 2.2. Structures of D-aspartyl and L-*iso*aspartyl residues in a peptide linkage. Note the similarity of the orientation of the side chains of the D-aspartyl and *iso*-L-aspartyl residues. Reprinted with permission from *Journal of Biological Chemistry* 259:10722(1984). Copyright 1984, American Society for Biochemistry and Molecular Biology.

stereospecifically; only one isomer was a substrate. Moreover, the proton abstraction was shown to occur at the same site as in the carboxylation reaction.

Chemical reagents also provide information about the residues involved in binding or catalysis. Labeling reagents specific for a particular amino acid are available. Moreover, amino acids can have different reactivities dependent upon the environment of the residue within the protein. These properties enable the specific labeling of a small number, possibly one, correlated to the inactivation of the enzymic activity. The phosphatase activity of calcineurin has been shown to be inactivated by reagents which label thiol[19] or arginyl[20] groups. In each case, inactivation was correlated to the labeling of a single residue. These results suggest that these residues may be crucial to the catalytic machinery of calcineurin. Unfortunately, no specific residues correlated to the loss of activity could be identified because of extensive modification, particularly following arginine labeling. This fact highlights one of the chief disadvantage of chemical labeling studies. A second disadvantage is that the overall structure of the protein may be disrupted because of extensive labeling. In general, these types of experiments yield equivocal results because of these problems.

A special class of labeling reagents, affinity labeling reagents, can overcome some of these difficulties. Affinity labeling reagents are synthesized to mimic all or a portion of the substrate to optimize specific binding. The labeling group is chosen to be selectively reactive; the reaction conditions may be modified from the initial incubation conditions to enable reactivity. Azido derivatives of nucleotides are often used to probe nucleotide binding sites. The ADP-ribosyltransferase activity of diptheria toxin was inactivated by the photolabeling of a specific tyrosine residue (Tyr_{65}) by 8-azidoadenine or azidoadenosine.[21] A novel azido compound, 3-azido-L-tyrosine, labels and inactivates tubulin:tyrosine ligase.[22] These compounds effectively bind to specific binding sites in the dark and react with the target residue when exposed to light. Controlling the light exposure allows specific control of the reaction. Other reagents, such as fluorosulfobenzoyl adenosine (FSBA), specifically react with target residues. FSBA reacts with the lysine residue present in the ATP binding site present in protein kinases.[23]

There are labeling reagents that only become reactive during the course of catalysis. These so-called suicide inactivators are designed to mimic the transition state or some intermediate and are reactive only as the reaction approaches this reaction coordinate. N-glycosyltransferase is irreversibly inhibited by a peptide substrate containing

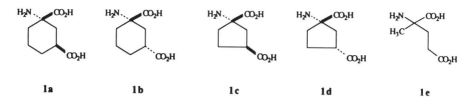

Figure 2.3. Diastereomeric inhibitors of Vitamin K dependent carboxylase. Note the relative positioning and orientation of the two carboxylic acid groups present in each molecule. Reprinted with permission from *International Journal of Peptide and Protein Research* 37:210 (1991). Copyright 1991, Manksgaard International Publisher's Ltd.

epoxyethylglycine (shown in Figure 2.4), a threonine analog, at the second residue from the acceptor asparagine.[24] With the typical glycosyl acceptor, the hydroxyl group of threonine is involved in hydrogen bonding with the side chain of asparagine, either to the amide[24] or carbonyl[25] group. In either situation, the nucleophilicity of the amide group is enhanced to facilitate attack on dolichol bound sugar. The glycosylated peptide is subsequently released by the enzyme. With the epoxy derivative, the enzyme is covalently linked to the threonyl derivative during the concerted reaction and cannot turnover the product even though the peptide is apparently glycosylated. Effectively, the enzyme can catalyze only a single cycle of the reaction.

Application of model compounds containing the modification may also be exploited to investigate regulatory phenomena. For example, the tyrosine kinase, pp60[c-src], is a substrate for myristoylation. The function of the myristoyl group is not completely clear, but seems to be involved in targeting the kinase to the membrane.[26,27] Supporting evidence for this model includes the altered localization of mutated kinase, which lacks the myristoylation site and will not associate with the membrane.[26,27] More recently, membrane proteins that associate with myristoylated pp60[c-src] have been identified.[28] The association of pp60[c-src] with these membrane proteins can be blocked by the presence of myristoylated peptides from the amino terminus of pp60[c-src]. The same peptide sequences lacking the myristoyl group fail to inhibit the association of the kinase with the membrane fraction. These data strongly support a role for myristoylation in regulating the interaction of pp60[c-src] with membrane proteins, if not directly with the membrane phospholipid components. Other examples are discussed in Chapter 7.

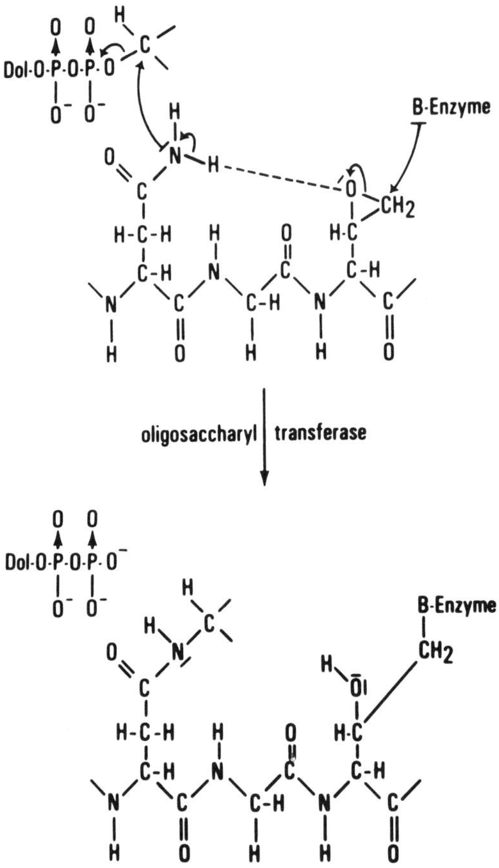

Figure 2.4. Suicide inactivation of protein glycosyltransferase by an epoxide peptide analog. A single catalytic cycle is possible; the glycosylated protein (peptide) product cannot dissociate off the enzyme. Reprinted with permission from *Biochimica Biophysica Acta* 906:161 (1987). Copyright 1987, Elsevier Science Publishers.

Primary Sequence Information

Structural analysis of modifying enzymes yields clues about the structure-function relationships of various domains of the protein. The protein primary structure determined either directly from sequencing or predicted from the nucleotide sequence can aid the assignment of unique regions. For example, the ATP-binding site of protein kinase has the conserved motif, GXGXXG-(18-32 AAs)-K.[29] This sequence, in fact, serves as a marker for new kinases described from nucleotide sequences.

Sequence information may enable the deduction of function when compared with other enzymes with similar properties. Methyltransferases for protein side chains, metabolites, and nucleic acids all share a requirement for S-adenosylmethionine (SAM) as the methyl donor. The amino acid sequences for protein carboxyl methyltransferases have been recently elucidated[30,31,32] and compared with other SAM utilizing enzymes.[31] Three regions of common structural features were identified; two of which were present in both the aspartyl and glutamyl methyltransferases. One sequence, LKPXGXL(L/I)L is the most highly conserved in these two and the other small molecule methyltransferases. An initial hypothesis may be that this region will correspond to a common function, possibly the binding site for SAM. A recent study indicated that this sequence was not the site covalently labeled by a photoactivated SAM derivative.[33] Still, this conserved sequence may prove functionally important to methyltransferase activity. Although the sequences for arginyl and lysyl methyltransferases are not known, differences may be expected, particularly for the arginyl modification. The specificity of carboxyl and arginyl methyltransferases for SAM and SAM analogs demonstrate significant differences.[34,35] In particular, it seems that different faces of the adenine moiety are important for binding to the two classes of enzymes. Interestingly, the apparent affinity for SAM of the two enzymes is the same. Other structural features of higher order structure may be similar. Nevertheless, comparison of primary structure information suggests the possible identification of a functional domain of the carboxyl methyltransferases.

With the advent of site directed mutagenesis, much of the information obtained from specific labeling reagents and sequencing can be tested. The specific amino acid identified from labeling experiments can be replaced and the effects of protein function can be analyzed. Pertussis toxin is composed of 5 subunits; the S1 subunit has both ADP-ribosyltransferase and NAD glycohydrolase activities. The ADP-ribosylation activity of pertussis requires the presence of thiol agents presumably to maintain the protein thiol groups in the appropriate oxidation state.[36] This result suggests a dependence of activity on a cysteine residue in the enzyme moiety. Indeed, alkylation of cysteine groups is correlated with the inactivation of the enzyme activity.[37] Only two cysteine groups, Cys_{41} and Cys_{200}, are present in pertussis S1. Mutagenesis was performed to yield truncated forms of the protein missing 16, 40, or 55 amino acids from the carboxyl terminal end.[38] In all cases, the ADP-ribosyl transferase activity of the mutant protein was diminished; the amount of activity lost increased with increased amount of the carboxyl terminus lost. Clearly, a segment of the carboxyl terminus was crucial for transferase activity. The most active mutant, missing only 16 amino acids, was the only mutant protein which retained cysteine$_{200}$. This finding suggests that this residue may be necessary for activity. Interestingly, the NAD-glycohydrolase activity of the recombinant S1 proteins was not affected by these deletions to the same extent.[37] This observation may be an indication that the carboxyl terminus is important in the interaction with the acceptor group, not the catalytic site. A series of site directed mutagenesis experiments focused on replacing Cys$_{41}$ with other amino acids.[39] Although ADP-ribosyltransferase activity was diminished by the substitution, the activity was not completely abolished. Moreover, NAD-glycohydrolase activity was not significantly changed by the point mutations. Deletion of either Cys$_{41}$ or Ser$_{40}$ did, however, completely abolish transferase activity. It seems that Cys$_{41}$ is not involved in catalysis,[38] but that disruption of the local structure in this part of the molecule can have deleterious effects on the global function of the protein.

REACTION MECHANISMS

As already discussed, modifying enzymes catalyze the same types of reactions as enzymes with small molecule substrates. These enzymes also utilize similar types of reaction mechanisms, including similar intermediates. Application of kinetic, chemical, and structural probes provide different insights into the reaction and enable the elucidation of the reaction mechanism.

Covalent Catalysis

Covalent catalysis is predicated on the formation of a covalent enzyme linked reaction intermediate. N-myristoyltransferase from yeast is specifically radiolabeled when [^{14}C]-myristoyl-CoA is used as the substrate.[40] Kinetic studies have indicated an ordered Bi-Bi mechanism with myristoyl-CoA the first substrate bound and myristoyl-peptide the last product released.[41] This kinetic mechanism is supported by results which demonstrated the formation of a ternary complex of the enzyme with myristoyl-CoA and a peptide substrate.[42] Together with the identification of a covalent reaction intermediate, such a mechanism requires the binding of acceptor and transfer of the myristoyl group prior to the release of the product, CoA. The reverse, release of CoA before acceptor binding and transfer, would be manifested as ping-pong kinetics. Kinetics have also demonstrated the primary influence of the myristoyl group in the interaction with the enzyme[42,43,44] and is consistent with the absolute specificity of the transferase for myristic acid compared to other acyl groups.[44,45]

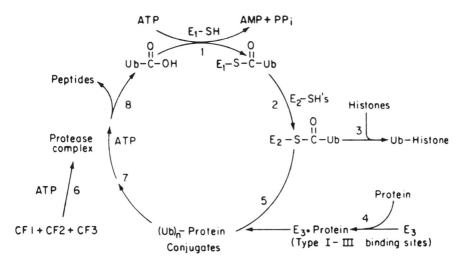

Figure 2.5. Ubiquitin dependent degradation pathway. There are multiple steps involving covalent enzyme intermediates. Particularly note reactions 1 and 2 in which ubiquitin conjugates are linked to enzymes E_1 and E_2 through thioester bonds. Reprinted with permission from *Journal of Biological Chemistry* 263:15237 (1988). Copyright 1988, American Society for Biochemistry and Molecular Biology.

Heteroatom derivatives of myristoyl-CoA have been examined as alternate donors.[46,47] Generally, singly substituted derivatives (thio-, or oxo-) completely replace myristoyl-CoA. In some cases, an increase in K_m is offset by an increase in V_{max}. Double replacement derivatives or changes in acyl chain length, however, produce significant deleterious changes in the kinetic parameters. These results illustrate the primary influence of chain length, not hydrophobicity or steric effects, on substrate (donor) utilization.[46] This may be related to the bent structure indentified in the donor molecule.[48]

The ubiquitin-linked protein degradation pathway illustrates an exceptional use of covalent intermediates. Ubiquitin (Ub) mediated degradation utilizes three protein components to accomplish the conjugation of ubiquitin to the substrate and the subsequent degradation of the substrate[49] as shown in Figure 2.5. The protein components are designated E_1 Ub-activating protein, E_2 carrier protein, and E_3 ligase. Crucial to the process is the initial activation of Ub by E_1 via the conjugation of Ub to a specific cysteine residue.[50,51] The reaction requires ATP for the formation of an activated adenylated Ub in a prereaction step. The E_1–Ub conjugate then transfers the Ub to a thiol group on E_2. The E_2–Ub conjugate serves as the donor for transfer of the ubiquitin to the target protein. The ubiquitinated protein is the substrate for proteolysis. Application of a nonhydrolyzable ubiquitin derivative, adenosyl-phospho-ubiquitinol, has provided support for part of this proposed mechanism, the formation of adenylated Ub and subsequent conjugation to E_1.[52] Two sites for ubiquitin binding on E_1 have been implicated; a noncovalent site and the site for conjugation to E_1. The ubiquitin derivative inhibits conjugate formation catalyzed by E_1 by binding at the noncovalent site with no conjugation to the thiol residue of E_1 can occur because the inhibitor is not hydrolyzable. The subsequent reaction pathway will then be terminated.

Concerted Reactions

More typically, modifying enzymes use a direct on-line transfer reaction pathway. Included in this group are protein kinases and ADP-ribosyltransferases. Choleragen catalyzed ADP-ribosylation has been studied with a combination of kinetic and chemical approaches. Because β-NAD+ is specifically utilized in this modification, analysis of the product stereochemistry should reveal information about the transfer mechanism. Examination of the stereospecificity of choleragen catalyzed ADP-ribosyl transfer implicates a direct transfer reaction.[53] Additional explanation of the role of the substrate is provided in Chapter 3. Recent kinetic studies[54] have clarified earlier reports[55,56] and have implicated a random sequential kinetic mechanism for the ADP-ribosylation reaction. Such a kinetic mechanism is consistent with the proposed chemistry because all of the reactive species are collected on the enzyme without the need for formation of a covalent intermediate.

Protein phosphorylation reactions also use direct on-line transfer. Detailed analysis of the reaction mechanism for the cAMP-dependent protein kinase[57,58] has been done with a combination of kinetic, chemical, and physical probes. The spatial relationship of a peptide substrate, Leu-Arg-Arg-Ala-Ser-Leu-Gly, to an inert Cr+3 analog of ATP has been elucidated with nuclear magnetic resonance spectroscopy;[57,58,59] Cr+3 serves as a paramagnetic reporting group (see Chapter 3). The γ-phosphoryl group of ATP is subject to nucleophilic attack by the proximal hydroxy amino acid, the target residue (see Figure 2.6). An electrophile may also be involved, but has not been identified. Kinetic study[58] of the pH dependence implicates enzymic with pK_as of 6.2 and 8.5 in the binding of the same peptide. The 6.5 group seems to serve as a catalytic base and may be an acidic residue identified by chemical modification studies.[60] Indeed, site directed mutagenesis of the catalytic subunit have implicated three glutamatic acid residues in catalysis.[61] Replacement of

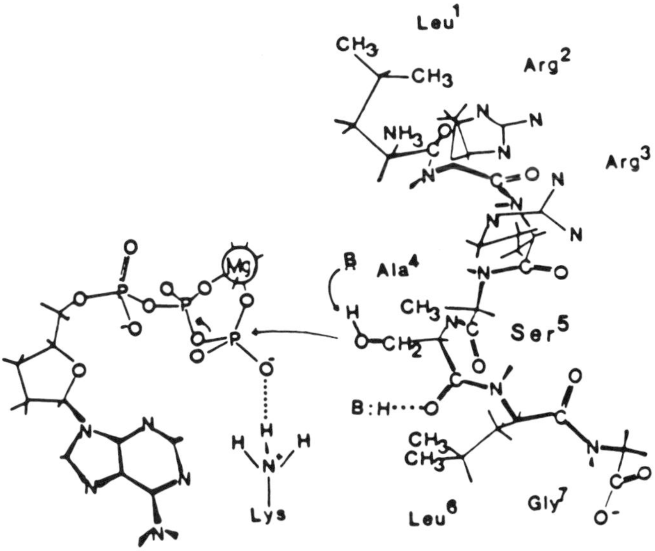

Figure 2.6. Chemical mechanism for the cAMP dependent protein kinase for the phosphorylation of a peptide substrate, kemptide. Shown in the figure is the nucleophilic attack of the serine hydroxyl on the terminal phosphate group of the donor ATP. Note the orientation of the e-amino group of the lysyl residue and the divalent magnesium ion to the phosphate groups of ATP. Reprinted with permission from *Biochemistry* 26:4118 (1987). Copyright 1987, American Chemical Society.

each of these glutamyl groups dramatically decreased kinase activity with a peptide substrate. Interestingly, alterations in the peptide substrate sequence demonstrated a clear requirement for basic residues (arginyl groups) for efficient phosphorylation.[62,63,64] Corresponding experiments with mutant enzyme and peptide substrate analogs suggest that specific electrostatic interactions occur between the glutamyl residues of the enzyme and the arginyl groups of the substrate.[61] The specific interaction of the protein kinase inhibitor protein (PKI) also was shown to be partially influenced by arginyl groups.[65]

Labeling studies have also identified a lysine residue critical for nucleotide binding,[66] typical for kinase nucleotide binding domains. A direct role for this lysine has not been assigned, but the positive charge may neutralize the phosphoryl group of the ATP. This lysine group is conserved in all protein kinases that have been sequenced, regardless if kinase is specific for serine/threonine or tyrosine residues.

Solution of the structure of the cAMP-dependent protein kinase catalytic subunit crystallized with an inhibitor peptide (PKI 5-24) has been accomplished[67] (Figure 2.7A). The solved structure supports the data obtained using chemical modifications. The invariant residues, Lys_{72}, Glu_{91}, and Asp_{184} identified as critical residues from chemical labeling, were shown to be in close proximity (Figure 2.7B). Evidence for a second triad including Asp_{166}, Asn_{171}, and Asp_{184} was also established. Asp_{166} is directed toward an alanine residue in the inhibitor peptide. The position of this alanine corresponds to the serine substrate site with Asp_{166} as a good candidate to function as a general base. The conserved Asp_{184} is involved in two triads. The position of Asp_{184} reduces the possibility that this residue functions as a catalytic base. The authors of this study suggest another possibility: Asp_{184} is involved in the chelation of Mg^{+2}. Such a role suggests that the orientation of Asp_{184} may change subsequent to the binding of MgATP. Reorientation of Asp_{184} may enable maximization of the nucleophilic potential of the serine hydroxyl for the transferase reaction. An update of the structure of the protein in the presence of MgATP has now been reported.[68] This structure confirmed that Asp_{184} interacts with the metal ion. Asn_{171} was shown to be involved in anchoring the γ-phosphate of MgATP and Asp_{166} was confirmed as the catalytic base for the transfer reaction.

As with protein kinases, a protein phosphatase, calcineurin, has also been proposed[69,70] to use direct transfer in the transfer of a phosphoryl group from substrate to water. Although not definitive, no evidence currently exists to support an covalent enzyme intermediate. A series of low molecular weight substrates has been exploited to probe the chemical mechanism of calcineurin.[69] Kinetic results indicate a dependence on substrate structure that can be used to evaluate β_{LG}, the Bronsted constant for the leaving group. Comparison of the β_{LG} value for the enzyme reaction with values for model chemical reactions provides insight into the chemical mechanism. A reaction pathway was proposed in which the enzyme supplies electrophilic assistance to the direct nucleophilic attack on the phosphate ester bond.

Not all protein phosphatases utilize a direct on-line transfer mechanism. A different protein phosphatase, a tyrosine phosphatase, has been shown to form a covalent phosphoryl-enzyme intermediate as an intermediate during catalysis[71] occurring as a phosphocysteine group. By using site directed mutagenesis, the cysteine residue was changed to serine with complete loss of enzyme activity. It seems that serine was not able to function as the appropriate nucleophile in this reaction. Hydrolysis of a radiolabeled substrate yielded a radiolabeled protein containing a chemical bond with properties similar to that of butylthiophosphate. Use of NMR spectroscopy has confirmed the presence of a phosphocysteine

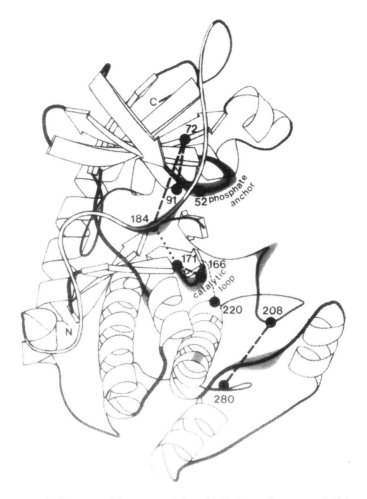

Figure 2.7A. Structural features of the cAMP dependent protein kinase. An overview of the structure illustrating the conserved phosphate binding regions (ATP binding) and the conserved catalytic loop. Other aspects are discussed in the text. Reprinted with permission from *Science* 253:414 (1991). Copyright 1991 by the AAAS.

residue for a protein tyrosine phosphatase.[72]

Transient Intermediate

In addition to the discussed mechanisms, there are also modifications which proceed with the formation of a transient free radical intermediate. For example, the primary intermediate in the peroxide dependent

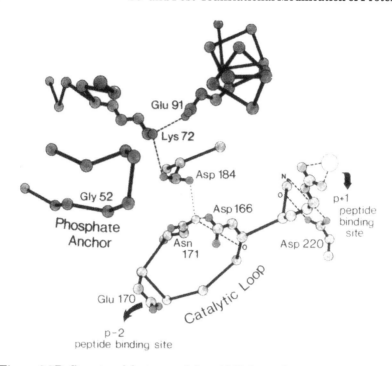

Figure 2.7B. Structural features of the cAMP dependent protein kinase. An expanded view of the active site region showing the relative orientation of functional residues. Note the potential for Asp-184 to participate in two triad units. Other aspects are discussed in the text. Reprinted with permission from *Science* 253:414 (1991). Copyright 1991 by the AAAS.

iodination reaction has been postulated to be a free radical with additional free radical intermediates present in the catalytic scheme.[73,74] Interestingly, it has been speculated that thyroid peroxidase utilizes both two electron and single electron chemistry for the oxidation of diiodotyrosine.[74] Another modification for which mechanistic information remains unclear is vitamin K dependent carboxylation.

The mechanism of vitamin K dependent carboxylation of glutamate residues has remained elusive.[75] Coupled to glutamyl carboxylation is the generation of vitamin K epoxide catalyzed by a vitamin K epoxidase.[75,76] The epoxidase shares the same cellular localization as the carboxylase activity. Vitamin K epoxide has been postulated as a reaction intermediate because (1) peroxide has been shown to block carboxylation;

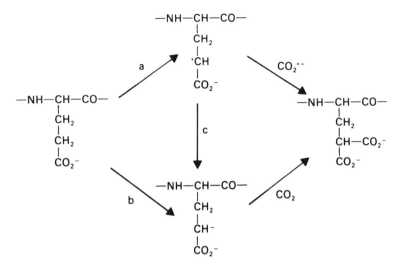

Figure 2.8. Alternate mechanisms for glutamyl carboxylation. Depicted are (a) free radical pathway; (b) carbon ion pathway. Also possible is the conversion of a free radical to a carbon ion (step c). Supporting data for each mechanism is presented in the text. Much data available is consistent with either mechanism. Reprinted with permission from *The Biochemical Journal* 266:749 (1990). Copyright 1990, The Biochemical Society and Portland Press.

(2) chemical oxidation of vitamin K also blocks carboxylation; and (3) hydroxyvitamin K has not been identified. These observations suggest that vitamin K epoxidation precedes glutamyl carboxylation. The involvement of peroxide and epoxide intermediates suggests free radical intermediates at least in the utilization of vitamin K. Moreover, carboxylation proceeds readily in the absence of molecular oxygen if vitamin K hydroquinone is used instead of the quinone form.[77] Tritium exchange reactions[78,79] and studies with stereospecific fluorinated glutamate acceptors[80] clearly demonstrate the generation of a carbanion intermediate of the target residue. As discussed by Dowd, et al., formation of a carbanion intermediate requires the action of a sufficiently strong base to abstract the g-proton of glutamate.[81] Initial formation of a free radical intermediate before yielding a carbanion cannot be excluded and, in fact, has been hypothesized.[75] Figure 2.8 illustrates a reaction scheme incorporating both anion and free radical intermediates.[75,82] Dowd et al.[81] infer, conversely, that vitamin K dependent carboxylation probably does

proceed via a carbanion intermediate. They have proposed a reaction scheme that posits that vitamin K has the potential for base strength amplification provided by the predicted exothermic formation of vitamin K epoxide in a model spontaneous reaction. A thermodynamic study has confirmed that epoxide formation can provide sufficient energy for subsequent amplification of the base.[83] An investigation[84] using isotopic oxygen ($^{18}O_2$) provided data consistent with the model of Dowd and associates. The exact reaction pathway, however, remains to be confirmed.

Both iodination and carboxylation show characteristics supportive of a free radical intermediate, but do not completely preclude a carbanion intermediate. A free radical intermediate is indicated in the amidation of peptides. Amidation of neuropeptides occurs exclusively at carboxyl terminal glycine residues.[85] The reaction is catalyzed by the peptide glycine α-amidating monooxygenase complex.[85,86] Interestingly, the reaction is a two-step process[87,88,89] with a hydroxylated glycine intermediate synthesized by a specific enzyme, a monooxygenase. Hydroxylation occurs at the α-carbon of the terminal glycine and is subsequently dealkylated[90,91,92] to yield the final amidated product. The α-amino group from the terminal glycine provides the amide function for the product peptide. Although the hydroxylated glycine is labile to alkali,[88] there is a specific enzyme that catalyzes the dealkylation reaction, peptidylamidoglycolate lyase.[91,92] Both enzymes are generated from a single precursor[93,94] providing assurance that the reaction pathway can be modulated as a single process. These two enzymes together form the monooxygenase complex.

The amidation reaction requires both Cu^{+2} and ascorbic acid and shares catalytic properties[85,95] with dopamine hydroxylase, also a monooxygenase. The ascorbate requirement seems to be necessary for only the initial hydroxylation of the α-carbon of glycine because subsequent dealkylation of a hydroxylglycine peptide no longer demonstrates this requirement.[88–92,95] Indeed, previously hydroxylated substrate can yield the amidated product in the absence of ascorbate and copper.[87,88] Consistent with these data, results with a model chemical reaction suggest that Cu^{+2} coordination to the amido nitrogen may be involved in the initial dehydrogenation reaction producing the substrate for hydroxylation.[96]

The enzymatic reaction mechanism shows many characteristics of a free radical mechanism. First, there are many similarities to the reaction mechanism of dopamine hydroxylase which generates the semidehydro-ascorbate free radical[97,98,99] directly identified by EPR spectroscopy.[94]

$$\text{Ascorbate} + O_2 \xrightarrow[\text{Cu(II)}]{} \text{dehydroascorbate} + H_2O_2 \quad \text{(NET)} \quad (1)$$

$$\text{Ascorbate} + \text{Cu(II):peptide} \longrightarrow$$
$$\text{semidehydroascorbate} + \text{Cu(I):peptide} \quad (2)$$

$$\text{Cu(I):peptide} + H_2O_2 \rightarrow \text{Cu(++):peptide} + \bullet OH + {}^-OH \quad (3)$$

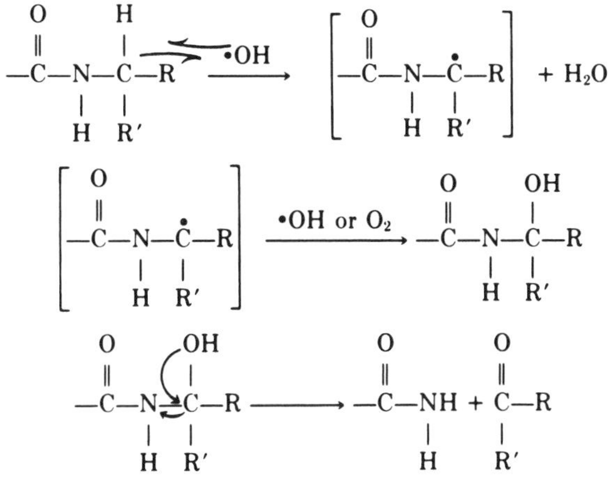

(4)

(5)

(6)

Figure 2.9. Model of free radical mechanism for peptide α-amidation. Shown is a reaction sheme proposed for the nonenzymatic amidation reaction. Although it does not illustrate the function of the enzyme, this scheme does depict the chemistry involved in the free radical mechanism proposed for the enzymatically catalyzed reaction. Reprinted with permission from *Journal of Biological Chemistry* 260:9088 (1985). Copyright 1985, American Society for Biochemistry and Molecular Biology.

Indeed, the amidation reaction was shown to generate semihydroascorbate during catalysis confirming the classification as a monooxygenase.[100] Second, primary isotope effects for the enzymatic amidation indicate that breakage of a C-H bond is the rate limiting step consistent with the formation of a free radical intermediate.[85] Third, the chemistry of the reduction of Cu^{+2} to Cu^{+1} by ascorbate has been shown to yield semihydroascorbate.[95,101] Semihydroascorbate self-reacts to yield ascorbate and dehydroascorbate. Fourth, the single electron donor, $FeCN_4^-$, replaced ascorbate in supporting activity.[86] These observations all support a free radical mechanism (Figure 2.9) as developed for a nonenzymatic model reaction.[102] An alternate mechanism[103] was suggested utilizing a dehydration reaction with an imino acid derivative as an intermediate. This mechanism requires a dehydration reaction prior to hydrolysis. Recent studies demonstrated that O_2, not H_2O, was the source of ^{18}O incorporated into the product.[104]

REGULATORY FEATURES OF MODIFYING ENZYMES

Compartmentalization

The specific localization of individual modifying enzymes defines many of the options available for reactivity. Obviously, the subcellular localization of an enzyme reflects its role in cellular regulation. Protein kinases and protein phosphatases are present in both the cytosol and nucleus and are actively involved in the regulation of metabolism and cell division. Additional levels of regulation are possible dependent upon the relationship of specific enzymes located within the same cellular organelles. Protein synthesis, particularly of proteins destined for the membrane or for secretion, is extensively regulated by modification reactions. Among the modifications are N- and O-linked glycosylation, fatty acylation, peptide amidation, and tyrosine sulfation. Not all of these reactions will occur in a representative protein. For example, fatty acylation and N-glycosylation seem to be independent events[105] as are glycosylation and tyrosine sulfation.[106]

Acetylation

The acetylation of histones is an interesting case of compartmentalized regulation. In contrast to the side chain lysine modification, which is predominantly nuclear,[107] the acetylation of the amino terminus of histones is a cytoplasmic modification.[108] Additionally, there is evidence for the temporal regulation of the different acetylation reaction types, with amino terminal acetylation being an early event during histone synthesis.[109]

Fatty Acylation

A variety of acyl transfer reactions, including myristoylation, palmitoylation, and isoprenylation have been identified.[110,111,112,113] Interestingly, these modifications are temporally and spatially distinct.[114,115,116,117] Myristoylation seems to be an early co-translational event and is not correlated with the migration of proteins to the membrane. Palmitoylation, in contrast, is a distinct post-translational reac-

tion evident only in membrane proteins. In fact, palmitoylation may be catalyzed at the membrane. These properties suggest that different populations[116,117,118] of proteins are substrates for myristoylation and palmitoylation. Estimates of turnover rates for these modifications confirm that different populations do exist.[117]
Partial purification of these acyltransferases has demonstrated that myristoylation can occur on tRNA bound polypeptides[119] implicating the direct association of the transferase with ribosomal structures. Palmitoylation is reportedly localized to within the ER-Golgi secretory structure[118,119,120] although it is suggested to be exposed to the cytosol. Not unexpectedly, cycloheximide has a more dramatic effect on myristoylation than on palmitoylation.[121] Evidence suggests no coupling of palmitoylation with protein synthesis, but a long term influence of cycloheximide on palmitoylation is possible because this modification is a dynamic process with turnover of the acyl group at a greater rate than of protein.[122] Clearly compartmentalization of enzymes catalyzing related reactions functions to discriminate substrates and to provide additional control to the reaction.

Glycosylation

The initial step in protein glycosylation, the primary modification of protein by the transfer from dolichol linked carbohydrate, occurs in the endoplasmic reticulum (RER). The initial modification event[123] and subsequent processing of the carbohydrate tree all happen in the RER. Affinity labeling of the oligosaccharyl transferase using peptide substrate analogs[124] showed that the enzyme is oriented both toward the lumen of the RER and toward the cytosol. The synthesis of the dolichol donor structure has been shown to be external to the RER.[123,125] The site of synthesis of the lipid-linked carbohydrate is postulated to be the cytosolic side of the ER membrane. Access of the transferase to its target protein or the carbohydrate donor is separate. The distinct sites of substrate and donor availability provide an additional mechanism for controlling glycosylation; accessibility to two compartments is modulated.

Phosphorylation

A novel system for controlling the regional compartmentalization of the cAMP dependent protein kinase has been proposed based on the identification of specific binding proteins for the type II cAMP dependent protein kinase, more specifically the RII regulatory subunit.[126] Designated A-Kinase Anchoring Proteins, or AKAPs, these proteins contain an amphipathic helix which may serve as the binding site for the RII subunit. These AKAPs seemingly are specific for the type II regulatory subunit and do not interact with the regulatory subunit of the type I cAMP dependent protein kinase. Based on gel shift binding studies and compeptition experiments, a 23 residue peptide has been shown to comprise the minimum sequence necessary for the interaction of the RII subunit.[127] The peptide identified has an elevated level of acidic residues.[127] Mutagenesis of arginine residues in the amino terminal region of the RII subunit has established that the acidic residues directly interact with the target arginine residues in the RII subunit.[128] Thus, the specific interaction of the protein kinase with these anchoring proteins may serve as a convenient cellular mechanism for the localization of the kinase. The specificity for the type-II kinase illustrates the potential for different regulatory pathways specific to each class of the cAMP dependent protein kinase. An important research area, then, is distinguishing the role of the two classes of cAMP dependent kinases. Which member functions in a particular cellular control process? How are they coordinately regulated? Because the two classes share the same catalytic subunit, how are the regulatory subunits regulated?

Availability of Substrates and Cofactors

Ligand Control

A ligand can influence an enzymatic reaction with a protein by binding (1) to the substrate, (2) to the enzyme, or (3) to both. By doing kinetic studies with the natural protein and an alternative substrate that does not bind the ligand—for example, a peptide, an amino acid derivative, or a molecule containing related functional groups to the reactive site in the protein—evaluation of rate data obtained in the presence and the absence of ligand allows the first two cases to be readily distinguished. The case in which the modifier binds to both is the most complex but may be treated by kinetic analysis.[129] If the reaction is influenced by ligand-binding to the protein substrate, further kinetic studies at different concentrations of substrate may be done to define how the ligand influences the reaction.

Consider the action of AMP on the dephosphorylation of phosphorylase *a*. AMP inhibits the reaction with the natural protein but has no effect on reactions with a small phosphorylated peptide, which suggests that the AMP-phosphorylase *a* complex is not recognized by the phosphatase.[122] X-ray analysis of phosphorylase *a* in the presence of AMP supports this view and shows that the bound ligand induces a conformational change in the protein shielding the seryl phosphate so that it is not available for binding to the phosphatase.[130,131]

Substrate Competition

The competition of various enzymes for required substrates and cofactors involved in the reactions is another mechanism by which these reactions are modulated. Protein synthesis again is a useful example. As noted, a number of modifications are found in newly synthesized proteins. Some proteins contain a single modification; others are mutiply modified. For example, a subpopulation of palmitoylated proteins are also glycosylated,[132] but there is no evidence for the concurrence of glycosylation and myristoylation. Even the overlap of palmitoylation and glycosylation may not be functionally significant because the two processes seem to be independent.[132,133] Regulation may also be simply caused by the availability of metabolites. The isoprenylation of proteins can be modulated by changes in mevalonate levels induced by an inhibitor of mevalonate synthesis.[113]

Methylation. The methylation of proteins can be classified by the target residue modified.[134,135,136,137,138] Enzymes specific for the methylation of arginine,[134,135] lysine,[136] histidine,[137] or carboxyl[138,139] residues have been isolated and characterized. All methyltransferases seem to have quite narrow substrate specificity, both for residue type and for the small number of physiological substrates identified.[140,141,142,143,144] One property that is shared is the absolute requirement for S-adenosyl-methionine as the donor molecule. Because there is overlap in the tissue distribution of the methyltransferases, competition for donor could be an important determinant of the activity of each enzyme. Interestingly, all methyltransferases have K_m values for SAM in the range 2–10 μM.[138,145,146,147,148,149] The relative amounts of each type of methylation will, therefore, depend less on SAM levels than on other properties of the enzymes, although the availability of the donor will be crucial.

Glycosylation. As stated, the acceptor and the donor for N-linked glycosylation are synthesized on opposite sides of the ER membrane.[124,125] The role of the separation is extended by the competition of various activities for the common pool of dolichol phosphate.[150] These enzymes all utilize dolichol phosphate to synthesize different sugar dolichol adducts. The distribution of the different adducts can modulate the level of glycosylation by altering the population of donor molecules. This is a different degree of competition for donor than described for methylation. The methyltransferases compete for donor molecules, but in all cases the substrate acceptor is a protein. Competition for dolichol derivatives is between enzymes with quite different acceptor substrates. The recognition requirements may be quite different.

Table 2.2. Hydroxylase K_m for 2–oxoglutarate

Hydroxylase	K_m (mM)
Pro-4	0.020
Pro-3	0.005
Lys	0.100
Asp	0.005

Hydroxylation. Hydroxylases[151,152,153,154] specific for different amino acid residues (proline, lysine, and aspartic acid) share a requirement for 2–oxoglutarate, ascorbate, and Fe^{+2}. Proline (both Pro-4-hydroxylase and Pro-3-hydroxylase) and lysine hydroxylases also share the same protein substrate, collagen.[151,152,153] These enzymes do have different requirements for 2–oxoglutarate as a cofactor.[151,152,153] In particular, lysine hydroxylase has a much higher K_m for this cofactor than do the other hydroxylases (see Table 2.2). These data predict that proline hydroxylation will predominate over lysine hydroxylation. Indeed, proline hydroxylation of collagen is more abundant than lysine hydroxylation.[155]

Each of these hydroxylases also requires ascorbate, which is consumed in the absence of the protein substrate albeit at substoichiometric amounts.[156] The vitamin seems to be required for reactivation of the hydroxylase.[157] As such, the net level of ascorbate available is crucial for maintaining the active state of the hydroxylase. Physiological levels of ascorbate may be modulated by the demand for the vitamin as a cofactor in peptide amidation.[85] Although the tissue distribution of these processes is distinct, there must be some degree of interdependence because of the demand for ascorbate in both reactions. In addition to proline and lysine residues, β-hydroxylaspartic acid is another product of hydroxylation;[154] asparagine modification has also been implicated.[158] Aspartyl modification has been isolated in some vitamin K dependent proteins involved in blood clotting.[159,160,161,162] Unlike the other hydroxylation reactions, aspartyl hydroxylation requires only 2–oxoglutarate and Fe^{+2} as cofactors; no involvement of ascorbate has been indicated. The absence of an ascorbate requirement suggests that there is a fundamental difference in the mechanisms for the various hydroxylations.

REGULATORY PROTEINS

As already discussed, a number of cofactors are involved in the post-translational modification of proteins. Additionally, some of these enzymes are modulated by other proteins, either bound enzyme subunits or specific inhibitor or activator proteins. Some of these activities may be monomeric,[163,164,165] or multimers of a single subunit type,[158,166,167] or multimers of different protein subunits.[168,169] Surprisingly little is currently known about the macromolecular state of many of the enzymes catalyzing these modifications. One particular example is the glycosyltransferase, which catalyzes the initial modification event in the N-glycosylation of proteins. An oligoglucosyl binding protein with molecular weight 57,000 Da was identified with a radiolabeled substrate analog, but recent evidence seems to refute a direct role for this protein in protein glycosylation.[170] Its role and the identity of a specific glycosyltransferase remain unclear.

Subunits

The diverse requirements of bacterial toxins require a complex of multiple proteins, each with a specific biological role. The holotoxin of cholera toxin has two subunits,[171,172] that of pertussis toxin six subunits of five proteins.[35] The complex of these subunits results in the complete function of the toxin. Similarly, eukaryotic enzymes often are composed of multiple subunits of distinct function. Proline-4-hydroxylase has two subunit types comprising a tetramer.[168,169]

The holomeric form of protein phosphatase-2A is composed of three subunits,[173,174] designated A (M_r 63,000 Da), B (M_r 55,000 Da), and C (M_r 38,000 Da). The C-subunit is the catalytic moiety and the other two serve regulatory functions. Interestingly, this enzyme can be purified as the ABC heterotrimer, the AC heterodimer, or the C subunit alone. Experiments examining various recombinations of the subunits clearly established that the regulatory subunits influence the substrate specificity of the catalytic subunit.[173,174,175,176] The A subunit has the peculiar property of stimulating the activity with some substrates and inhibiting activity with other substrates.[176] This characteristic is demonstrated in Figure 2.10 for the dephosphorylation of eIF-2a and 40S ribosomes. A conformational change induced in the catalytic subunit must modulate these interactions. Because the substrate specificity determinants for this, or any other, phosphatase are not known, the differential effects of the A subunit may reflect distinct subsets of substrates that interact differently with the AC heterodimer. The regions of the A and C subunits involved in these interactions, however, have not been identified. Possibly, there are unique, but distinct, substrate recognition regions on the catalytic subunit that overlap. Association of the A subunit may differentially modulate these two substrate sites. Regulating the interaction of noncatalytic subunits may be an important process.

Discrete Regulatory Proteins

Additional regulation may be achieved by nonsubunit, auxiliary proteins. Activator and inhibitor proteins are known. A general example is that of calmodulin, which is required for the activation of both a specific protein phosphatase[177,178] and a specific protein kinase.[179] The coordinate regulation of these opposing enzymes is an important, albeit poorly understood, signaling process. Another intriguing example is the putative heterodimer of protein phosphatase-1.[180,181] The complex of the PP-1 catalytic subunit with a regulatory protein, inhibitor-2, is postulated to be regulated by a cycle of phosphorylation-dephosphorylation.[181,182,183]

In contrast to calmodulin, which has multiple targets, there are activator proteins specific to a distinct enzyme or reaction. ADP-ribosylation of low molecular weight G-proteins by bacterial toxins is enhanced by an associated protein, the ADP-ribosylation factor or ARF.[184,185,186,187,188] ARF is itself a low molecular weight G-protein. In fact,

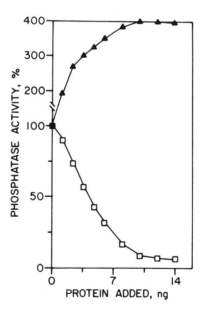

Figure 2.10. Influence of the A subunit (60 kDa) of protein phosphatase 2A on substrate selectivity. With isolated eIF-2 (closed symbols, upper curve), the addition of the regulatory A subunit enhances the phosphatase activity of the free catalytic subunit. With intact 40S ribosomal particles, the A subunit inhibits the activity of the free catalytic subunit. Reprinted with permission from *Journal of Biological Chemistry* 264:7267 (1989). Copyright 1989, American Society for Biochemistry and Molecular Biology.

currently there is evidence for four distinct proteins[178,179,180] in the ARF family with both soluble and membrane bound forms. All have guanine nucleotide binding activity.[189] Distinct forms have been described for the activation of cholera[184,165,186,187] and botulinum C3[188] toxins. The effect of ARF on cholera toxin-catalyzed modifications is complex. ARF and cholera toxin have been reported to form a set of complexes dependent upon the conditions. Sodium dodecylsulfate promotes the formation of ARF-CT complex, which has activity with a number of protein substrates, including auto-ADP-ribosylation. In contrast, deoxycholate prevents the formation of these complexes; instead an aggregate of ARF is formed, with a different influence on the ADP-ribosyltransferase activity of cholera toxin. The effect of ARF, therefore, is dependent upon the nature of the complex species formed in solution.[189,190]

A similarly specific activator protein has recently been reported for peptide amidation.[191,] This protein has been named stimulator of peptidyl glycine α-amidating monooxygenase, or SPAM. Although not well characterized, SPAM seems to activate PAM by both increasing V_{max} and decreasing K_m. The activity of SPAM is reported to require Cu^{+2}, as does the amidating enzyme, but no information is available suggesting that SPAM has catalytic activity. It is possible that SPAM functions to stimulate the oxidation reaction involving Cu^{+2} and ascorbate. An alternative role for SPAM is the metabolism (recycling) of semidehydroascorbate; semidehydroascorbate reductase has been shown to enhance the activity of dopamine β-monooxygenase.[192] In addition to sharing cofactors, PAM and SPAM are colocalized to the secretory pathway in rat atrium.[184]

Specific inhibitor proteins have also been identified. Only type-1 protein phosphatase of the cytosolic protein phosphatases is sensitive to inhibition by the proteins designated phosphatase inhibitor-1 and inhibitor-2.[193,194,195] The phosphatases specific for the mitochondrial pyruvate and α-keto acid dehdrogenase complexes are similarly modulated by unique inhibitor proteins.[196,197] Protein kinases are subject to regulation by specific inhibitors. The best characterized is the inhibitor protein of the cAMP-dependent protein kinase, designated protein kinase inhibitor, PKI.[198,199] Inhibitors for other modification reactions will be discussed in a later chapter.

The functional region of PKI has been defined by proteolytic digestion of the protein yielding an active fragment comprised of residues 5 to 24; subsequent analysis showed that residues 6 to 22 are the minimum required for inhibitor activity.[200,201,202] The active fragment has the sequence: Thr-Tyr-Ala-Asp-Phe-Ile-Ala-Ser-Gly-Arg-Thr-Gly-Arg-Arg-Asn-Ala-Ile-NH_2. Initial review of the sequence of this peptide suggested that the region between Arg_{15} and Arg_{19} would be important determinants analogous to studies with peptides as substrates.[200] These substrates have an absolute requirement for a cluster of basic residues on the amino terminal side of the serine that is phosphorylated.[62,63,64] Structure-function analysis[200,201,202] with synthetic analogs of the PKI fragment demonstrated a only partial requirement for this basic region (Arg_{15} to Arg_{19}). Secondary structure analysis of the different analogs indicated that the amino terminal ten residues formed an α-helical region that was necessary for good potency as an inhibitor.[201,202] Moreover, phenylalanine at position-10 proved to be the most crucial residue and required the α-helical region for the proper orientation relative to the kinase active site.[202] This phenylalanine residue could only be replaced by aromatic residues for retention of potent inhibitory activity. Interestingly, there is some evidence for participation of the tyrosine residue.[201] Removal of the peptide segment from the amino terminus through Tyr_7 causes a fivefold loss in inhibitor potency; changes in surrounding residues cause a two to threefold loss in activity.[201,202] Moreover, phosphorylation of this residue by the epidermal growth factor receptor kinase abolished inhibitory activity.[203] The hydroxyl group of the tyrosine residue may mimic the serine hydroxyl to help the binding interaction. Recent crystallographic analysis of the kinase-PKI complex reveal that the arginine residues interact with glutamyl residues on the kinase catalytic subunit.[204] Glutamyl residues in the active site of the kinase have been implicated from site directed mutagenesis.[61] These results are also consistent with the model proposed for the interaction of peptide substrates with the catalytic subunit.[57,59,61]

Previously discussed, the AKAPs also function to regulate the function of the cAMP dependent protein kinase.[126,127] There is no reported effect on the activity of the kinase by these proteins. Instead, these proteins serve to anchor the kinase and may specifically target the protein to specific regions within the cellular framework. The effect of targeting on the activity and substrate specificity of the kinase has not been established. This field of research may provide critical insights into the cellular control of enzymes involved in signal transduction.

Macromolecular Complexes

In addition to individual regulatory proteins functioning as discrete entities, modification reactions may also be regulated as part of a macromolecular complex. Lysine acetylation of histone proteins is related to transcriptional competence,[205,206] and can be enhanced by DNase treatment of the chromosomal material. Clearly, the structure of DNA in the nucleosomal material modulates protein acetylation. In fact, certain acetylases are only reactive with histone proteins; chromatin is resistant. Similarly, the extent of acetylation may influence the structure adopted by the chromatin material. It seems that the topological structure of chromatin may be related to the level of histone acetylation.[207] A hypothesis has been presented that the removal of the acetyl groups may be the crucial regulatory step in the organization of chromatin structure.[208] In this model, the deacetylase enzyme serves as the regulatory activity for controlling transcriptional competence.

The specific structural form of chromatin also influences poly(ADP)-ribosylation of histone proteins reflecting the role of modification in regulating DNA replication and repair.[209,210,211] The extent of modification, defined as polymer length, is differentially modulated by the different histone proteins.[209] The relative composition of histone proteins in the chromatin, therefore, will define the level of poly(ADP)-ribosylation. ADP-ribosylation of histone proteins is increased in previously acetylated chromatin protein.[212] Enhancement of acetylation by treatment with n-butyrate also causes an increase in the level of ADP-ribosylation.

The DNA component of chromatin can also enhance the level of ADP-ribosylation.[213] The optimal DNA requirement is not clear, but a small segment (octamer) seems to stimulate activity of the polymerase more greatly than a larger sequence of activated DNA, which is a heterogeneous mixture of DNA fragments. Polymeric ADP-ribosylation is induced by DNA strand breakage. The increased activation of the polymerase by a DNA fragment with a defined structure (the octamer has blunt dephosphorylated ends) may reflect the actual process that causes the induction of activity. For the example of the octamer, the absence of the phosphate group may be crucial; mono-ADP-ribosylation is greatly influenced by the presence of a phosphoryl group near the ADP-ribosylation site.

So-called signal transduction particles represent another situation in which protein modification is modulated by the aggregation of diverse proteins. Transmission of growth factor stimulation, particularly by epidermal growth factor and platelet-derived growth factor, involves the modulation of a number of enzyme activities[214,215,216,217] including growth factor autophosphorylation.[218,219] Many of the target enzymes are found associated in signal transduction particles. This collection of associated proteins may serve as a paradigm for a phosphorylation cascade following activation of the growth factor receptor/ kinase. These receptor/ kinases are known to phosphorylate and activate a series of serine/ threonine protein kinases.[220,221,222,223 224,225] The signal transduction particle is mimicked by the association of various enzymes with middle T antigen of polyoma virus.[226,227,228] Immunoprecipitation of middle T antigen specifically coprecipitates pp60[c-src,226] phosphatidylinositol-3-kinase,[227] and the heterodimer of protein phosphatase-2A.[228,229]

AUTOREGULATION

Besides regulatory mechanisms based on substrate and cosubstrate availability or exogenous factors, enzymes may be self-regulated by structural features, including the catalysis of self-modification.

Autoinhibitor Domain

A novel regulatory structure has been identified that is comprised of a specific primary sequence in the enzyme. These novel sequences have been found primarily in protein kinases and mimic the sequence of the kinase phosphorylation site. Because of this similarity these sequences inhibit activity by interacting with the active site, thereby preventing access of the substrate. These regions have been designated autoinhibitory domains,[230,231] or because they resemble phosphorylation sites, pseudosubstrate domains.[232] The inhibitory role of these autoinhibitor domains has been corroborated using synthetic peptides based on the identified sequence.[233,234,235] Essentially, these domains block catalytic activity by associating and masking the substrate binding region. Elevating the substrate can relieve the autoregulatory process.

For some of the kinases demonstrated to have an autoinhibitory domain, the domain is in close proximity to the kinase autophosphorylation site.[230,231,232,233] No evidence is available, however, directly addressing the relationship between autophosphorylation and the interaction of this site with the catalytic site. Hypothetically, autophosphorylation may modulate the effect of the autoinhibitor domain. Phosphorylation can have significant effects on the local conformation of sites in proteins as for the phosphorylated region in glycogen phosphorylase. Crystallographic analyses of phosphorylase b and phosphorylase a reveal significant changes in the local structure.[236,237] Clearly, with autoinhibitor domains near the termini of the kinases, it would not be unexpected that structural changes could be induced.

SH2 Domains

A novel type of autoregulatory structure is the phosphotyrosine binding sequences present in a variety of proteins. First identified in pp60[v-src], these regions have been designated Src Homology (SH) regions domain 2 or domain 3.[238] Subsequently, these noncatalytic regions were found in other oncogene nonreceptor tyrosine kinases of the src-family. Domain 2 (SH2) is of particular interest because it is present in proteins contained in signal transduction particles bound to growth factor receptors. These cytoplasmic proteins included phospholipase C,[239,240] GTPase activating protein (GAP),[239,240,241,242] and phosphatidylinositol-3-kinase.[243]

It seems the SH2 domain is important for the association with proteins that contain phosphotyrosine. Interestingly, many of these proteins are subsequently phosphorylated at a tyrosine residue.

This property for binding to regions containing phosphotyrosine implicates a possible role for SH2 domains in the regulation of tyrosine kinase activity. The cellular homolog, pp60[c-src], is activated following changes in the SH2 domain or by dephosphorylation of a specific tyrosine residue, Tyr$_{527}$.[244,245] Because SH2 domains associate with phosphotyrosine regions,[238] the SH2 domain in pp60[c-src] may serve to prevent the dephosphorylation of Tyr$_{527}$ by blocking the accessibility of a tyrosine phosphatase to this site. Blocking pp60[c-src] dephosphorylation will maintain the kinase in an inactive state. A corollary function is the enhanced tyrosine phosphorylation of cellular proteins as pp60[c-src] is activated and associated proteins become bound.

Further evidence for this type of role for SH2 domains was obtained from studies of the *crk* oncogene product, p47[gag-crk].[246,247,248] The crk gene product has an abundance of SH2–like sequences, which are crucial for its transformation activity.[246] Expression of p47[gag-crk] increases the level of phosphotyrosine in cellular proteins consistent with the apparent inhibition of tyrosine phosphatase activity. The increase in phosphotyrosine does not seem to be because of any tyrosine kinase activity associated with the *crk* oncogene product.

In this model, SH2 domains function as a unique autoinhibitory sequence with a mechanism different from the "pseudosubstrate" autoinhibitor sequences present in serine kinases. There is evidence, however, that a portion of the SH2 domain functions more like the typical autoinhibitory domain.[247] A peptide derived from the SH2 domain of pp60[c-src] was shown to modulate the activity of the enzyme. The peptide inhibited the phosphotransferase activity of pp60[c-src], but was shown to be dispensable with no loss of activity when blocked by a peptide specific antibody. Because the sequence incorporated the determinants for autophosphorylation, the authors speculated that this autoinhibitory region may directly associate with the SH2 domain present in the protein.[249] Moreover, SH2 domains may also regulate tyrosine phosphatase activity by preventing access to the substrate site. A novel role for SH2 domains has been implicated for the modulation of protein tyrosine phosphatase activity.[238] Such a sequence provides a convenient mechanism for regulating the interaction between the phosphatase and phosphotyrosyl containing proteins. This mechanism is consistent with the apparent inhibition of tyrosine phosphatase by SH2 domains. In this mechanism, SH2 domains serve as up regulating signals for tyrosine phosphorylation in activated oncogenes.

Automodification

Autoinhibitor domains are segments of the native protein structure that regulate activity. Enzymes catalyzing modification reactions may themselves be modified, particularly by automodification reactions. The process may occur within a single protein molecule, in a molecular complex of defined quaternary structure, or between separate but identical molecules. Consider three cases: (1) an enzyme modifies itself in an intramolecular reaction; (2) an enzyme reacts intermolecularly in a molecular complex such as a system with quaternary structure that does not dissociate upon dilution; and (3) modification occurs between two protein molecules that do dissociate upon dilution. Case 3 may be distinguished readily by analyzing the protein concentration dependence of the autocatalytic reaction. By using the van't Hoff equation in logarithmic form, $\log V = n(\log[C]) + \log(k)$, where V equals the initial rate of modification, n is the reaction order, C is the concentration of enzyme, and k is a rate constant, distinct patterns can be identified. Case 3 predicts a linear line with a slope of 2 with concentrations below the K_m value. Cases 1 and 2 predict unitary slopes. When these considerations were applied to the autophosphorylation of cyclic AMP-dependent protein kinase, it was conclusively demonstrated that the results did not fit case 3.[250] Kinetics do not distinguish between the first two cases, but definition of the product of phosphorylation does. With cyclic AMP-dependent protein kinase, the acceptor is a regulatory protein and fits with the description of case 2. Autophosphorylation occurs intramolecularly with Ca^{+2}/calmodulin kinase[251] and protein kinase C.[252] In the latter instance, a study of kinetics in mixed micelles, which may contain one molecule of protein kinase, provided convincing evidence that the process of autophosphorylation occurred by an intrapeptide mechanism.

In principle, autocatalysis may activate, inhibit, or have no effect on the rate of the enzymatic reaction with exogenous substrates. The reaction may occur slowly or rapidly in comparison with the typical catalytic reaction. Kinetic studies help determine the consequence of autocatalysis on enzyme activity. If the reaction occurs slowly in the presence of exogenous substrates, the progress curve for formation of exogenous product may resemble one of curves illustrated in Figure 2.11. In curve A, autocatalysis causes inactivation, as indicated by the slope of the curve, which decreases to zero. In curve B, autocatalysis causes activation, and the final slope is a measure of the activity of the modified enzyme. Inhibition of Ca^{+2}/calmodulin kinase and activation of phosphorylase kinase,[253] respectively, may be explained by rate-limiting autocatalytic reactions. Changing the conditions of the reaction mixture

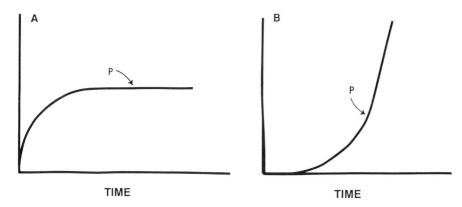

Figure 2.11. Theoretical plots for the generation of automodified enzyme. In plot A (left), automodification serves to decrease activity. In plot B (right), the automodification reaction is slow during the initial phases, but then an activated form of the enzyme is generated resulting in an enhanced automodification reaction.

helps to distinguish the possibilities. When the concentration of substrate, for example, ATP in autophosphorylation reactions is altered, the shape of the curves change. This change may provide information regarding the rates of modification and characteristics of the modified enzyme.

When modification occurs rapidly, the progress curve of the formation of exogenous product may be linear and may not reveal directly whether any autocatalytic reaction has occurred. If, however, the second product of autocatalysis is added, for example, ADP in autophosphorylation or nicotinamide in ADP-ribosylation, and if this product reverses the modification reaction, analysis of initial rates can reveal an autocatalytic event and its consequences. Consider an autocatalytic reaction that causes activation. Substrate A reacts in an autocatalytic reaction with E. The product, B, reverses the reaction. After modification, E* can react with exogenous substrate through E*A. Using equilibrium assumptions, we obtain the following rate equation. This equation predicts that a

$$\frac{1}{V} = \frac{B}{V_M K_1 K_2 (A)^2} + \frac{1}{V_M K_2 A} + \frac{1}{V_M}$$

linear double reciprocal plot will be obtained in the absence of B, but that in the presence of B, competitive inhibition will be found and a nonlinear plot will be generated. Even though the enzyme is present in catalytic amounts, a fast recycling reaction in the presence of product could cause the system to use 2 moles of the substrate, A. By analyzing rates with and without B, information about kinetic parameters and the equilibrium constant of the autocatalytic reaction may be obtained.

Effect of Automodification

The initial step in ubiquitination is the formation of a covalent conjugate of ubiquitin and the Ub-activating enzyme, E_1, via a thiolester linkage. There is now evidence that E_1 can also catalyze an autoubiquitination reaction yielding not a thiolester linkage, but a bond with properties similar to a peptide bond.[254] Autoubiquitination inhibited the typical ubiquitin activating activity of E_1. In contrast to the effect on activity as for autoubiquitination, other automodification reactions may have different influences. Substrate modification by mono-(ADP)-ribosyltransferase from hen liver nuclei serves as a template for elongation by a separate poly(ADP-ribose) synthetase.[255] In a similar reaction, the synthetase catalyzes the self-(ADP)-ribosylation as the initial priming reaction for additional automodification.[256,257] The automodification occurred at multiple sites in the protein and served to cause a decrease in activity from an increase in the K_m for NAD^+.[258]

A complex system of automodification serves to modulate the activity of the multifunctional calmodulin dependent protein kinase, also called calmodulin dependent protein kinase II. Both autophosphorylation and substrate phosphorylation are enhanced by calmodulin in a cooperative manner.[259] Interestingly, there is a bimodal effect (Figure 2.12) of the autophosphorylation induced by calmodulin.[260,261,262,263,264] The initial modification produces an enzyme form which no longer requires calmodulin for kinase activity.[260,261,262,263,264] Coupled with this autophosphorylation event is the activation of the kinase toward exogenous substrates. Further autophosphorylation, in the absence of Ca^{+2}, generates an inhibited form of the enzyme.[261,263,265] Distinction of these effects seems to be related to the phosphorylation site.[255,266,267,268,269,270,271,272] Activation accompanies the autophosphorylation of a specific threonyl residue, Thr_{286}.[264,265,266,267,268,269] The inhibitory autophosphorylation event has been suggested to be near the calmodulin binding domain[270,271] at Thr_{305} and Ser_{314} (α-type catalytic subunit) or Thr_{306} and Ser_{315} (β-subunit). The phosphorylation pattern may then define the properties of the phosphorylated protein. Selectivity for the different autophosphorylation sites may be the controlling feature; indeed autophosphorylation may be modulated by the ATP level used in the reaction.[273] Low ATP results in the inactivated kinase, whereas physiological levels of ATP result in an activated, calmodulin-independent

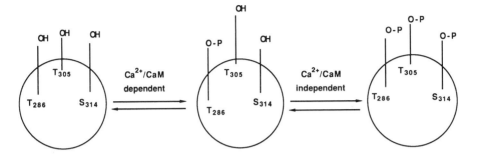

Figure 2.12. Bimodal effect of phosphorylation on the activity of the type II calmodulin activated protein kinase. The initial autophosphorylation at threonine-286 renders the kinase active in the absence of calmodulin. Subsequent autophosphorylation reactions at threonine-305 and serine-314 function to inactivate the enzyme. This two-step process provides an autoregulatory role.

kinase. The effect of different ATP levels may be determined by different K_m values for each of the two autophosphorylation sites. Definition at this level has not been accomplished.

In contrast to automodification, a different modification may influence the subsequent addition of modifying enzyme. For example, poly(ADP)-ribosylation of proline hydroxylase has been shown to inhibit its activity.[274] Another example is the recently reported regulation ofprotein ubiquitination by phosphorylation.[275] Two of the enzymes, E_1 and E_2 involved in protein ubiquitination were shown to be regulated by the action of two protein kinases. E_1 was specifically phosphorylated and activated by the cAMP dependent protein kinase. Likewise, subunit E_2 was phosphorylated and activated by protein kinase C. The phosphorylation reactions seem to be specific for the target protein.

SEQUENTIAL MODIFICATIONS

Sequential reactions

When several reactions can take place with one substrate, it may be important to determine whether the modifications occur randomly or in a particular order, and what the relative rates of these processes are. When a reaction of an enzyme with a substrate containing two modifiable sites occurs randomly, the rate of change of one site can be described by the equation shown, where v is the observed initial velocity and V_1 is the maximum velocity for modification of site 1. Because the intercept contains terms for interactions at two sites, only apparent K_m and V_{max} terms may be obtained; the real values depend upon the ratio of K_1/K_2.

$$\frac{1}{V} = \frac{K_1}{V_1}\left(\frac{1}{S}\right) + \frac{1}{V_1}\left(1 + \frac{K_1}{K_2}\right)$$

Thus, even though one process is being measured of two random events, the kinetics do not describe only just one event. Because a comparison of kinetic parameters is often made between reactions with a natural protein substrate and a peptide substrate, the interpretation of kinetic parameters needs to be guarded if the protein contains multiple modification sites and the peptide only one. In this instance it is possible that a lower K_m and higher V_{max} may be obtained with the peptide substrate.

If the reaction occurs in a sequential way, information can be derived about the relative rates of the processes by analyzing curves of product formation as a function of time of reaction. Two reaction paths (Figure 2.13) are possible where: B1 and B2 represent products with different modified sites. These paths may be distinguished by determining whether B1 or B2 is generated as an intermediate. Equations for evaluating k_1 and k_2 can be determined using the following relationships.

$$A = A_0 e^{-k_1 t}$$

$$B = k_1 A_0 \frac{e^{-k_1 t} - e^{k_2 t}}{k_2 - k_1}$$

$$C = A_0 \left[1 + \frac{1}{k_1 - k_2}\left(k_2 e^{-k_1 t} - k_1 e^{-k_2 t}\right) \right]$$

The apparent rate constants are related to the enzyme kinetic parameters V_{max}/K_m for the two steps if the reactions take place with concentrations below K_m values. Inasmuch the ratio of V_{max}/K_m is a measure of catalytic efficiency, the ratio of the apparent constants describes the effectiveness of the two processes.

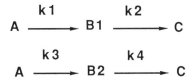

Figure 2.13. Scheme illustrating two possible reaction pathways for sequential reactions.

Kinetic constants for the reactions catalyzed by phosphorylase kinase (phosphorylase b to phosphorylase ab to phosphorylase a) have been measured.[276] A HPLC assay was devised to separate the starting material, phosphorylase ab (the half-phosphorylated species, a dimer in which only one monomer is phosphorylated), and the final product. By calculating how much of each form was present during the course of the reactions and by using the equations for a consecutive reaction, the values of k_1 and k_2 were determined. The rate constant k_2 was approximately 3.7 greater than k_1. Thus, in this instance the first phosphorylation event favors the second reaction.[276]

Sequential reactions may proceed independently or with some degree of interdependence. Glycogen synthase is extensively phosphorylated; the phosphorylation of some of the sites are interdependent and follows a definite order.[277,278,279] This phenomenom has been labeled hierarchal. There is a specific site that must be modified before additional regulatory sites can be phosphorylated (Figure 2.14). This is an example of synergistic phosphorylation of an individual protein. In this example, phosphorylation by glycogen synthase kinase 3 will not happen unless glycogen synthase has first been phosphorylated by casein kinase II. Phosphorylation of a serine site by CK-II generates the recognition site for glycogen synthase kinase 3. The first phosphorylation by GSK-3 generates another site for a second modification by GSK-3. The process continues up to three modifications by GSK-3. Although the phosphorylation sites are closely localized, the reactions do not proceed in a processive fashion. The phosphorylated protein seems to dissociate from the kinase prior to subsequent phosphorylation.

Another example is the methylation of the terminal carboxyl group in a series of low molecular weight G-proteins.[280,281,282] These proteins are also isoprenylated (farnesylated or geranylgeranylated) at a cysteine residue three residues from the carboxyl terminus. After isoprenylation,

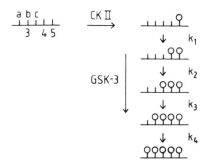

Figure 2.14. Sequential phosphorylation reactions catalyzed by glycogen synthase kinase-3 (GSK-3). Following the initial phosphorylation reaction at site 5 catalyzed by casein kinase II, a recognition site for GSK-3 is produced. This new structure is phosphorylated by GSK-3 at site 4 creating an additional recognition sequence. Sequentially, each phosphorylation at the three targets in site 3 by GSK-3 yields a recognition site for the subsequent phosphorylation reaction. Reprinted with permission from *Journal of Biological Chemistry* 262:14042 (1987). Copyright 1987, American Society for Biochemistry and Molecular Biology.

the terminal three residues are removed by proteolysis and the carboxyl terminus methylated.[280-282] The presence of the isoprenyl group is an absolute requirement for methylation.[283] This sequence of reactions has important biological implications, because proteolysis and methylation are necessary modifications for the localization and activity of these proteins. For example, p21K-ras(B) is not localized to the membrane after isoprenylation, but only after subsequent proteolysis and carboxyl methylation.[284]

Sequential modifications may also simply modulate subsequent modification events. Acetylation of histone H3 enhanced the phosphorylation of H3 by a calcium dependent protein kinase.[285] In contrast, ADP-ribosylation of arginyl groups in substrates for the cAMP-dependent protein kinase inhibited the of the substrate by the kinase.[286,287] The inhibitory effect of ADP-ribosylation was manifested in both protein and peptide substrates.[286,287,288] Prior phosphorylation prevented ADP-ribosylation of the peptide substrate.[286]

A complex system of sequential reactions is present in the regulation of adaptation to chemotactic signals.[289,290,291] Among the proteins involved in the chemotactic response are four membrane receptors,[289,292,293] a methyltransferase,[294,295,296] a methyl-esterase,[297,298] a protein kinase,[299,300,301,302] and other proteins with regulatory functions.[303,304,305,306] Overall, in response to external stimuli, membrane-bound proteins are methylated on glutamyl residues as the initial event.[307,308,309] The response of the bacteria is a change in its movement pattern between normal swimming and tumbling. Attractants cause an increase in the amount of methylation and a decrease in the level of tumbling.[289] Repellants decrease the level of methylation and an increase in tumbling. Extensive investigation has begun to decipher the role of the other proteins and the pathway for transmission of the initial stimulus. In *S. tymph.* the necessary proteins are the products of the genes designated CheR, CheB, CheA, CheW, CheY, and CheZ. CheR is the methyltransferase[294] and CheB is the methylesterase.[297] CheR directly modifies the methyl accepting chemotactic proteins (MCPs) and is antagonized by the methylesterase. CheW seems to be responsible for transducing the signal from the membrane bound MCPs to the cytoplasmic proteins, CheA, CheY, and CheZ.[310] CheA is a protein kinase that autophosphorylates[299,311] on histidine (His_{48}) and can transfer[299,312,313,314] its phosphoryl group to CheY (Asp_{57}) and CheB. CheW and CheA can directly form a complex *in vitro* suggesting that CheW can regulate the level of CheA activity[304] and this complex was shown to enhance the autophosphorylation of the CheY protein.[315] CheY is the response element directly initiating the motility response.[305] The transient phosphorylation of CheY is a necessary step for this activity. Methylation of MCPs is a crucial step for transmission of the signal to CheW; CheW is the connection between the level of methylation caused by a stimulus and the level of cytoplasmic phosphorylation caused by the stimulus.[310,316] Partial reconstitution of receptor with CheW and CheA demonstrated that the complex was stable to changing environmental conditions.[317] CheB is also transiently phosphorylated by CheA; the phosphorylation is reported to cause an activation of the esterase activity of the enzyme. This crossover provides a convenient mechanism for feedback regulation to achieve an adaptation response to a new environment. No catalytic activity has been ascribed to the CheZ product, but it is reported to enhance the dephosphorylation of the CheY protein.[299] As such, it serves as an antagonist to the action of CheY and could be a regulatory protein to control the tumbling response. A hypothetical model[289] has been developed for the interconnectedness of these various functions and is presented in Figure 2.15. The reconstitution experiments already described provided some insight into the mechanism of the transduction pathway. The authors[317] demonstrated that association-dissociation of the ternary complex—of receptor, CheW, and CheA—was not affected by the transduction of signal and concluded that signalling was accomplished by a conformational change in the complex, not an association phenomenom.

This model in Figure 2.15 only explains the available data concerning the direct response through a single receptor to external signals. This model does not agree with the data observed from the indirect response of a receptor to stimulation of a different receptor.[318] Stimulation of the aspartate receptor causes increased methylation of both the aspartate and serine receptors. The indirect response of the serine receptor

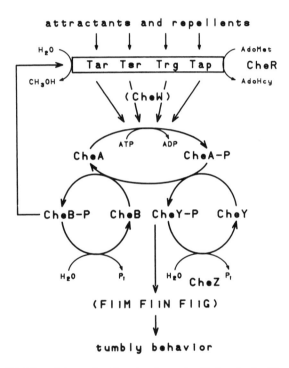

Figure 2.15. Signal transduction in chemotactic bacteria. The same cascade systems respond to both attractants and repellants via the membrane proteins, tar, tsr, trg, and tap. A series of enzymatic reactions is required for effective signaling: methylation by CheR; autophosphorylation by CheA; and the transient phosphorylation of CheY and CheB. Note also the site of action of the methylesterase activity of CheB. Reprinted with permission from *Microbiological Reviews* 53:450 (1989). Copyright 1989, American Society for Microbiology.

required the expression of the CheA and the CheW gene products. The model in Figure 2.15 suggests that activation of CheA will ultimately result in decreased methylation of MCPs, not an increase.

Cascades

A particular arrangement of sequential reactions provides an unique system for the transfer of biological information. Amplification of biological signals is commonly accomplished by cascades of enzymatic reactions; for example, in blood clotting a signal leads to an activation of an enzyme, which in turn modifies another sequence of reactions, and so on, leading to the effective conversion of fibrinogen to fibrin. A second type of cascade, a cyclic process, is possible in reversible covalent modification reactions such as reactions of phosphorylation-dephosphorylation, ADP-ribosylation-deADP-ribosylation, acetylation-deacetylation, carboxymethylation-demethylation, and adenylation-deadenylation. These cyclic cascades enable tight metabolic control because they provide: (1) increased signal amplification to a primary stimulus; (2) increased sites for allosteric interactions with effector molecules; (3) rate amplification for a quickly needed response with the ultimate enzymatic rate bearing a multiplicative function of all preceding enzymatic rates; and (4) a sigmoidal response of the interconvertible enzyme to changing metabolite concentration. Theoretical considerations of these cascades pioneered by Stadtman, Chock, and associates have shown how these cases occur.[319,320] Figure 2.16 shows a monocyclic-cascade that is affected by two converter enzymes, E1 and E2. The effector molecules, a and b, regulate these respective enzymes. The reaction catalyzed by E1 utilizes the reactant, B, to modify the protein, W, to generate the active form, W*. Enzyme E2 reverses the process. The amplification of a signal response to a change, for example in concentration of the effector, a, of E1 engendered by the cyclic cascade, may be additionally bolstered if one or both of the converting enzymes operate when their protein substrates are present at saturating concentration. An equation describing the steady state, in which the total concentration of W is in great excess over E1 and E2, a zero-order condition, has been generated.[321] The maximal

$$W^* = \frac{\left(\frac{V_1}{V_2} - 1\right) - K_2\left(\frac{K_1}{K_2} + \frac{V_1}{V_2}\right) + \left\{\left[\frac{V_1}{V_2} - 1 - K_2\left(\frac{K_1}{K_2} + \frac{V_1}{V_2}\right)^2 + 4K_2\right]\left(\frac{V_1}{V_2} - 1\right)\left(\frac{V_1}{V_2}\right)\right\}^{1/2}}{2\left(\frac{V_1}{V_2} - 1\right)}$$

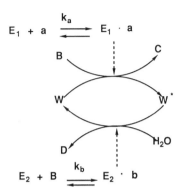

Figure 2.16. Representation of a monocyclic cascade. Enzymes E1 and E2 both act on protein substrate W. Shown are the unmodified and modified forms of W (W and W*, respectively). Also illustrated are ligands a and b which influence the activity of the interconverting enzymes, E1 and E2.

velocities for the two processes are designated V_1 and V_2. The constants, K_1 and K_2, are the Michaelis constants divided by the total concentration of W. The symbol [W*] is the mole fraction, $W*/W_T$.

When $K_1 = K_2 = 0.01$, the substrates W and W* are saturating, and when $K_1 = K_2 = 1$, W and W* are in a condition of half-saturation. A change in one of the velocities, for example, V_1, caused by an increase in the concentration of the effector, a, increases the amount of W* by positive terms in the equation but is counterbalanced by negative terms depending upon the magnitude of K_1 and K_2. At low values of these constants, the zero-order condition (case 1), the negative terms are insignificant and the transition for forming W* is steep. At high values, a condition approaching first order, (case 2), the transition is more gradual because the negative terms dampen the response. Conceptually, this means that the reverse process, which is going as fast as it can under zero-order conditions, is not influenced by an increase in W* caused by an increase in V_1. But when W* is not saturating, any increase in W* can cause an increase in the reverse direction and decrease the effectiveness of the change in V_1. Figure 2.17 shows how fractions of modified (W*) and unmodified (W) can change in these two cases.

Two experimental systems, glycogen phosphorylase[322] and iso-citrate dehydrogenase,[323] have been demonstrated to be subject to zero-order sensitivity. The effect of varying the ratio of kinase-phosphatase on the mole fraction of active phosphorylase a under zero- and first-order conditions illustrates nicely the relation between theoretical and experimental conditions.[322] As the amount of kinase is increased over the level of phosphatase, the steady state concentration of phosphorylase a is increased. The dependency of phosphorylase a on the kinase/phosphatase ratio is more sensitive at a higher phosphorylase concentration.

Increase in sensitivity resulting from an incremental change through E_1, E_2, or both under zero-order conditions is obviously coupled to the turnover of the reactant, for example, B in Figure 2.16. If this reactant is ATP as it is in phosphorylation-dephosphorylation reactions, the increased sensitivity is paid for by the energy expenditure of ATP hydrolysis. Different mechanisms can influence the process, and a common effector activating one pathway and inhibiting the second can maximize sensitivity and minimize energy expenditure.[324,325]

The kinetics of interconversion of glycogen phosphorylase follows zero-order ultrasensitivity in the presence of glycogen, and this model system may describe what occurs in the glycogen particle.[326] Thus, the kinetic study and modeling may help explain how glycogenolysis is regulated in intact muscle.

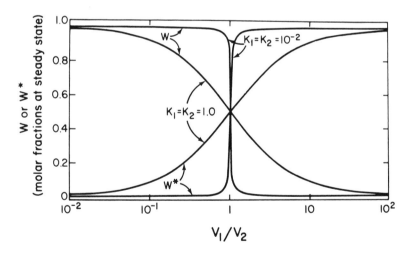

Figure 2.17. Distribution of Modified and Unmodified substrate. The steady state distribution of the two forms of the substrate protein (W and W*) calculated as a function of the ratio of interconversion rates. Reprinted with permission from *Proceedings of the National Academy of Science USA* 78:6840 (1981).

REVERSIBILITY

A key feature of any covalent modification involved in cellular regulation is the role of reversibility. To achieve a precise role in cellular regulation, the effect of the modification must be able to be reversed. The inability to reverse the effect of the modification will generate a system nonresponsive to future environmental signals. Enzymes responsible for catalyzing the removal of a modifying group have been identified; these include fatty acylesterase,[327] aminocyl-peptide hydrolase,[328] ubiquitin hydrolase,[329] carboxyl methylesterase,[330,331] protein phosphatases,[332,333,334] and protein-N-glycosyl amidase.[335,336] In each instance, the final products are the same as for the direct hydrolytic reaction.

The reversibility of ADP-ribosylation presents a more complex system. Reversibility of both mono- and poly-ADP-ribosylation is known. Removal of an ADP-ribose group from mono-ADP-ribosylated proteins is accomplished by direct hydrolysis of the linkage.[337,338,339] Removal of the complete poly-ADP-ribosyl chain is more involved. Poly-ADP-ribose can be degraded by three activities: (1) a glycohydrolase;[340,341,342] (2) a pyrophosphatase;[343,344] and (3) a lyase.[345,346] Glycohydrolase activity cleaves at ribose-ribose bonds to simply remove an ADP-ribose group with processivity involving exoglycosidic cleavage proceeding from the adenine end to the ribose terminus. Figure 2.18 demonstrates the cleavage catalyzed by the pyrophosphatase and the lyase activities. ADP-ribose protein lyase may be the most intriguing because it does not yield ADP-ribose as the product. Instead, a derivative, 5'-ADP-3"-deoxypent-2"-enofuranose, is produced.[346,347,347] This compound has a modified ribose, 3-deoxy-D-glycero-2–ulose, as the sugar.

Bacterial Systems

The reversibility of covalent modification in bacterial systems has important differences. For two well characterized cases, nucleotidylation of glutamine synthetase[348,349] and phosphorylation of isocitrate dehydrogenase,[350] activities for the forward and reverse reactions are both present on the same protein molecule.

The cyclic phosphorylation-dephosphorylation of isocitrate dehydrogenase is regulated by a single protein with both kinase and phosphatase activities.[350] The control of IDH activity by phosphorylation is important for the response to environmental conditions; growth on acetate requires the activation of the glyoxalate bypass.[350,351] As part of this need, IDH is phosphorylated and inactivated. Inactivation of IDH prevents further metabolism via the tricarboxylic acid cycle. Instead, isocitrate is metabolized by the glyoxalate bypass. The directional activity of the kinase/

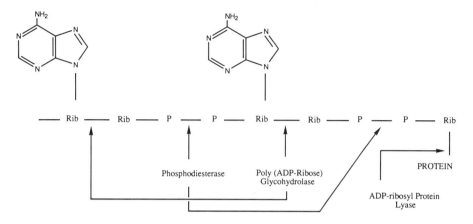

Figure 2.18. Multiple pathways for the metabolism of poly-(ADP)-ribosyl groups in proteins. Shown are the hydrolytic reaction (glycohydrolase) and two other metabolic fates. Note the sites within the polymer which are targets for the different enzymes. The reaction product of the lyase reaction contains a modified carbohydrate structure; see text for discussion and references.

phosphatase is modulated by various metabolites, including isocitrate, 3-phosphoglycerate, pyruvate, oxaloacetate, and 2–oxoglutarate.[352] All of these compounds activate IDH phosphatase and inhibit IDH kinase such that changes in these metabolites will influence the balance between kinase and phosphatase activities. This system represents a simple mechanism for regulating the proportion of modified protein in response to external signals.

Similarly, nucleotidylation functions to coordinate the response to external signals.[348,349] The nucleotidylation control pathway is a bicyclic cascade involving both adenylylation and uridylylation to regulate the activity of glutamine synthetase (GS). GS is specifically adenylylated by a specific adenylyl transferase that transfers the AMP moiety of ATP to the phenolic hydroxyl group of the tyrosine.[353,354,355] The same protein also catalyzes the deadenylylation of glutamine synthetase.[354,355] The two reactions are catalyzed by the same protein at two different sites. The selectivity of adenylylation vs. deadenylylation is modulated by a regulatory protein, designated P_{II}.[356] This protein is, in turn, regulated by uridylylation.[357] Uridylylated P_{II} stimulates the adenylyl-transferase. The modification of P_{II} is catalyzed by a unique uridylyl transferase.that transfers the UMP moiety of UTP to a tyrosine residue. The same protein catalyzes the deuridylylation reaction.

A unique result of adenylylation is the conversion of glutamine synthetase from a Mg^{+2} dependent enzyme to a Mn^{+2} activated enzyme.[358] Adenylylation also induces the sensitivity of GS to feedback inhibition by the end products of glutamine metabolism[358]. Because Mg^{+2} is normally an abundant divalent metal, adenylylation would lower the activity of glutamine synthetase.

Both transferases are subject to metabolite control (Figure 2.19). The adenylylation state of GS is correlated with the intracellular ratio of glutamine to 2–oxoglutarate,[348,349,359] higher when the ratio is higher. The reverse is true for uridylylation. Adenylylation is also closely related to the growth rate[360,361] of cells (*E.coli*). The relationship between

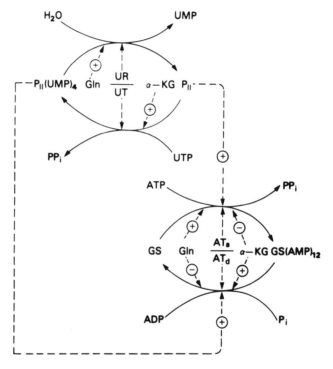

Figure 2.19. Nucleotidylation. Represented is the bicyclic cascade system using adenylylation-deadenylylation and uridylylation-deuridylylation to control glutamine synthetase activity. Shown are both of the connecting reactions for the regulation of adenylylation by protein P_{II}. Indicated are the sites for postive and negative metabolite control of each reaction by various ligands involved in glutamine metabolism. Reprinted with permission from *Current Topics in Cellular Regulation* 27:3 (1985). Copyright 1985, Academic Press.

adenylyltransferase and growth does not manifest any effect on the synthesis of glutamine synthetase.[348,361,362] Loss of uridylyltransferase activity dramatically influences growth[360] and induces a strict glutamine requirement for growth because of the reduction of GS activity. The uridylylated modulator protein, P_{II}, provides a link between the regulation of GS activity and GS synthesis via its effect on GS and another protein, NR_I.[363] This protein directly stimulates GS synthesis when kept in the active state by uridylylated P_{II}.

These bacterial systems demonstrate dual effects of a single protein with opposing activities.[350,354,355,364] It seems a simple mechanism to achieve efficient regulation in response to the environment. Advancing through evolutionary stages shows the divergence of opposing activities to two different proteins. Generation of distinct proteins having the opposing activities will provide additional sensitivity to the system by the development of additional levels of regulation beyond metabolite control evident in the bacterial control pathways. In bacteria, the relative amounts of modified protein and unmodified protein are determined by the relative distribution of the states of the dual-functional enzyme. With separate proteins providing the opposing activities, it is the relative levels of activities that define the flux through the cyclic modification-demodification pathway. A more responsive, more sensitive pathway is possible.

REFERENCES

1. Stadtman, T.C., *J. Biol. Chem.* 266, 16257-16260 (1991).
2. Shamberger, R.J., *Biochemistry of Selenium*, pp. 77-78, New York: Plenum Press, (1983).
3. Berry, M.J., Banu, L., Chen, Y., Mandel, S.J., Kieffer, J.D., Harney, J.W., and Larsen, P.R., *Nature* 353, 273-276 (1991).
4. Safran, M., Farwell, A.P., and Leonard, J.L., *J. Biol. Chem.* 266, 13477-13480 (1991).
5. Berry, M.J., Kieffer, J.D., Harney, J.W., and Larsen, P.R., *J. Biol. Chem.* 266, 14155-14158 (1991).
6. Tonks, N.K., Diltz, C.D., and Fischer, E.H., *J. Biol. Chem.* 6731-6737 (1988).
7. Folk, J.E., *Adv. Enzymol.* 54, 1-56 (1983).
8. Barber, J.R., and Clarke, S., *J. Biol. Chem.* 258, 1189-1196 (1983).
9. Brunauer, L.S., and Clarke, S., *J. Biol. Chem.* 261, 12538-12543 (1986).
10. Johnson, B.A., Freitag, N.E., and Aswad, D.W., *J. Biol. Chem.* 260, 10913-10916 (1985).
11. Stephenson, R.C., and Clarke, S., *J. Biol. Chem.* 264, 6164-6170 (1989).
12. Luo, S., Liao, C.-X, McClelland, J.F., and Graves, D.J., *Int. J. Peptide Protein Res.* 29, 728-733 (1987).

13. Martin, B.L., Wu, D., Tabatabai, L., and Graves, D.J., *Arch. Biochem. Biophys.* 276, 94-101 (1990).
14. Chazin, W.J., Kordel, J., Thulin, E., Hofmann, T., Drakenberg, T., and Forsen, S., *Biochemistry* 28, 8646-8653 (1989).
15. Murray, Jr., E.D., and Clarke, S., *J. Biol. Chem.* 259, 10722-10732 (1984).
16. Lowenson, J.D., and Clarke, S., *J. Biol. Chem.* 265, 3106-3110 (1990).
17. Acher, F., and Azerad, R., *Int. J. Peptide Protein Res.* 37, 210-219 (1991).
18. Ducrocq, C., Righini-Tapie, A., Azerad, R., Green, J.F., Friedman, P.A., Beaucourt, J.-P., and Rousseau, B. J., *J. Chem. Soc., Perkins Trans.* I, 1323-1328 (1986).
19. King, M.M., *J. Biol. Chem.* 261, 4081-4084 (1986).
20. King, M.M., and Heiny, L.P., *J. Biol. Chem.* 262, 10658-10662 (1987).
21. Papini, E., Santucci, A., Schiavo, G., Domenighini, M., Neri, P., Rappuoli, R., and Montecucco, C., *J. Biol. Chem.* 266, 2494-2498 (1991).
22. Coudijzer, K., and Joniau, M., *F.E.B.S. Lett.* 268, 95-98 (1990).
23. Taylor, S.S., Kerlanage, A.R., and Zoller, M. J., *Methods Enzymol.*, 99, 140-153 (1983).
24. Bause, E., *Biochem. J.*, 209, 323-330 (1983).
25. Selton, B.M., Trowbridge, I.S., Cooper, J.A., and Scolnick, E.M., *Cell* 31, 465-474 (1982).
26. Resh, M.D., *Molec. Cell. Biol.* 8, 1896-1905 (1988).
27. Garber, E.A., Cross, F.R., and Hanafusa, H., *Molec. Cell. Biol.* 5, 2781-2788 (1985).
28. Feder, D., and Bishop, J.M., *J. Biol. Chem.* 266, 19040-19046 (1991).
29. Kemp, B.E., and Pearson, R.B., *Trends in Biochem. Sci.* 15, 342-346 (1990).
30. Gilbert, J.M., Fowler, A., Bleibaum, J., and Clarke, S., *Biochemistry* 27, 5227-5233 (1988).
31. Ingrosso, D., Fowler, A.V., Bleibaum, J., and Clarke, S., *J. Biol. Chem.* 264, 20131-20139 (1989).
32. Henzel, W.J., Sultis, J.T., Hsu, C.-A., and Aswad, D.A., *J. Biol. Chem.* 264, 15905-15911 (1989).
33. Syed, S. K., Kim, S., Paik, W. K., *Biochemistry* 32, 2242-2247 (1993).
34. Oliva, A., Galletti, P., Zappia, V., Paik, W.K., and Kim, S., *Eur. J. Biochem.* 104, 595-602 (1980).
35. Casellas, P., and Jeanteur, P., *Biochim. Biophys. Acta* 519, 25-268 (1978).
36. Sekura, R.D., Fish, F., Manclark, C.R., Meade, B., and Zhang, Y.-l., *J. Biol. Chem.* 258, 14647-14651 (1983).
37. Kaslow, H.R., Schlotterbeck, J.D., Mar, V.L., and Burnette, W.N., *J. Biol. Chem.* 264, 6386-6390 (1989).
38. Cortina, G., and Barbieri, J.T., *J. Biol. Chem.* 266, 3022-3030 (1991).
39. Locht, C., Lobet, Y., Feron, C., Cieplak, W., and Keith, J.M., *J. Biol. Chem.* 265, 4552-4559 (1990).
40. Rudnick, D.A., McWherter, C.A., Adams, S.P., Ropson, I.J., Duronio, R.J., and Gordon, J.I., *J. Biol. Chem.* 265, 13370-13378 (1990).
41. Rudnick, D.A., McWherter, C.A., Rocque, W.J., Lennon, P.J., Getman, D.P., and Gordon, J.I., *J. Biol. Chem.* 266, 9732-9739 (1991).

42. Rudnick, D. A., Rocque, W. J., McWherter, C. A., Toth, M. V., Jackson-Machelski, E., and Gordon, J. I., *Proc. Natl. Acad. Sci., USA* 90, 1087- 1091 (1993).
43. Heuckeroth, R.O., Towler, D.A., Adams, S.P., Glaser, L., and Gordon, J.I., *J. Biol. Chem.* 263, 2127-2133 (1988).
44. Towler, D.A., Adams, S.P., Eubanks, S.R., Towery, D.S., Jackson- Machelski, E., Glaser, L., and Gordon, J.I., *Proc. Natl. Acad. Sci., USA* 84, 2708-2712 (1987).
45. Towler, D.A., Eubanks, S.R., Towery, D.S., Adams, S.P., and Glaser L., *J. Biol. Chem.* 262, 1030-1036 (1988).
46. Heuckeroth, R.O., Glaser, L., and Gordon, J.I., *Proc. Natl. Acad. Sci., USA* 85, 8795-8799 (1988).
47. Kishore, N.S., Lu, T., Knoll, L.J., Katoh, A., Rudnick, D.A., Mehta, P.P., Devadas, B., Huhn, M., Atwood, J.L., Adams, S.P., Gokel, G.W., and Gordon, J.I., *J. Biol. Chem.* 266, 8835-8855 (191).
48. Kishore, N. S., Wood, D. C., Mehta, P. P., Wade, A. C., Lu, T., Gokel, G. W., Gordon, J. I., *J. Biol. Chem.* 268, 4889-4902 (1993).
49. Ciechanover, A., Gonen, H., Elias, S., and Mayer, A., *New Biologist* 2, 227-234 (1990).
50. Haas, A.L., Warms, J.V.B., Hershko, A., and Rose, I.A., *J. Biol. Chem.* 257, 2543-2548 (1982).
51. Haas, A.L., and Rose, I.A., *J. Biol. Chem.* 257, 10329-10337. (1982)
52. Wilkinson, K.D., Smith, S.E., O'Connor, L., Sternberg, E., Taggart, J.J., Berges, D.A., and Butt, T., *Biochemistry* 29, 7373-7380 (1989).
53. Oppenheimer, N.J., *J. Biol. Chem.* 253, 4907-4910 (1978).
54. Larew, J.S.-A., Peterson, J.E., and Graves, D.J., *J. Biol. Chem.* 266, 52-57 (1991).
55. Osborne, J.C., Stanley, S.J., and Moss, J., *Biochemistry* 24, 5235-5240 (1985).
56. Mekalanos, J.J., Collier, R.J., and Romig, W.R., *J. Biol. Chem.* 254, 5849-5854 (1979).
57. Mildvan, A.S., Rosevear, P.R., Fry, D.C., Bramson, H.N., and Kaiser, E.T., *Curr. Top. Cell. Reg.* 27, 133-144 (1985).
58. Yoon, M.-Y., and Cook, P.F., *Biochemistry* 26, 4118-4125 (1987).
59. Granot, J., Mildvan, A.S., Bramson, H.N., Thomas, N., and Kaiser, E.T., *Biochemistry* 20, 602-(1981)
60. Buechler, J.A., and Taylor, S.S., *Biochemistry* 29, 1937-1943 (1990).
61. Gibbs, C.S., and Zoller, M.J., *Biochemistry* 30, 5329-5334 (1991).
62. Kemp, B.E., Graves, D.J., Benjamini, E., and Krebs, E.G., *J. Biol. Chem.* 252, 4888-4894 (1977).
63. Feramisco, J.R., Glass, D.B., and Krebs, E.G., *J. Biol. Chem.* 255, 4240-4245 (1980).
64. Prorok, M., and Lawrence, D.S., *Biochem. Biophys. Res. Commun.* 165, 368-371 (1989).
65. Cheng, H.-C., Kemp, B.E., Pearson, R.B., Smith, A.J., Musconi, L., Van Patten, S.M., and Walsh, D.A., *J. Biol. Chem.* 261, 989-992 (1986).
66. Zoller, M.J., Nelson, N.C., and Taylor, S., *J. Biol. Chem.* 256, 10837- 10842 (1981).

67. Knighton, D.R., Zheng, J., Ten Eyck, L.F., Ashford, V.A., Xuong, N.-H., Taylor, S.S., and Sowadski, J.M., *Science* 253, 407-414 (1991).
68. Zheng, J., Knighton, D. R., Ten Eyck, L. F., Karlsson, R., Xuong, N.-H, Taylor, S. S., and Sowadowski, J. M., *Biochemistry* 32, 2154-2161 (1993).
69. Martin, B., Pallen, C.J., Wang, J.H., and Graves, D.J., *J. Biol. Chem.* 260, 14932-14937 (1985).
70. Martin, B.L., and Graves, D.J., *J. Biol. Chem.* 261, 14545-14550 (1986).
71. Guan, K.L., and Dixon, J.E., *J. Biol. Chem.* 266, 17026-17030 (1991).
72. Pot, D.A., and Dixon, J.E., *J. Biol. Chem.* 267, 140-143 (1992).
73. Magnusson, R.P., Taurog, A., and Dorris, M.L., *J. Biol. Chem.* 259, 197- 205 (1984).
74. Ohtaki, S., Nakagawa, H., Kimura, S., and Yamazaki, I., *J. Biol. Chem.* 256, 805-810 (1981).
75. Vermeer, C., *Biochem. J.* 266, 625-636 (1990).
76. de Metz, M., Soute, B.A.M., Hemker, H.C., Fokkens, R., Lugtenburg, J., and Vermeer, C., *J. Biol. Chem.* 257, 5326-5329 (1982).
77. Larson, A.E., Friedman, P.A., and Suttie, J.W., *J. Biol. Chem.*, 256, 11210-11212 (1981).
78. McTigue, J.J., and Suttie, J.W., *J. Biol. Chem.* 258, 12129-12131 (1983).
79. Anton, D.L., and Friedman, P.A., *J. Biol. Chem.* 258, 14084-14087 (1983)
80. Vidal-Cros, A., Gaudry, M., and Marquet, A., *Biochem. J.* 266, 749-755 (1990).
81. Dowd, P., Ham, S.W., and Geib, S.J., *J. Am. Chem. Soc.* 113, 7734-7743 (1991).
82. Gallop, P.M., Friedman, P.A., and Henson, E. in "Vitamin K Metabolism and Vitamin K dependent Proteins" Suttie, J.W. Ed., University Park Press: Baltimore. pp. 408-412 (1980).
83. Arnett, E.M., Dowd, P., Flowers III, R.A., Ham, S.W., and Naganathan, S., *J. Am. Chem. Soc.* 114, 9209-9210 (1992).
84. Kuliopolis, A., Hubbard, B.R., Lam, Z., Koski, I., Furie, B., Furie, B.C., and Walsh, C.T., *Biochemistry* 31, 7722-7728 (1992)
85. Kizer, J.S., Bateman, R.C., Miller, C.R., Humm, J., Busby, W.H., and Youngblood, W.W. Endocrinology 118, 2262-2267 (1986).
86. Dickinson, C.J., and Yamada, T., *J. Biol. Chem.* 266, 334-338 (1991).
87. Young, S.D., and Tamburini, P.P., *J. Am. Chem. Soc.* 111, 1933-1934 (1989)
88. Tajima, M., Iida, T., Yoshida, S., Komatsu, K., Namba, R., Yanagi, M., Noguchi, M., and Okamoto, H., *J. Biol. Chem.* 265, 9602-9605 (1990).
89. Suzuki, K., Shimoi, H., Kawahara, T., Matsuura, Y., and Nishikawa, Y., *EMBO J.* 9, 4259-4265 (1990).
90. Katopodis, A.G., Ping, D., and May, S.W., *Biochemistry* 29, 6115-6120 (1990).
91. Eipper, B.A., Perkins, S.N., Husten, E.J., Johnson, R.C., Keutmann, H.T., and Mains, R.E., *J. Biol. Chem.* 266, 7827-7833 (1991).
92. Katopodis, A.G., Ping, D., Smith, C.E., and May, S.W., *Biochemistry* 30, 6189-6194 (1991).
93. Kato, I., Yonekura, H., Tajima, M., Yanagi, M., Yamamoto, H., and Okamoto, H., *Biochem. Biophys. Res Commun.* 172, 197-203 (1990).

94. Perkins, S.N., Husten, E.J., and Eipper, B.A., *Biochem. Biophys. Res. Commun.* 171, 926-932 (1990).
95. Merkler, D.J., and Young, S.D., *Arch. Biochem. Biophys.* 289, 192-196 (1991).
96. Reddy, K.V., Jin, S.-J., Arora, P.K., Sfeir, D.S., Maloney, S.C.F., Urbach, F.L., and Sayre, L.M., *J. Am. Chem. Soc.* 112, 2332-2340 (1990).
97. Skotland, T., and Ljones, T., *Biochim. Biophys. Acta* 630, 30-35 (1980).
98. Diliberto, E.J., Jr., and Allen, P.L., *J. Biol. Chem.* 256, 3385-3393 (1981).
99. Dhariwal, K.R., Black, C.D.V., and Levine, M., *J. Biol. Chem.* 266, 12908-12914 (1991)
100. Merkler, D.J., Kalathila, R., Consaluo, A.P., Young, S.D., and Ash, D.E., *Biochemistry* 31, 7282-7288 (1992).
101. Shinar, E., Navok, T., and Chevion, M., *J. Biol. Chem.* 258, 14778- 14783 (1983).
102. Bateman, R.C., Jr., Youngblood, W.W., Busby, W.H., Jr., and Kizer, J.S., *J. Biol. Chem.* 260, 9088-9091 (1985).
103. Bradbury, A.F., Finnie, M.D.A., and Smyth, D.G., *Nature* 298, 686-688 (1982)
104. Noguchi, M., Seino, H., Kochi, H., Okamoto, H., Tanaka, T., and Hirama, M., *Biochem. J.* 283, 883-888 (1992).
105. Appukuttan, P.S., and Wu, H.C., *F.E.B.S. Lett.* 255, 139-142 (1989).
106. Aratani, Y., and Kitagawa, Y., *F.E.B.S. Lett.*, 235, 129-132 (1988).
107. Lopez-Rodas, G., Tordera, V., Sanchez del Pino, M.M., and Franco, L., *Biochemistry* 30, 3728-3732 (1991).
108. Yamada, R., and Bradshaw, R.A., *Biochemistry* 30, 1017-1021 (1991).
109. Allfrey, V.G. (1977) in "Chromatin and Chromosome Structure" Li. H.J., and Eckhardt, R., Eds. New York: Academic. pp. 167-191.
110. Magee, A.I., and Schlesinger, M.J., *Biochim. Biophys. Acta* 694, 279-289(1982).
111. Carr, S.A., Biemann, K., Shoji, S., Parmelee, D.C., Titani, K., *Proc. Natl. Acad. Sci., USA*, 79, 6128-6131 (1982).
112. Aitken, A., Cohen, P., Santikarn, S., Williams, D.H., Calder, A.G., Smith, A., Klee, C.B., *F.E.B.S. Lett.*, 150, 314-318 (1982).
113. Repko, E.M., and Maltese, W.A., *J. Biol. Chem.* 264, 9945-9952 (1989).
114. Olson, E.N., and Spizz, G., *J.Biol. Chem.* 261, 2458-2466 (1986).
115. Olson, E.N., Towler, D.A., and Glaser, L., *J. Biol. Chem.* 260, 3784-3790 (1985).
116. Magee, A.I., and Courtneidge, S.A., *EMBO J.* 4, 1137-1144 (1985).
117. McIlhinney, R.A.J., Pelly, S.J., Chadwick, J.K., and Cowley, G.P., *EMBO J.* 4, 1145-1152 (1985).
118. Kasinathan, C., Grzelinka, E., Okazaki, K., Slomiany, B.L., and Slomiany, A., *J. Biol. Chem.* 265, 5139-5144 (1990).
119. Wilcox, C., Hu, J.-S., and Olson, E.N., *Science* 238, 1275-1278 (1987).
120. Slomiany, A., Liau, Y.H., Takagi, A., Laszewicz, W., and Slomiany, B. J. Biol. Chem. 259, 13304-13308 (1984).
121. Wilcox, C., Hu, J.-S., and Olson, E.N., *Science* 238, 1275-1278 (1987).
122. Bizzozero, O.A., and Good, L.K., *J. Biol Chem.* 266, 17092-17098 (1991).

123. Abeijon, C., and Hirschberg, C.B., *J. Biol. Chem.* 265, 14691-14695 (1990).
124. Kaplan, H.A., Naider, F., and Lennarz, W.J., *J. Biol. Chem.* 263,7814- 7820 (1988).
125. Welpley, J.P., Kaplan, H.A., and Shenbagamurthi, P., *Arch. Biochem. Biophys.* 246, 808-819 (1986).
126. Carr, D.W., Stofko-Hahn, R.E., Fraser, I.D.C., Bishop, S.M., Ascott, T.S., Brennan, R.G., and Scott, J.D., *J. Biol. Chem.* 266, 14188-14192 (1991).
127. Carr, D.W., Hausken, Z.E., Fraser, I.D.C., Stofko-Hahn, R.E., and Scott, J.D., *J. Biol. Chem.* 267, 13376-13382 (1991).
128. Scott, J.D. Personal Communication.
129. Graves, D.J., and Martenson, T.M., *Methods Enzymol.* 64, 325-341 (1980)
130. Madsen, N.B., Kavinsky, P.J., and Fletterick, R.J., *J. Biol. Chem.* 253,9097-9101 (1978).
131. Sprang, S.R., Withers, S.G., Goldsmith, E.J., Fletterick, R.J., and Madsen, N.B., *Science* 254, 1367-1371 (1991).
132. Agrawal, H.C., and Agrawal, D., *Biochem. J.* 263, 173-177 (1989).
133. Rose, J.K., Adams, G.A., and Gallione, C.J., *Proc. Natl. Acad. Sci., USA* 81, 2050-2054 (1984).
134. Brostoff, S., and Fylar, E.H., *Proc. Natl. Acad. Sci., USA* 68, 765-769 (1971).
135. Carnegie, P.R., *Nature* 229, 25- 28 (1971)
136. Kakimoto, Y., *Biochim. Biophys. Acta* 243, 31- 37 (1971)
137. Vijayasarathy, C., and Rao, B.S.N., *Biochim. Biophys. Acta* 923, 156- 165 (1987).
138. Van der Werf, P., and Koshland, D.E., Jr., *J. Biol. Chem.* 252, 2793- 2795 (1977).
139. Kleene, S.J., Toews, M.L., and Adler, J., *J. Biol. Chem.* 252, 3214-3218 (1977).
140. Janson, C.A., and Clarke, S., *J. Biol. Chem.* 255, 11640-11643 (1980).
141. Ghosh, S.K., Paik, W.K., and Kim, S., *J. Biol. Chem.* 263, 19024-19033 (1988).
142. Venkatesan, M., and McManus, I.R., *Biochemistry* 18, 5365-5371 (1979).
143. Honda, B.M., Dixon, G.H., and Candido, E.P.M., *J. Biol. Chem.* 250, 8681-8685 (1975).
144. Ghosh, S.K., Syed, S.K., Jung, S., Paik, W.K., and Kim, S., *Biochim. Biophys. Acta* 1039, 142-148 (1990).
145. Gupta, A., Jensen, D., Kim, S., and Paik, W.K., *J. Biol. Chem.* 257, 9677-9683 (1982).
146. Chang, F.N., Cohen, L.B., Naickas, I.J., and Chang, C.N., *Biochemistry* 14, 4994-4998 (1975).
147. Simms, S.A., and Subbaramaiah, K., *J. Biol. Chem.* 266, 12741-12746 (1991).
148. Ullah, A.H.J., and Ordal, G.W., *Biochem. J.* 194, 795-801 (1981).
149. Jamaluddin, M., Kim, S., and Paik. W.K., *Biochemistry* 14, 694-698 (1975).
150. Rosenwald, A.G., Stoll, J., and Krag, S.S., *J. Biol. Chem.* 265, 14544- 14553 (1990).
151. Majamaa, K., Hanauske-Abel, H.M., Gunzler, V., and Kivirikko, K.I., *Eur. J. Biochem.* 138, 239-245 (1984).

152. Majamaa, K., Turpeenniemi-Hujanen, T.M., Latipaa, P., Gunzler, V., Hanauske-Abel, H.M., Hassinen, I.E., and Kivirikko, K.I., *Biochem. J.* 229, 127-133 (1985).
153. Kwirikka, K.I., and Prockop, D.J., *Biochim. Biophys. Acta*, 258, 366- 379 (1972).
154. Gronke, R.S., Welsch, D.J., VanDusen, W.J., Garsky, V.M., Sardana, M.K., Stern, A.M., and Friedman, P.A., *J. Biol. Chem.* 265, 8558-8565 (1990).
155. Kivirikko, K.I., and Myllyla, R. in "The Posttranslational Modification of Proteins" Vol. 1, Freedman, R.B., and Hawkins, H.C., Eds. New York: Academic Press. pp. 54-104 (1980).
156. Myllyla, R., Majamaa, K., Gunzler, V., Hanauske-Abel, H.M., Kivirikko, K.I., *J. Biol. Chem.* 259, 5403-5405 (1984).
157. Myllyla, R., Kuutti-Savolainen, E.-R., and Kivirikko, K.I., *Biochem. Biophys. Res. Commun.* 83, 441-448 (1978).
158. Journet, A., and Tosi, M., *Biochem. J.* 240, 783-787 (1986).
159. McMullen, B.A., Fujikawa, K., Kisiel, W., Sasagawa, T., Howald, W.N., Kwa, E.Y., and Weinstein, B., *Biochemistry* 22, 2875-2884 (1983).
160. Lundwall, A., Dackowski, W., Cohen, E., Schaffer, M., Mahr, A., Dahlback, B., Stenflo, J., and Wydro, R., *Proc. Natl. Acad. Sci., USA* 83, 6716-6720 (1986).
161. Fernlund, P., and Stenflo, J., *J. Biol. Chem.*, 258, 12509-12512 (1983).
162. Stenflo, J., Lundwall, A., and Dahlback, B., *Proc. Natl. Acad. Sci., USA* 84, 368-372 (1987).
163. Tanuma, S.-I., Kawashima, K., and Endo, H., *J. Biol. Chem.* 263, 5485- 5489 (1988).
164. Towler, D.A., Adams, S.P., Eubanks, S.R., Towery, D.S., Jackson-Machelski, E., Glaser, L., and Gordon, J.I., *Proc. Natl. Acad. Sci., USA* 84, 2708-2712(1987).
165. Lee, F.-J.S., Lin, L.-W., and Smith, J.A., *J. Biol. Chem.* 263, 14948- 14955 (1988).
166. Yamada, R., and Bradshaw, R.A., *Biochemistry* 30, 1010-1016 (1991).
167. Ghosh, S.K., Paik, W.K., and Kim, S., *J. Biol. Chem.* 263, 19024-19033 (1988).
168. Bassuk, J.A., Kao, W.W.-Y., Herzer, P., Kedersha, N.L., Seyer, J., DeMartino, J.A., Daugherty, Mark, G.E., and Berg, R.A., *Proc. Natl. Acad. Sci., USA* 86, 7382-7386 (1989).
169. Myllyla, R., Pihlajaniemi, T., Pajunen, L., Turpeenniemi-Hujanen, T., and Kivirikko, K.I., *J. Biol. Chem.* 266, 2805-2810 (1991).
170. Noiva, R., Kaplan, H.A., and Lennarz, W.J., *Proc. Natl. Acad. Sci., USA* 88, 1986-1990(1991).
171. Holmgren, J., and Lonnroth, I., *J. Gen. Microbiol.* 86, 49-65 (1975).
172. Gill, D.M., *Adv. Cyclic Nucleotid. Res.* 8, 85-118 (1977).
173. Tung, H.Y.L., Alemany, S., and Cohen, P., *Eur. J. Biochem.* 148, 253-263 (1985).
174. Usui, H., Imazu, K., Meata, K., Tsukamoto, H., Azuma, K., and Takeda, M., *J. Biol. Chem.* 263, 3752-3761 (1988).

175. Waulkens, P., Goris, J., Merlevede, W., *J. Biol. Chem.*, 262, 1049-1059 (1987).
176. Chen, S.-C., Kramer, G., and Hardesty, B., *J. Biol. Chem.* 264, 7267- 7275 (1989).
177. Stewart, A.A., Ingebritsen, T.S., Manalan, A., Klee, C.B., and Cohen, P., *F.E.B.S. Lett.* 137, 80-84 (1982).
178. Yang, S.-D., Tallant, E.A., and Cheung, W.Y., *Biochem. Biophys. Res. Commun.* 106, 1419-1425 (1982).
179. Colbran, R.J., and Soderling, T.R., *Curr. Topics Cell. Regul.* 31, 181-221 (1990).
180. Resink, T.J., Hemmings, B.A., Tung, H.Y.L., Cohen, P., *Eur. J. Biochem.*, 133, 455-461 (1983).
181. Villa-Moruzzi, E., Ballou, L.M., and Fischer, E.H., *J. Biol. Chem.* 259, 5857-5863 (1984).
182. DePaoli-Roach, A.A., *J. Biol. Chem.* 259, 12144-12152 (1984).
183. Li, H.-C., Price, D.L., and Tabarini, D., *J. Biol. Chem.* 260, 6416-6426 (1985).
184. Monaco, L., Murtaugh, J.J., Newman, K.B., Tsai, S.-C., Moss, J., and Vaughan, M., *Proc. Natl. Acad. Sci., USA* 87, 2206-2210 (1990).
185. Tsuchiya, M., Price, S.R., Nightingale, M.S., Moss, J., and Vaughan, M., *Biochemistry* 28, 9668-9673 (1989).
186. Bobak, D.A., Nightingale, M.S., Murtaugh, J.J., Price, S.R., Moss, J., and Vaughan, M., *Proc. Natl. Acad. Sci., USA* 86, 6101-6105 (1989).
187. Kahn, R.A., Kern, F.G., Clark, J., Gelmann, E.P., and Rulka, C., *J. Biol. Chem.* 266, 2606-2614 (1991).
188. Ohtsuka, T., Nagata, K.-i., Iiri, T., Nozawa, Y., Ueno, K., Ui, M., and Katada, T., *J. Biol. Chem.* 264, 15000-15005 (1989).
189. Weiss, O., Holden, J., Rulka, C., and Kahn, R.A., *J. Biol. Chem.* 264, 21066-21072 (1989).
190. Tsai, S.-C., Adamik, R., Moss, J., and Vaughan, M., *Biochemistry* 30, 3697-3703 (1991).
191. Perkins, S.N., Husten, E.J., Mains, R.E., and Eipper. B.A., *Endocrinology* 127, 2771-2778 (1990).
192. Diliberto, E.J., Jr., and Allen, P.L. Molec. Pharmacol. 17, 421-426 (1980).
193. Ingebritsen, T.S., and Cohen, P., *Eur. J. Biochem.* 132, 255-261 (1983).
194. Pato, M.D., Adelstein, R.D., Crouch, D., Safer, B., Ingebritsen, T.S., and Cohen, P., *Eur. J. Biochem.* 132, 283-287 (1983).
195. Brautigan, D.L., Gruppuso, P.A., and Mumby, M., *J. Biol. Chem.* 261, 14924-14928 (1986).
196. Damuni, Z., Tung, H.Y.L., and Reed, L.J., *Biochem. Biophys. Res. Commun.* 133, 878-883 (1986).
197. Damuni, Z., Humphreys, J.S., and Reed, L.J., *Proc. Natl. Acad. Sci., USA* 83, 285-289 (1986).
198. Walsh, D.A., Ashby, C.D., Gonzalez, C., Calkins, D., Fischer, E.H., and Krebs, E.G., *J. Biol. Chem.* 246, 1977-1985 (1971).
199. DeMaille, J.G., Peters, K.A., and Fischer, E.H., *Biochemistry* 16, 3080- 3086 (1977).

200. Scott, J.D., Glaccum, M.B., Fischer, E.H., and Krebs, E.G., *Proc. Natl. Acad. Sci., USA* 83, 1613-1616 (1986).
201. Reed, J., DeRopp, J.S., Trewhella, J., Glass, D.B., Liddle, W.K., Bradbury, E.M., Kinzel, V., and Walsh, D.A., *Biochem. J.* 264, 371-380 (1989).
202. Glass, D.B., Cheng, H., Mendo-Mueller, L., Reed, J., and Walsh, D.A., *J. Biol. Chem.* 264, 8802-8810 (1989).
203. Van Patten, S.M., Heisermann, G.J., Cheng, H.-C., and Walsh, D.A., *J. Biol. Chem.* 262, 3398-3403 (1987).
204. Knighton, D.R., Zheng, J., Eyck, L.F.T., Xuong, N.-H., Taylor, S.S., and Sowadski, J.M., *Science* 253, 414-420 (1991).
205. Ridsdale, J.A., Hendzel, M.J., Deleuve, G.P., and Davie, J.R., *J. Biol. Chem.* 265, 5150-5156 (1991).
206. Csordas, A., *Biochem. J.* 265, 23-38 (1990).
207. Boulikas, T., *Proc. Natl. Acad. Sci., USA* 86, 3499-3503 (1989).
208. Boulikas, T., *J. Biol. Chem.* 265, 14638-14647 (1990).
209. Lutter, L. C., Juds, Paretti, R. F., *Molec. Cell Biol.* 12, 5004-5014 (1992).
210. Lopez-Rodas, G., Brosch, G., Georgieva, E.I., Sendra, R., Franeg, L., Loidl, P., *F.E.B.S. Lett.* 317, 175-180 (1993).
211. Naegeli, H., and Althaus, F.R., *J. Biol. Chem.* 266, 10596-10601 (1991).
212. Golderer, G., and Grobner, P., *Biochem. J.* 277, 607-610 (1991).
213. Hengartner, C., Laguex, J., and Poirier, G.G., *Biochem. Cell Biol.* 69, 577-580 (1991).
214. Ullrich, A., and Schlessinger, J. Cell 61, 203-212 (1990).
215. Margolis, B., Li, N., Koch, A., Mohammadi, M., Hurwitz, D.R., Zilberstein, A., Pawson, T., and Schlessinger, J., *EMBO J.* 9, 4375-4380 (1990).
216. Escobedo, J.A., Kaplan, D.R., Kavanaugh, W.M., Turck, C.W., and Williams, L.T., *Molec. Cell. Biol.*, 11, 1125-1132 (1991).
217. Anderson, D., Koch, C.A., Grey, L., Ellis, C., Moran, M.F., and Pawson, T., *Science*, 250, 979-982 (1990).
218. Ushiro, H., and Cohen S., *J. Biol. Chem.* 255, 8363-8365 (1980).
219. Pike, L.J., and Krebs, E.G. in "The Receptors" Vol. III, Conn, M.P., Ed., New York: Academic Press. pp. 93-134 (1986).
220. Ahn, N.G., Seger, R., Bratlein, R.L., Diltz, C.D., Tonks, N.K., and Krebs, E.G., *J. Biol. Chem.*, 266, 4220-4227 (1991).
221. Price, D.J., Gunsalus, J.R., and Avruch, J., *Proc. Natl. Acad. Sci. USA*, 87, 7944-7948 1990).
222. Boulton, T.G., and Cobb, M.H., *Cell. Regul.*, 2, 357-371 (1991).
223. Lee, R.-m., Rapp, U.R., and Blackshear, P.J., *J. Biol. Chem.*, 266, 10351-10357 (1991).
224. Czech, M.P., Klarlund, J.K., Yagaloff, K.A., Bradford, A.D., and Lewis, R.E., *J. Biol. Chem.*, 263, 11017-11020 (1988).
225. Sturgill, T.W., and Wu, J., *Biochim. Biophys. Acta*, 1092, 350-357 (1991).
226. Courtneidge, S.A., and Smith, A.E., *Nature*, 303, 435-439 (1983).
227. Whitman, G., Kaplan, D.P., Schaffhausen, B., Cantley, L., and Roberts, T.M., *Nature*, 315, 239-242 (1985).
228. Pallas, D.C., Shahrik, L.K., Martin, B.L., Jaspers, S., Miller, T.B., Jr., Brautigan, D.L., and Roberts, T.M., *Cell,* 60, 167-176 (1990).

229. Walter, G., Ruediger, R., Slaughter, C., and Mumby, M., *Proc. Natl. Acad. Sci., USA*, 87, 2521-2525 (1990).
230. Soderling, T.R., *J. Biol. Chem.* 265, 1823-1826 (1990).
231. Smith, M.K., Colbran, R.J., and Soderling, T.R., *J. Biol. Chem.* 265, 1837-1840 (1990).
232. Kemp, B.E., Pearson, R.B., House, C., Robinson, P.J., and Means, A.R., *Cellular Signalling* 1, 303-311 (1989).
233. Hardie, G., *Nature* 335, 592-593 (1988).
234. House, C., and Kemp, B.E., *Science* 238, 1726-1728 (1987).
235. Pearson, R.B., Wettenhall, R.E.H., Means, A.R., Hartshorne, D.J., and Kemp, B.E., *Science* 241, 970-973 (1988).
236. Sprang, S.R., Archarage, K.R., Goldsmith, E.J., Stuart, D.I., Varviell, K., Fletterick, R.J., Madsen, N.B., and Johnson, L.N., *Nature* 336, 215-221 (1988).
237. Goldsmith, E.J., Sprang, S.R., Hamlin, R., Xuong, N.-H., and Fletterick, R.J., *Science* 245, 528-532 (1989).
238. Koch, C.A., Anderson, D., Moran, M.F., Ellis, C., and Pawson, T., *Science* 252, 668-674 (1991).
239. Anderson, D., Koch, C.A., Grey, L., Ellis, C., Moran, M.F., and Pawson, T., *Science* 250, 979-982 (1990) .
240. Fazioli, F., Kim, U.-H., Rhee, S.G., Molloy, C.J., Segatto, O., and Di Fiore, P.P., *Molec. Cell. Biol.* 11, 2040-2048 (1991).
241. Moran, M.F., Polakis, P., McCormick, F., Pawson, T., and Ellis, C., *Molec. Cell. Biol.* 11, 1804-1812 (1991).
242. Brott, B.K., Decker, S., Shafer, J., Gibbs, J.B., Jove, R., *Proc. Natl. Acad. Sci., USA* 88, 755-759 (1991).
243. Fukui, Y., and Hanafusa, H., *Molec. Cell. Biol.* 11, 1972-1979 (1991).
244. MacAuley, and Cooper, J.A., *New Biologist*, 2, 828-840 (1990).
245. Cartwright, C.A., Kaplan, D.L., Cooper, J.A., Hunter, T., and Eckhart, W., *Molec. Cell. Biol.*, 6, 1562-1570 (1986).
246. Matsuda, M., Mayer, B.J., Fukui, Y., and Hanafusa, H., *Science* 248, 1537-1540 (1990).
247. Shen, S.-H., Bastien, L., Posner, B.I., and Chretien, P., *Nature* 352, 736-739 (1991).
248. Matsuda, M., Mayer, B.J., and Hanafusa, H., *Molec. Cell. Biol.* 11, 1607-1613 (1991).
249. Fukami, Y., Sato, K-i., Ikeda, K., Kamisango, K., Koizumi, K., and Matsuno, T., *J. Biol. Chem.* 268, 1132-1140, 1993.
250. Todhunter, J.A., and Purich, D.L., *Biochim. Biophys. Acta* 485, 87-94 (1977).
251. Kuret, J., and Schulman, H., *J. Biol. Chem.* 260, 6427-6433 (1985).
252. Newton, A.C., and Koshland, D.E., Jr., *J. Biol. Chem.* 262, 10185-10188 (1987).
253. King, M.M., Fitzgerald, T.J., and Carlson, G.M., *J. Biol. Chem.* 257, 9925-9930 (1982).
254. Arnold, J.E., and Gevers, W., *Biochem. J.* 267, 751-757 (1990).

255. Tanigawa, Y., Tsuchiya, M., Imai, Y., and Shimoyama, M., *J. Biol. Chem.* 259, 2022-2029 (1984).
256. Naegeli, H., Loetscher, P., and Althaus, F.R., *J. Biol. Chem.*, 264, 14382-14385 (1989).
257. Taniguchi, T., *Biochem. Biophys. Res. Commun.* 147, 1008-1012 (1987).
258. Kawaichi, M., Ueda, K., and Hayaishi, O., *J. Biol. Chem.* 256, 9483-9489 (1981).
259. Le Vine III, H., Sayhoun, N.E., and Cuatrecasa. P., *Proc. Natl. Acad. Sci., USA* 83, 2253-2257 (1986).
260. Kwiatkowski, A.P., Shell, D.J., and King, M.M., *J. Biol. Chem.* 263, 6484-6486 (1988).
261. Lai, Y., Nairn, A.C., and Greengard, P., *Proc. Natl. Acad. Sci., USA* 83, 4253-4257 (1986).
262. Miller, S.G., and Kennedy, M.B., *Cell* 44, 861-870 (1986).
263. Kaitoh, T., and Fujisawa, H., *J. Biol. Chem.* 266, 3039-3044 (1991).
264. Lou, L.L., and Schulman, H. J., *Neurosci.* 9, 2020-2032 (1989).
265. Colbran, R.J., Smith, M.K., Schworer, C.M., Fong, Y.-L., and Soderling, T.R., *J. Biol. Chem.* 264, 4800-4804 (1989).
266. Hanson, P.I., Kapiloff, M.S., Lou, L.L., Rosenfeld, M.G., and Schulman, H., *Neuron* 3, 59-70 (1989).
267. Waldmann, R., Hanson, P.I., and Schulman, H., *Biochemistry* 29, 1679-1684 (1990).
268. Fong, Y.-L., Taylor, W.L., Means, A.R., and Soderling, T.R., *J. Biol. Chem.* 264, 16759-16763 (1989).
269. Waxham, M.N., Aronowski, J., Westgate, S.A., and Kelly, P.T., *Proc. Natl. Acad. Sci., USA* 87, 1273-1277 (1990).
270. Ohsako, S., Nakazawa, H., Sekihara, S.-i., Ikai, A., and Yamauchi, T., *J. Biochem.* 109, 137-143 (1991).
271. Patton, B.L., Miller, S.G., and Kennedy, M.B., *J. Biol. Chem.* 265, 11204-11212 (1990).
272. Colbran, R.J., and Soderling, T.R., *J. Biol. Chem.* 265, 11213-11219 (1990).
273. Lou, L.L., Lloyd, S.J., and Schulman, H., *Proc. Natl. Acad. Sci., USA* 83, 9497-9501 (1986).
274. Hussain, M.Z., Ghani, Q.P., and Hunt, T.K., *J. Biol. Chem.* 265, 7850-7855 (1989).
275. Kong, S.-K., and Chock, P.B., *J. Biol. Chem.* 267, 14189-14192 (1990).
276. Harris, W.R., and Graves, D.J., *Arch. Biochem. Biophys.* 276, 102-108 (1990).
277. Fiol, C.J., Mahrenholz, A.M., Wang, Y., Roeske, R.W., and Roach, P.J., *J. Biol. Chem.* 262, 14042-14048 (1987).
278. Roach, P.J., *FASEB J.* 4, 2961-2968 (1990).
279. Roach, P.J., *J. Biol. Chem.* 266, 14139-14142 (1991).
280. Clarke, S., Vogel, J.P., Deschenes, R.J., and Stock, J., *Proc. Natl. Acad. Sci., USA* 85, 4643-4647 (1988).
281. Hrycyna, C.A., and Clarke, S., *Molec. Cell. Biol.* 10, 5071-5076 (1990)
282. Tan, E.W., Perez-Sala, D., Canada, F.J., and Rando, R.R., *J. Biol. Chem.* 266, 10719-10722 (1991).

283. Yamane, H.K., and Fung, B.K.-K., *J. Biol. Chem.* 264, 20100-20105 (1989).
284. Hancock, J.F., Cadwaller, K., Marshall, C.J., *EMBO J.* 10, 641-646 (1991).
285. Whitlock, Jr., J.P., Galeazzi, D., and Schulman, H., *J. Biol. Chem.* 258, 1299-1304 (1983).
286. Kharadia, S.V., and Graves, D.J., *J. Biol. Chem.* 262, 17379-17383 (1987).
287. Matsuura, R., Tanigawa, Y., Tsuchiya, M., Mishima, K., Yoshimura, Y., and Shimoyama, M., *Biochem. J.* 253, 923-926 (1988).
288. Tsuchiya, M., Tanigawa, Y., Ushiroyama, T., Matsuura, R., and Shimoyama, M., *Eur. J. Biochem.* 147, 33-40 (1985).
289. Stock, J.B., Ninfa, A.J., and Stock, A.M., *Microbiol. Rev.* 53, 450-490 (1989).
290. Sanders, D.A., and Koshland, D.E., Jr., *Proc. Natl. Acad. Sci., USA* 85, 8425-8429. (1988)
291. Borkovich, K.A., and Simon, M.I., *Cell* 63, 1339-1348 (1990).
292. Goy, M.F., Springer, M.S., amd Adler, J., *Cell* 15, 1231-1240 (1978).
293. Springer, M.S., Goy, M.F., and Adler, J., *Proc. Natl. Acad. Sci., USA* 74, 3312-3316 (1977).
294. Springer, W. R., and Koshland, D.E., Jr., *Proc. Natl. Acad. Sci., USA* 74, 533-537 (1977).
295. Burgess-Cassler, A., Ullah, A.H.J., and Ordal, G.W., *J. Biol. Chem.* 257, 8412-8417 (1982).
296. Simms, S.A., Stock, A.M., and Stock, J.B., *J. Biol. Chem.* 262, 8537-8543 (1987).
297. Simms, S.A., Keane, M.G., and Stock, J.B., *J. Biol. Chem.* 261, 10161-10168 (1985).
298. Simms, S.A., Cornman, E.W., Mottonen, J., and Stock, J., *J. Biol. Chem.* 262, 29-31 (1987).
299. Hess, J.F., Oosawa, K., Kaplan, N., and Simon, M.I., *Cell* 53, 79-87 (1988).
300. Oosawa, K., Hess, J.F., and Simon, M.I., *Cell* 53, 89-96 (1988).
301. Wylie, D., Stock, A., Wong, C.-Y., and Stock, J., *Biochem. Biophys. Res. Commun.* 151, 891-896 (1988).
302. Stock, A., Chen, T., Welsh, D., and Stock, J., *Proc. Natl. Acad. Sci., USA* 85, 1403-1407 (1988).
303. Stock, A., Mottonen, J., Chen, T., and Stock, J., *J. Biol. Chem.* 262, 535-537 (1987).
304. Gegner, J.A., and Dahlquist, F.W., *Proc. Natl. Acad. Sci., USA* 88, 750-754 (1991).
305. Silverman, M., and Simon, M., *Proc. Natl. Acad. Sci., USA* 74, 3317-3321 (1977).
306. Silverman, M., and Simon, M., *J. Bacteriol.* 130, 1317-1325 (1977).
307. Terwilliger, T.C., and Koshland, Jr., D.E., *J. Biol. Chem.* 259, 7719-7725 (1984).
308. Rice, M.S., and Dahlquist, F.W., *J. Biol. Chem.* 266, 9746-9753 (1991).
309. Weiss, R.M., Chasalow, S., and Koshland, Jr., D.E., *J. Biol. Chem.* 265, 6817-6826 (1990).
310. Liu, J., and Parkinson, J.S., *Proc. Natl. Acad. Sci., USA* 86, 8703-8707 (1989).
311. Hess, J.F., Bourret, R.B., and Simon, M.I., *Nature* 336, 139-143 (1988).

312. Sanders, D.A., Gillee-Castro, B.L., Stock, A.M., Burlingame, A.L., and Koshland, Jr., D.E., *J. Biol. Chem.* 264, 21770-21778 (1989).
313. Lupas, A., and Stock, J., *J. Biol. Chem.* 264, 17337-17342 (1989).
314. Stewart, R.C., Roth, A.F., and Dahlquist, F.W.J., *Bacteriol.* 172, 3388-3399 (1990).
315. McNally, D. F., and Matsumura, P., *Proc. Natl. Acad. Sci., USA*, 88, 6269-6273, (1991).
316. Borkovich, K.A., Kaplan, N., Hess, J.F., and Simon, M.I., *Proc. Natl. Acad. Sci., USA*, 86, 1208-1212 (1989).
317. Gegner, J.A., Graham, D.R., Roth, A.F., and Dahlquist, F.W., *Cell* 70, 975-982 (1992)
318. Sanders, D.A., and Koshland, D.E., Jr., *Proc. Natl. Acad. Sci., USA* 85, 8425-8429 (1988).
319. Chock, P.B., and Stadtman, E.R., *Methods Enzymol.* 64, 297-325 (1980).
320. Cardenas, M.L., and Cornish-Bowden, A., *Biochem. J.* 257, 339-345 (1989).
321. Goldbetter, A., and Koshland, D.E., Jr., *Proc. Natl. Acad. Sci., USA* 78, 6840-6844 (1981).
322. Meinke, M.H., Bishop, J.S., and Edstrom, R.D., *Proc. Natl. Acad. Sci., USA* 83, 2865-2868 (1986).
323. LaPorte, D.C. and Koshland, Jr., D.E., *Nature* 305, 286-290 (1983).
324. Shacter, E., Chock, P.B., and Stadtman, E.R., *J. Biol. Chem.*, 259, 12260-12264 (1984).
325. Goldbetter, A., and Koshland, D.E., Jr., *J. Biol. Chem.* 262, 4460-4471 (1987).
326. Meinke, M.H., and Edstrom, R.D., *J. Biol. Chem.* 266, 2259-2266 (1991).
327. Berger, M., and Schmidt, M.F.G., *J. Biol. Chem.* 261, 14912-14918 (1986).
328. Radhakrishna, G., and Wold, F., *J. Biol. Chem.* 264, 11076-11081 (1989)
329. Agell, N., Ryan, C., and Schlesinger, M.J., *Biochem. J.* 273, 615-620 (1991).
330. Gagnon, C., Harbour, D., and Camato, R., *J. Biol. Chem.* 259, 10212-10215 (1984).
331. Veeraragavan, R., and Gagnon, C., *Biochem. J.* 260, 11-17 (1989).
332. Ingebritsen, T.S., and Cohen, P., *Science* 221, 331-338 (1983).
333. Cohen, P., and Cohen, P.T.W., *J. Biol. Chem.* 264, 21435-21438 (1989).
334. Alexander, D.A., *New Biologist* 2, 1049-1062 (1990).
335. Ridley, J.M., and Van Etten, R.L., *J. Biol. Chem.* 260, 15488-15494 (1985).
336. Fisher, K.J., Tollersud, O.K., and Aronson, Jr., N.N., *F.E.B.S. Lett.* 269, 440-444 (1990).
337. Chang, Y.-C., Soman, G., and Graves, D.J., *Biochem. Biophys. Res. Commun.* 139, 932-939 (1986).
338. Moss, J., Tsai, S.-C., Adamik, R., Chen, H.-C., and Stanley, S.J., *Biochemistry* 27, 5819-5823 (1988).
339. Tanuma, S.-i., and Endo, H., *F.E.B.S. Lett.* 261, 381-384 (1990).
340. Hatakeyama, K., Nemoto, Y., Ueda, K., and Hayaishi, O., *J. Biol. Chem.* 261, 14902-14911 (1986).
341. Tanuma, S.-i., and Endo, H., *Eur. J. Biochem.* 191, 57-63 (1990).
342. Maruta, H., Inageda, K., Acki, T., Nishima, H., and Tanuma, S.-I., *Biochemistry* 30, 5907-5912 (1991).

343. Tanuma, S.-i., *Biochem. Biophys. Res. Commun.* 163, 1047-1055 (1989).
344. Miro, A., Costas, M.J., Garcia-Diaz, M., Hernandez, M.T., and Cameselle, J.C., *F.E.B.S. Lett.* 244, 123-126 (1989).
345. Oka, J., Ueda, K., Hayaishi, O., Komura, H., and Nakanishi, K., *J. Biol. Chem.* 259, 986-995 (1984).
346. Ueda, K., Hayaishi, O., Oka, J., Komura, H., and Nakanishi, K. in "ADP-Ribosylation of Proteins" Althaus, F.R., Hilz, H., and Shall, S., Eds. Berlin:Springer-Verlag. pp.159-166 (1985).
347. Yan, S.-C.B., *Trends Biochem. Sci.* 9, 331-332 (1984).
348. Shapiro, B.M., Kingden, H.S., and Stadtman, E.R., *Proc. Natl. Acad. Sci., USA* 58, 642-649 (1967).
349. Brown, M.S., Segal, S., and Stadtman, E.R., *Proc. Natl. Acad. Sci., USA* 68, 2949-2953 (1971).
350. LaPorte, D.C., and Koshland, D.E., Jr., *Nature* 300, 458-460 (1982).
351. LaPorte, D.C., Thorsness, P.E., and Koshland, D.E., Jr., *J. Biol. Chem.* 260, 10563-10568 (1985).
352. Nimmo, G.A., and Nimmo, H.G., *Eur. J. Biochem.* 141, 409-414 (1984).
353. Kingden, H.S., Shapiro, B.M., and Stadtman, E.R., *Proc. Natl. Acad. Sci., USA* 58, 1703-1710 (1967).
354. Caban, C.E., and Ginsburg, A., *Biochemistry* 15, 1569-1580 (1976).
355. Shapiro, B.M., and Stadtman, E.R., *J. Biol. Chem.* 244, 3769-3771 (1968).
356. Adler, S.D., Purich, D., and Stadtman, E.R., *J. Biol. Chem.* 250, 6264-6272 (1975).
357. Mangum, J.H., Magni, G., and Stadtman, E.R., *Arch. Biochem. Biophys.* 158, 514-525 (1973).
358. Segal, A., Brown, M.S., and Stadtman, E.R., *Arch. Biochem. Biophys.* 161, 319-327 (1974).
359. Senior, P.J., *J. Bacteriol.* 123, 407-418 (1975).
360. Foor, F., Cedergren, R.J., Streichei, S.L., Rhee, S.G., and Magasink, B., *J. Bacteriol.* 134, 562-568 (1978).
361. Magasink, B., and Bueno, R., *Curr. Top. Cell Reg.* 27, 215-220 (1985).
362. Bancroft, S., Rhee, S.G., Neumann, C., and Kustu, S., *J. Bacteriol.* 134, 1046-1055 (1978).
363. Bueno, R., Pahel, G., and Magasink, B., *J. Bacteriol.* 164, 816-822 (1985).
364. Weiss, V., and Magasink, B., *Proc. Natl. Acad. Sci., USA* 85, 8919-8923 (1988).

3

The Protein Substrate

FACTORS INFLUENCING THE CHEMICAL REACTIVITY OF THE AMINO ACID RESIDUES

Extensive studies of chemical modification reactions have provided key insights about the structure of proteins and the properties of their functional groups. Specific chemical reactions depend on the reagent used and the protein being investigated. For example, phenylglyoxal or related reagents react with the guanidinyl side chain of arginyl residues, but proteins show unequal reactivity and only a few residues in a given protein are reactive. In nucleotide binding proteins, an arginyl residue in the binding site often is very reactive, but arginyl residues oriented toward the surface of the protein surprisingly aren't modified.[1] Reagents reacting with cysteinyl residues can modify residues at the surface readily, but residues in the interior of proteins often are sluggishly reactive, although a thiol in the active site of ficin is unusually reactive. An ε-amino group of a lysyl residue in the active site of acetoacetyl carboxylase is much more reactive than other amino groups in the protein. Differential reactivities depend on the structure of the protein and are changed when the native structure is denatured. But what underlying chemical principles influence the reactivities of functional groups in proteins? How can we use these concepts to better understand how an enzyme can specifically modify a protein substrate?

$$\Delta E = \underbrace{-\frac{Q_{nuc.}\, Q_{elec.}}{\varepsilon R}}_{\substack{\textit{The} \\ \textit{Coulombic} \\ \textit{term}}} + \underbrace{\frac{2(c_{nuc.}\, c_{elec.}\, \beta)^2}{E_{HOMO\,(nuc.)} - E_{LUMO\,(elec.)}}}_{\textit{The frontier orbital term}}$$

Figure 3.1. Equation for energy change for a reaction of a nucleophile and an electrophile. Reprinted with permission from *Frontier Orbitals and Organic Chemical Reactions,* p. 37 (1976). Copyright 1976, John Wiley & Sons, Ltd.

Chemical Reactivity and Nucleophilicity

Consider the example of the reaction of a nucleophilic group in a protein with a molecule possessing electrophilic characteristics. When molecules containing nucleophilic and electrophilic groups collide in the process of bond formation, energy is gained and lost when the molecular orbitals of the reactive portions overlap. The change in energy (ΔE) has been described mathematically[2,3] and the equation consists of three parts; core, electrostatic, and overlap terms.

$$\Delta E = \Delta E_1 \,(\text{core}) + \Delta E_2 \,(\text{electrostatic}) + \Delta E_3 \,(\text{overlap})$$

The core term is due to the repulsive interaction occurring between filled orbitals of the two species and is positive representing lost energy. The electrostatic term is negative and depends on the charge of the reactive species; it is important when ions or polar molecules react. The overlap term, also negative, is associated with the interaction of occupied orbitals with unoccupied orbitals. A major factor contributing to this term is the interaction of the occupied molecular orbital of the highest energy (HOMO) of the nucleophile with the unoccupied molecular orbital of lowest energy (LUMO) of the electrophile. The last two terms of the equation are of particular importance in explaining differential reactivity. For example, the reaction of related nucleophiles with a common electrophile. The modified equation is shown in Figure 3.1.[4]

Nucleophiles and electrophiles can be described as hard or soft.[5] A qualitative definition is linked to polarizability. High polarizability suggests that the distribution of electrons can easily change in a "soft" molecule, whereas changes don't occur as readily in "hard" molecules (low polarizability). Table 3.1 is a list of different nucleophiles and electrophiles with soft, intermediate, and hard characteristics.

Table 3.1. Some hard and soft acids (electrophiles) and bases (nucleophiles)

Bases (Nucleophiles)	Acids (Electrophiles)
Hard	Hard
H_2O, OH^-, F^-	H^+, Li^+, Na^+, K^+
$CH_3CO_2^-$, PO_4^{3-}, SO_4^{2-}	Be^{2+}, Mg^{2+}, Ca^{2+}
Cl^-, CO_3^{2-}, ClO_4^-, NO_3^-	Al^{3+}, Ga^{3+}
ROH, RO^-, R_2O	Cr^{3+}, Co^{3+}, Fe^{3+}
NH_3, RNH_2, N_2H_4	CH_3Sn^{3+}
	Si^{4+}, Ti^{4+}
	Ce^{3+}, Sn^{4+}
	$(CH_3)_2Sn^{2+}$
	$BeMe_2$, BF_3, $B(OR)_3$
	$Al(CH_3)_3$, $AlCl_3$, AlH_3
	RPO_2^+, $ROSO_2^+$, SO_3
	I^{7+}, I^{5+}, Cl^{7+}, Cr^{6+}
	RCO^+, CO_2, NC^+
	HX (H-bonding mol.)
Borderline	Borderline
$C_6H_5NH_2$, C_5H_5N, N_3^-, Br^-, NO_2^-,	Fe^{2+}, Co^{2+}, Ni^{2+}, Cu^{2+}, Zn^{2+}, Pb^{2+}
SO_3^{2-}, N_2	Sn^{2+}, $B(CH_3)_3$, SO_2, NO^+, R_3C^+,
	$C_6H_5^+$
Soft	Soft
R_2S, RSH, RS^-	Cu^+, Ag^+, Au^+, Tl^+, Hg^+
I^-, SCN^-, $S_2O_3^{2-}$	Pd^{2+}, Cd^{2+}, Pt^{2+}, Hg^{2+}, CH_3Hg^+,
R_3P, R_3As, $(RO)_3P$	$Co(CN)_5^{2-}$
CN^-, RNC, CO	Tl^{3+}, $Tl(CH_3)_3$, BH_3
C_2H_4, C_6H_6	RS^+, RSe^+, RTe^+
H^-, R^-	I^+, Br^+, HO^+, RO^+
	I_2, Br_2, ICN, etc.
	trinitrobenzene, etc.
	chloranil, quinones, etc.
	tetracyanoethylene, etc.
	O, Cl, Br, I, N, RO, RO_2
	M^0 (metal atoms)
	bulk metals
	CH_2, carbenes

Frontier Orbitals and Organic Chemical Reactions, p. 35, I. Fleming, 1976. Reprinted by permission of John Wiley & Sons, Ltd.

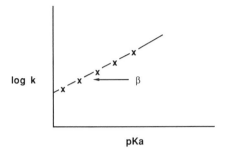

Figure 3.2. Bronsted plot—relationship between reactivity and pK_a of the attacking nucleophile.

Various scales have been used to describe hardness and softness and to predict reactivity.[6] Two soft molecules, a nucleophile and an electrophile, with a low energy gap $[E_{HOMO(nuc.)} - E_{LUMO(elec.)}]$ react favorably as described by the overlap term in the equation illustrated in 3.1.[4] Two hard molecules with corresponding higher energy gaps of the molecular orbitals can react effectively. In this case the second term, the electrostatic term, dominates. In many cases then, strong interactions occur between nucleophiles and electrophiles of like character, hard-hard, or soft-soft. Weaker interactions are found between soft and hard types. The order of reactivity of a series of nucleophiles depends on the relative contributions of the two terms. For example, in some ionic reactions; for example, Hg^+, a soft electrophile, reacts more favorably with soft ions in the order $HS^- > CN^- > Br^- > Cl^- > OH^- > F^-$. Ca^{+2}, a hard electrophile, reacts more favorably with hard ions in the order $OH^- > CN^- > HS^- > F^- > Cl^- > Br^- > I^-$. The reaction of S-adenosyl methionine and a thiol group is typical of a reaction of soft centers. Nucleophiles containing N or O at the reactive position have low polarizability (hardness) and are less reactive with S-adenosyl methionine.[7] Acylation and phosphorylation reactions involving the carbonyl and phosphoryl groups (hard centers) occur readily with hard nucleophiles such as O and N. For reactions at P(V), sulfur-containing nucleophiles may be less reactive. For example, the order of reactivity of O- and S-containing nucleophiles with diisopropyl phosphorochloridate is $CH_3CH_2O^- > PhO^- > PhS^-$.[8] How might this information help us understand phosphoryl transfer by an enzyme? Alkaline phosphatase, which catalyzes the dephosphorylation of a wide variety of phosphomonoesters, forms a phosphorylated covalent intermediate with a seryl residue as part of the reaction sequence. Under acidic conditions, the rate of the reaction is limited by the hydrolysis of the intermediate and, under basic conditions, by the dissociation of the enzyme phosphate complex. Changing the reactive seryl residue to a cysteinyl residue by site-directed mutagenesis alters the rate-limiting step of the reaction to the formation of the enzyme phosphoryl intermediate.[9] The differences in the stabilization of transition states at hard centers by S- and O-nucleophiles may contribute to the change in the rate-limiting step. Such differences in chemical features of O- and S-centers may be important in explaining phosphorylation of seryl residues by protein kinases and the paucity of reactions with cysteinyl residues.

The reaction rates of closely related nucleophiles, for example, those containing N, with a common electrophile often can be related to differences in basicity of the nucleophiles. A similar relationship might be found for a series of oxygen- or sulfur-containing nucleophiles, but a simple relationship between the two classes based on basicity alone is inadequate to explain their different reactivities. A Bronsted plot, log k versus the pK_a of the nucleophile, can be used to evaluate differences in reactivity of various nucleophiles. A relation can be obtained as shown in Figure 3.2.

This is typical of what is known as a *linear free energy relationship* because the logarithm of the rate constant is related to the free energy of activation of the reaction, and the logarithm of the dissociation constant (pK_a) is proportional to the free energy of dissociation of the nucleophile. The slope of the line (β) gives information about the transfer of charge in the transition state and may indicate the extent of bond formation. Sometimes nonlinear plots are obtained, and analysis of the curve can give decisive information about the transition state.[10]

Consider the plot is from a series of reactions of phenols with an acetylating reagent such as acetyl imidazole where the phenoxide ion is the reactive species. The linear plot with a positive slope suggests that dissociated phenols with higher electron density (higher pK_a's) are more reactive, that is, the higher the pK_a, the greater the nucleophilicity. Because charge and electron density are important in the reaction with different phenols, the order of reactivity with different phenols could depend on the pH of the reaction mixture. The phenolic group of O-fluorotyrosine has a pK_a value of 9.21, and the corresponding ionization in tyrosine occurs with a pK_a value of 10.07. At a pH value of 7.5, it was found that fluorotyrosine is approximately 7.2-fold more reactive than tyrosine is with acetyl imidazole.[11] But as the pH was raised, the order of reactivity changed, and it was calculated, that under pH conditions where both phenols are completely ionized, tyrosine would be 17 times more reactive than the fluoroderivative. Here, the basicity of the nucleophile dominates reactivity. Consider further the phosphorylation of tyrosyl residues in proteins, a type of acylation reaction, catalyzed by tyrosyl kinases. By using the insulin receptor, it was found that the V_{max} for fluorotyrosyl-containing gastrin is approximately ten-fold lower than that observed for the parent peptide. One explanation is that both tyrosyl derivatives are deprotonated or H-bonded in the active site, with the relative rates of reaction being partly determined by the nucleophilicity of the phenolate ions. In this case, the species with the higher pK_a (higher electron density), the tyrosyl residue, would be more reactive, as predicted by the Bronsted relationship. Although many factors are involved in different types of phosphorylation reactions, the basicity of the attacking nucleophile is likely very important. The alkoxide ion (of a seryl or a threonyl residue) is more basic and a more potent nucleophile than a phenoxide ion (a tyrosyl residue), and these differences may help explain why seryl kinases have generally higher V_{max} values than tyrosyl kinases.

The reaction rates or equilibrium constants for a series of related nucleophile attacking an electrophile also depend on the nature of the leaving group. This is termed nucleofagacity. The Bronsted coefficient (β) for the reaction of different tertiary amines with esters is +1.5 for the nucleophile. A plot of the logarithm of the equilibrium constant versus the logarithm of the pK_a of very basic leaving groups gave a slope of -1.5, showing the reaction is favored by decreasing the pK_a of the alcohol. With more activated leaving groups, phenols, the two corresponding Bronsted coefficients were found to be +0.1–0.2 for the nucleophile and -0.1–0.2 for the leaving group, respectively.[12] Hence, the characteristics of the leaving groups change the response to different nucleophiles and vice versa. A less reactive nucleophile may show more sensitivity to the nature of the departing leaving group than a more reactive nucleophile. These properties could be important in determining the specificity of some post-translational modification reactions. The phenolic group of tyrosine with a pK_a of 10 would be expected to be a better leaving group than the alcoholic group of serine, which has a pK_a value greater than 14.0. The differences in properties of the leaving groups of the protein substrates could contribute to the observed rate differences of different protein phosphatases. Calcineurin, a protein phosphatase, acts on phosphorylated seryl, threonyl, and tyrosyl residues in proteins, but with free amino acids, tyrosyl phosphate is highly favored. Using a series of phosphorylated esters with different leaving group potential, it was found that a linear relationship existed between the pK_a of the leaving group and the logarithm of the kinetic parameters, V_{max}/K_m.[13] The reaction is favored by increasing nucleofagacity. Alkaline phosphatase with a cysteinyl residue in place of a seryl residue at the active site shows a similar pattern of reactivity. A sulfur anion would be less reactive than an oxyanion and may explain why the mutant enzyme is particularly sensitive to the nucleofagacity of the leaving group.

By examining rate data and applying various free-energy relationships, the role of the nucleophilic reactivity of substrates has been evaluated in various enzyme-catalyzed processes. The application to post-translational modification reactions with proteins has been compromised because it has been difficult to chemically prepare variants of specific residues of changed nucleophilicity in the protein substrate. Two variants could be used to learn the importance of nucleophilicity of the protein substrate. One, analogs of the reactive portion can be chemically synthesized. An analog could be a peptide, amino acid, or some other derivative that is chemically related to the side chain. Two, the protein could be specifically changed by genetic methods to change the reactive residue. Some clues have been derived about the importance of nucleophilicity in ADP-ribosylation reactions catalyzed by cholera toxin. Cholera toxin normally modifies arginyl groups in proteins but can cause reactions with free arginine and various guanidine derivatives. A series of aromatic derivatives, benzylamino-guanidines, with different nucleophilicity were synthesized and used in kinetic experiments.[14] The reaction constant, k_{cat}, and k_{cat}/K_m were fit with parameters from the linear free-energy Hammett equation. The value of the parameter, ρ, was <0. If an electron-deficient center occurs in the transition state, the presence of electron-donating groups could enhance the formation of the transition state. It was suggested that this accounts for the observed increase in k_{cat} and fits a scheme of a S_N2 reaction where the guanidino group makes a nucleophilic attack on the C_1 carbon of ribose in β-NAD, causing an inversion in configuration producing the α-N-glycosidic derivative.[15] A deprotonated guanidinyl group found in the free protein or generated in the enzyme-substrate would make a potent nucleophile. Deprotonation or reduction of the pK_a of the guanidino group has been suggested to explain the enhanced chemical reactivity of arginyl groups in phosphate-binding sites in proteins. Similarly, the high reactivity of an ϵ-amino group of a lysyl residue in acetoacetyl decarboxylase is due to a reduction of its pK_a from the normal value of 10 to 5.9. This perturbation of the lysyl group then provides a potent nucleophile for the catalytic reaction at pH 5.9.[16]

STRUCTURAL AND CHEMICAL FEATURES OF THE SUBSTRATE

If an enzyme is going to catalyze the modification of some specific side-chain groups in the protein, it is clear that the residue to be reacted must be accessible to the active site of the enzyme. Hence, it seems unlikely that enzymatic reactions, in contrast to chemical modification, can occur

with residues deep in the interior of a protein. Reactions could occur in
a crevice or on the surface of the protein substrate. But are there specific
structural requirements for these types of reactions? Is a β-turn, an α-
helix, a β-sheet, or some other structure involved? These or other
structures creating a microenvironment about the modifiable residue
could cause the functional group to have certain chemical characteristics
important for enzyme recognition, for example, have a perturbed pK_a
value or be H-bonded. The use of synthetic peptides to model reactive
sites of protein substrates can give valuable information about features
influencing specificity and is particularly revealing when both
enzymological and physical studies are done. To illustrate different
principles and approaches used to probe the importance of structural and
chemical features of the protein substrate in post-translational modification reactions, various examples are presented.

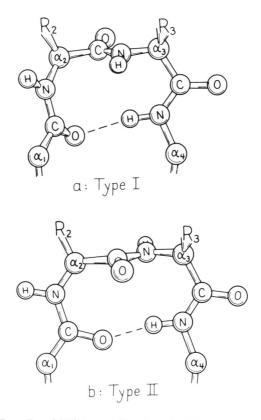

Figure 3.3. Type I and II β-turns. Reprinted with permission from *Adv.
Protein Chemistry* 34:167 (1981). Copyright 1981, Academic Press.

Table 3.2. Glycosyl-acceptor capabilities of various hexapeptides with a modified 'marker sequence'. Relative N-glycosyltransferase activities (V_{max}/K_m) are based on the activity with the serine containing peptide. Abbreviations used: n.d., not determined.

Amino acid sequence	K_m (mM)	V_{max} (c.p.m./10 min.)	Relative Activity
Tyr-Asn-Gly-Thr-Ser-Val	0.16	2080	40
Tyr-Asn-Gly-Ser-Ser-Val	1.6	550	1
Tyr-Asn- Gly-Cys-Ser-Val	n.d.	n.d.	0.4
Tyr-Asn-Gly-Val-Ser-Val	—	—	0
Tyr-Asn-Gly-Thr(OCH$_3$)-Val	—	—	0

Reprinted with permission from *Biochemistry Journal* 195, 639. Copyright 1981, The Biochemical Society and Portland Press.

β-Turns and Related Structures

The β-turn is an organized structural element in proteins that causes a reversal in direction of the peptide chain. A turn may consist of four amino acid residues where the carbonyl oxygen of residue i may be H-bonded to the NH residue of the i+3 residue in the turn. Figure 3.3 illustrates structures for two major turn types.

Specific amino acids sequences favor formation of turns, and certain amino acids are more likely found than others at specific sites, for example, glycine is the preferred residue at the third position of a type II β-turn because other residues with bulky side chains cause steric hindrance. Another type turn, γ, consists of three residues with possible H-bonding between residue i and residue i+2. A review on β-turns is useful for further study.[17] Turns have been suggested as recognition sites for certain types of post-translational modification reactions—for example, glycosylation, hydroxylation, phosphorylation.

Glycosylation

The amide nitrogen of an asparagine residue is the acceptor of a glucosamine-linked sugar, $Glc_3Man_9GlcNAc_2$ from a dolichol pyrophosphate derivative in enzymatic N-glycosylation reactions. Commonly, a sequence of three amino acids, Asn -X- (Ser, Thr), is found in the acceptor site of glycosylated proteins[18] where X is any amino acid except aspartic acid, glutamic acid, or proline.[19,20,21] Work based on the method of Chou and Fasman[22] for predicting secondary structures in proteins suggested that these sequences have a high probability of occurring in a β-turn or other loop structures[23,19] and that these structures could serve as recognition sites for enzymes involved in N-glycosylation, but the exact structure of the substrate is not yet proven.

The acceptor capabilities of different peptides with different amino acids at the hydroxamino acid position show how important this residue is in glycosylation (Table 3.2). Note the consequence of methylation of threonine on the enzymatic reaction.

One proposed structure to explain assistance of the hydroxy-amino acid (Model A shown in Figure 3.4) involves a hydrogen bond of the hydroxyamino acid with the carbonyl carbon of the amide.[24]

A second model, based on extensive kinetic data,[25,26,27] suggests that serine or threonine serves as a H-bond acceptor from the amide N of asparagine, enhancing the nucleophilic reactivity of the amide group for its reaction with the dolichol pyrophosphate sugar derivative. A scheme illustrating these features is shown as Model B in Figure 3.4. To test these two models, peptides containing threo- β- and erythro-β-fluoroasparagine were made, and kinetic studies were compared with the glycosylation of an asparagine-containing peptide. The fluoroderivatives gave lower V_{max} values and higher K_m values.[28] It was suggested from an earlier study that if Model B were correct, a reduced rate should be seen in glycosylation reactions because the electron withdrawing character of the fluoroatom would decrease the nucleophilicity of the amide N.[29] If Model A were correct, it was suggested that the fluoroatom in asparagine would enhance glycosylation.

Hence, an important point suggested by these studies is that the protein substrate itself might contribute to the catalytic process by supplying some group that activates the amino acid side chain to be modified.

Still, the exact structure of the substrate is not established. Circular dichroism measurements of peptides, containing the sequence Asn-X-Ser(Thr), in aqueous-organic mixtures suggest that a β-turn can be generated in small peptides.[30] DMSO at 10 percent promotes both formation of structure and glycosylation.[31] Further support of the view that organized structure is important for enzyme recognition is provided by results of affinity chromatography in the purification of an oligosaccharyl transferase from hen oviducts.[32] In 10 percent DMSO, binding of the enzyme to bound peptide is promoted. By removing the DMSO, the conformation of bound peptide is thought to change, and the dissolution of the enzyme occurs. This approach to purification may be quite useful for the preparation of other post-translational modifying enzymes which recognize organized structural elements in their substrates.

Another possible structural element is the asparagine turn.[33] This is a secondary structural motif in which H-bonding occurs between the carbonyl group of an aparaginyl-carboxamide side chain with an amide

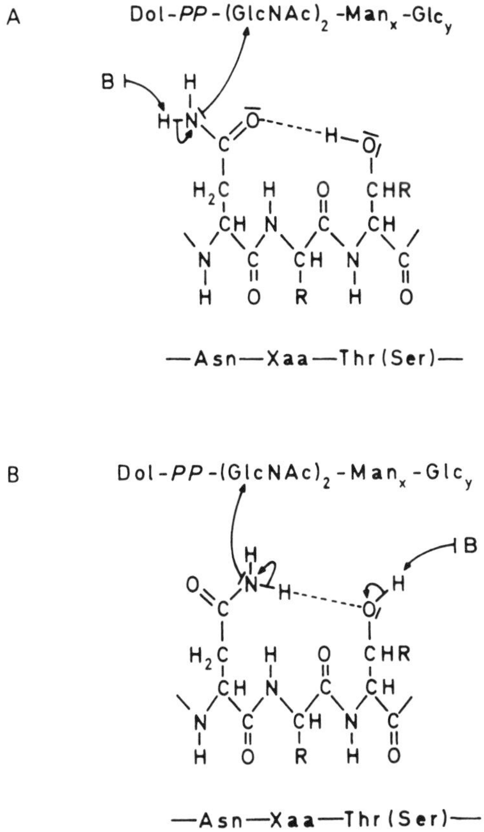

Figure 3.4. Models representing possible hydrogen-bonded interactions between the side chains of asparagine and the hydroxy amino acid in the "marker sequence" Asn-Xaa-Thr(Ser). Structure A (proposed by Marshall, 1972) incorporates a hydrogen bond between the hydroxy function of the hydroxy amino acid and the carbonyl-group oxygen atom of the asparagine residue. In structure B, the amide group of asparagine functions as hydrogen-bond donor and the hydroxy-group oxygen atom of the hydroxy amino acid as hydrogen-bond acceptor. The significance of both types of interactions for the catalytic mechanism of enzymic glycosyl transfer is discussed in the text. Reprinted with permission from *Biochemical Journal* 195:639 (1981). Copyright 1981, The Biochemical Society and Portland Press.

H of the polypeptide backbone. DMSO, a protic solvent thought to mimic solvent conditions at membrane glycosylation sites, induces a conformational state of a tripeptide, Ac-Asn-Ala-Thr-NH$_2$, characteristic of an

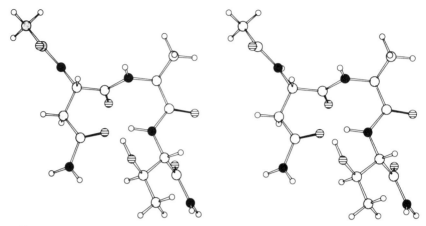

Figure 3.5. Stereoview of proposed conformational model for Ac-Asn-Ala-Thr-NH$_2$. Reprinted with permission from *Biochemistry* 30:4374 (1991). Copyright 1991, American Chemical Society.

asparagine turn (Figure 3.5). This peptide is a good substrate for N-glycosylation, but tripeptides used not possessing this conformational state are not.[34]

A mechanism has been proposed suggesting that proton transfer can occur effectively from the hydroxylic side chain of serine or threonine to the oxygen atom of the carboxyamido function of asparagine in the asparagine turn. This protonation could influence stabilization of a transition state in which the amide group is converted to an imidate or its reaction rate with the dolichol pyrophosphate oligosaccharide.[35] Various structures in substrates may help orient an exposed asparaginyl residue, change its chemistry, and influence reactivity.

N-linked glycosylation of β-protein C presents another feature of modification; that is, the rate of translation of the protein may determine in part which potential sites in the protein substrate are glycosylated.[36] The β-form is closely related to the α-form but differs in that no glycosylation is found at Asn$_{329}$ in the sequence Asn-Ser-Cys. This cysteinyl residue may promote glycosylation according to the results presented in Table 3.2. But further down the chain is Cys$_{345}$, which reacts to form a disulfide bond with Cys$_{331}$. Hence, it has been suggested that if a disulfide bond is formed first, then the activating effect of Cys$_{331}$ would disappear. If translation is slow, the protein may spend sufficient time in the endoplasmic reticulum allowing glycosylation before Cys$_{345}$ emerges to influence the reaction process.

Reactions involving O-glycosylation of seryl and threonyl residues also have been analyzed by the Chou and Fasman procedure, and the results are consistent with the placement of the reacting residue in a β-turn.[37] When synthetic peptides of the reactive threonyl residue of myelin basic protein were used, the minimal sequence Thr-Pro-Pro-Pro was required for glycosylation. The reaction was improved by an extension on the amino terminal side. In fact, the peptide Val-Thr-Pro-Arg-Thr-Pro-Pro-Pro reacted more favorably than the protein substrate.[38] But, whether a particular organized structural element in the peptide and the protein is needed for O-glycosylation is uncertain.

Hydroxylation

Several modification reactions are involved in the formation of a mature collagen molecule with a triple helix. Hydroxylation of prolyl residues occurs both as co- and post-translational events before the individual chains are intertwined.[39] But what structures might be recognized by the prolyl-4-hydroxylase that modifies procollagen? Again, using peptides as models for portions of modifiable proteins, insight was gained about the natural process. Structural analysis of pentapeptide substrates of the sequence Gly-X-Pro-Gly-Y suggests that the Gly-X-Pro portion of the sequence is in the poly proline type II structure, and the X-Pro-Gly-Y fragment is in a β-bend conformation.[40] The amino acid at position X is suggested to influence the orientation of the prolyl residue needed for effective hydroxylation by changing the psi_1 and psi_2 dihedral angles.[41] As hydroxylation proceeds, structural reorganization is thought to occur, leading to the extended conformation for formation of the triple helix.[42] Hence, hydroxylation is thought to have an important role in the kinetics of folding as well as its effect on thermodynamic stability of the triple helix of collagen.

Infrared spectroscopy of "collagen like peptides" and circular dichroic measurements of peptides in water and trifluorethanol suggest that good peptide substrates can have the structure of a combined polyproline II helix followed by a β-turn. As a polyproline helix is not stabilized by hydrogen bonding, even small peptides can show some organized structure. A structure for the binding of peptide to prolyl-4-hydroxylase has been proposed (Figure 3.6).[43]

Phosphorylation

An analysis of seryl and threonyl phosphorylation sites suggested to Small, et al.[44] that a turn might be a recognition point for certain protein kinases. But what types of experiments have been or can be done to test the idea that a specific structural element is needed for efficient phosphorylation by a specific kinase? Consider two cases: (1) casein kinase II, and (2) cyclic AMP-dependent protein kinase.

Casein kinase II phosphorylates substrates in which a seryl or a threonyl residue is clustered by acidic amino acids on the carboxyl end of the reactive site.[45] Tyrosine is never phosphorylated. Small peptides containing natural sequences of protein substrates can give V_{max} values equivalent to values obtained with the intact protein. The K_m's are somewhat higher, but overall the results suggest that the prominent features required for efficient catalysis and recognition are found in a small segment of the polypeptide chain. An acidic amino acid, three residues away from serine or threonine, is of central importance. Positions 1 and 2 from the serine or threonine can be substituted by alanine, but the best substrates contain multiple acidic amino acids. A β-turn has been suggested for casein kinase II substrates.[46] A turn potential for the tetrapeptide sequence ser-glu-glu-glu is 3.7-fold greater than the average probability for a type I β-turn.[47,48] It has been suggested that the preferences for amino acids in turns reflect local interactions and that predictive measures can also apply to peptides.[48] A turn may bring the side chain groups into position so that they may interact favorably with corresponding groups on the enzyme (negative carboxylates in the substrate with cationic groups in the enzyme). Another consideration is that the acidic amino acids may alter the microenvironment of the seryl or threonyl residue and change its reactivity. But structural information is needed for peptides in solution and when bound to the enzyme.

A synthetic octapetide peptide, glu-ser-leu-ser-ser-ser-glu-glu, of a phosphorylatable site in casein is acted upon readily by casein kinase II. On the basis of structural predictions, abnormal binding on a C_{18} reverse phase column and ^{13}C-NMR spectroscopy, a β-turn was proposed for the leu-ser-ser-leu segment with the carbonyl oxygen of ser 6 involved in hydrogen bonding.[46]

Cyclic AMP-dependent protein kinase also phosphorylates seryl or threonyl residues, but in this case the catalyst prefers basic amino acids on the amino terminal side of the phosphorylatable site. A phosphorylatable sequence may contain the residues arg-arg-x-ser. Such a sequence is found in pyruvate kinase and a peptide of this segment, leu-arg-arg-ala-ser-leu-gly, is phosphorylated nearly as effectively as the native protein. Primary structural requirements for phosphorylation have been defined from the use of synthetic peptides in kinetic studies, but what these variations in sequence mean in terms of conformation of the substrate is not obvious. Here, NMR spectroscopy has been extremely useful, particularly in determining peptide structures bound to the enzyme.

An analysis of the structure of bound peptide is based on the fact that paramagnetic ions bound to the protein perturb the ^{1}H-NMR of nearby hydrogen atoms. With certain assumptions, distances of hydrogen

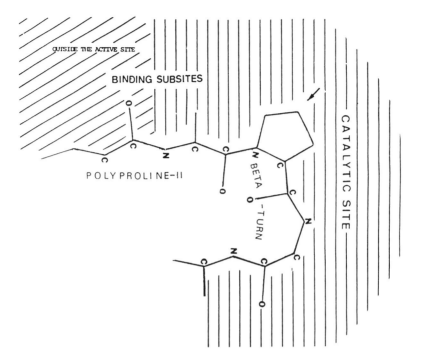

Figure 3.6. Conformational model for proline hydroxylation. The backbone of the peptide substrate and the proline ring are sketched in the PP-II + β-turn conformation inside the active site of proline 4-hydroxylase. The site of hydroxylation is indicated by the arrow. The peptide is shown with the PP-II arm interacting with the enzyme at the binding subsites situated in and outside the active site. The β-turn segment of the peptide is shown interacting with the enzyme at the catalytic site. Reprinted with permission from *Journal of Biological Chemistry* 264:2852 (1991). Copyright 1991, The American Society for Biochemistry & Molecular Biology.

atoms from various probes may be deduced and information derived about the structure of the bound peptide. Distances of carbon-bound protons of bound peptide substrates to Mn^{+2} in an inhibitory site and to Cr^{+3} in the bound complex b,g-bidentate,Cr^{+3}-b,g-methylene ATP ruled out that the bound peptides had an a-helical or b-sheet conformation.[49] Determination of distances to backbone C a protons and backbone amide (NH) protons of bound peptide substrates to Mn^{+2} with bound $Co(NH_3)_4AMPPCP$ in the ATP site eliminated b-bulge and b-turn conformations[50] and suggested that the peptide was bound as an extended coil. The results of kinetic studies with synthetic peptides of restricted conformational states due to the presence of peptide bonds containing N-methylated amino acids helped show that one of two coiled conformations (A) was recognized by cyclic AMP-dependent protein kinase.[50,51]

An active site that accommodates a coiled peptide substrate can have extensive contacts with the substrate and may be necessary for specificity and effective catalysis. The combined approach of physical measurements and detailed kinetic studies illustrated in these studies provided answers about specificity and suggests a means for study of other post-translational modification reactions.

Proteolytic processing

The release of hormones from prohormone precursors, the removal of signal peptides in protein translocation, and the activation of zymogens in a blood coagulation cascade are well documented cases of how limited proteolysis can influence important physiological events. What are some structural features in protein substrates that may influence the specificity of the reactions?

Prohormonal processing occurs often at sites containing multiple basic amino acids, for example, arg-arg, lys-arg, or lys-lys; cleavages may occur differentially and not all sequences are acted upon. A specific structure in the protein substrate that is nearby or part of the site to be cleaved may contribute to the specificity of the process. A reverse turn or an omega (Ω) loop may facilitate recognition. An Ω loop is a continuous stretch of the polypeptide chain, 6–16 amino acid residues long that forms a neck region, with the chains no more than 1 nm apart; thus, the structure actually resembles the Greek symbol Ω.[52] The loop may contain no regular secondary structure and may have been regarded as "random coils". The coil, which is surface oriented in proteins, could bind to a modifying enzyme. An approach taken to evaluate the significance of coils in prohormone processing was to determine if a relationship existed between predicted structure and enzymatic reactions. On this basis, it has been suggested that prohormonal processing may be associated with the recognition of Ω loops and that cleavage sites occur in the neck region.[53]

An amino terminal segment (15–30 residues long), the signal peptide, is found in pre forms of most secretory proteins; during co-translation and translocation, this segment of the protein, involved in transport, is cleaved off as it passes through the luminal side of the endoplasmic reticulum membrane by a signal peptidase. The signal peptide region usually consists of a positively charged n-terminal region, a hydrophobic stretch, a region containing residues that interrupt the organized structure of the hydrophobic section, and the cleavage region. It has been suggested that the region next to the cleavage region may contain a β-turn.[54] Results of site-directed mutagenesis show that the larger the accessible surface area is of the amino acid at the cleavage site, the worse is the cleavage. Also, for a given area, an increase in polarity of the side chain will cause a decrease in processing.[55] In a study of proteolytic processing of human pre(pro)apolipoprotein A-II, the position of proline, a residue thought to induce a β-turn, influenced processing.[56] The position of a turn could affect the way the portion of the polypeptide is oriented into the lumen for cleavage by the signal peptidase.

Helical and Sheet Structures

Carboxylmethylation

The esterification of specific carboxyl groups in proteins is catalyzed by various carboxyl methyltransferases with S-adenosylmethionine as the methyl donor. Reactions are well documented in prokaryotes and eukaryotes, but interesting differences exist in these processes and in the nature of the protein substrates. In the case of prokaroytes, esterification of the γ-carboxyl group of glutamyl residues occurs. But not any protein can be modified and only select glutamyl side chains in certain proteins can be modified. Four membrane-bound transducer proteins, Tsr, Tar, Trg, Tap, involved in adaptation mechanisms in chemotaxis, are specific targets for methylation reactions.[57,58] Multiple modification occurs on each receptor and an analysis of *in vivo* methylation of aspartate receptors in *E. coli* indicates about 0.5 moles of methyl groups are transferred. The methyl groups are distributed between four sites; the sequences and relative rates of modification are shown in Table 3.3.[59]

Table 3.3. Comparison of sequences with the relative rates of methylation and demethyulation at each site in RP 4372 pGK2 after adaptation to aspartate. A "consensus" sequence obtained is shown below the sequences from sites 1-4. The second residue in each sequence shown is methyl-esterified. For purposes of this comparison, serine and threonine residues were treated as if they were equivalent.

Site	Sequence	Agreements with consensus	Relative rates of modification	
			Methylation	Demethylation
3	Glu-Glu-X-X-Ala-Thr	4	100	100
2	Glu-Glu-X-X-Ala-Ser	4	63	46
1	GluGlu-X-X-Ser-Ala	2	9	29
4	Gln-Glu-X-X-Ala-Ala	2	2	23

Glu-Glu-X-X-Ala-thr/Ser = consensus sequence. Reprinted with permission from *Journal of Biological Chemistry* 261: 10814. Copyright 1986, The American Society for Biochemistry and Molecular Biology.

A consensus sequence of Glu-Glu*-X-X-Ala-Thr/Ser was suggested as a favorable target site with methylation occurring at Glu*. Results with Trg and other transducer proteins led to the suggestion of Ala/Ser-X-X-Glu-Glu*-X-Ala/OH-Ala-OH/Ala for a more extended consensus sequence for methylation.[60] To explain the specificity of methylation it was suggested that regions of the primary structure having a consensus sequence exist in a α-helix, depicted for aspartate and serine receptors in Figure 3.7.

Note that the glutamyl residues are on one face of the a-helix placing them in a similar position for an enzymatic reaction. In site 3 of the Tar receptor (Figure 3.7) modification occurs at residue. $_{309}$ Alanine and threonine at positions $_{312}$ and $_{313}$ are on an adjacent turn of the helix. Changing the order of these two residues by site-directed mutagenesis causes a four-fold decrease in reactivity of site 3.[61] Hence, it is believed these residues brought together in a spatial relationship on one side of an a-helix are important features in the protein substrate which could contribute to the specificity of modification. The receptor contains a large degree of a-helical character,[62] but direct proof that the modification site is an a-helix is needed. Alanyl residues, helical forming residues, are often found on the carboxyl side of the modification site.

Another important facet of carboxymethylation of these receptor proteins is that not all of the glutamyl sites are translated from their respective mRNAs as glutamyl residues. In fact, glutamate$_{309}$ in site 3 is made as glutamine and then deamidated by an enzyme known to catalyze the hydrolysis of the carboxymethyl esters of glutamyl residues of the receptor proteins.[63] Whether an α-helical structure is important for specific deamidation also is not known. The modification sites of sensory-transducer proteins often involve a pair of glutamyl or glutaminyl residues with the second residue being the favored site for modification.[64]

Hydroxylation

Results of the hydroxylation of prolyl containing substrates by a plant prolyl-4-hydroxylase provides convincing evidence that helical structure in a substrate is important for the activity of this enzyme. Polyproline is known to form a type II left-handed helix with three residues per turn in aqueous solutions and these poly(L-proline)-like helical structures are found in natural proteins such as collagen, cuticulin, and bovine pancreatic ribonuclease. By using a series of Boc-Pro$_n$ and Pro$_n$ oligomers as models for portions of natural proteins, it was found effective hydroxylation could only occur with derivatives containing organized structure. The optimal temperature for the reaction of Boc Pro$_8$ is at 32°, but the optimal temperature for Pro$_5$ is only at 15° (Figure 3.8), and this differ-

WILD TYPE   MUTANT

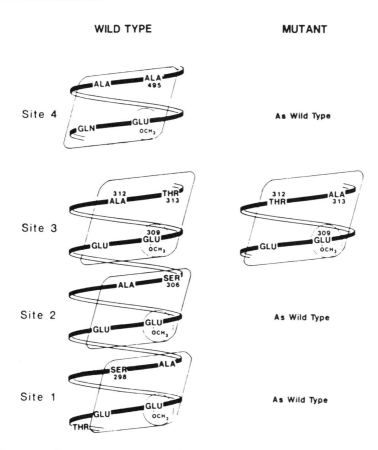

Figure 3.7. Recognition sequences for sites of methylation. The amino acid sequences of residues 293-313 and residues 490-495 of the S. typhimirium aspartate receptor are depicted as α-helices. Sites 1 and 3 are synthesized as glutamines but are rapidly deamidated by the chemotaxis methylesterase. Reprinted with permission from *Journal of Biological Chemistry* 259:7719 (1984). Copyright 1984, The American Society for Biochemistry & Molecular Biology.

ence is attributed to the stabilities of the two helical structures.

Thus, the temperature dependence of the reactions and the temperature effects on the CD spectra of Pro_n oligomers support the view that prolyl oligomers in a type II helix are preferred substrates for the plant prolyl-4-hydroxylase.[65] Hydroxyproline containing proteins in plant cell walls are known to have a polyprolyl type II helical conformation,[66,67] but the proof of structure of the unhydroxylated form awaits further study.

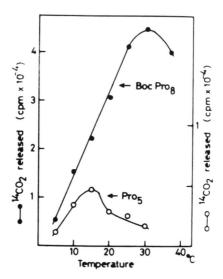

Figure 3.8. Optimum temperatures of prolyl hydroxylase toward Pro_5 and Boc-Pro_8. The enzyme and bovine serum albumin were preincubated with Pro_5 (open circle) or Boc-Pro_8 (closed circle) at a temperature illustrated on the abscissa, and then the separately preincubated mixture containing cofactors and co-substrate was added at the same temperature. The enzymic reactions were done for 30 min at the respective temperatures. Reprinted with permission from *Journal of Biological Chemistry* 256:11397 (1981). Copyright 1981, The American Society for Biochemistry & Molecular Biology.

Limited proteolysis

Several peptides possessing antibiotic activity are cosecreted with an endopeptidase from granular glands in frog skin. This peptidase, called "magaininase", acts on peptides of the magainin family,[68] producing half-peptides that no longer have antibiotic activity. Magaininase can cleave these structures at lys-lys, gly-lys, and leu-lys peptide bonds, but specificity cannot be explained by simple recognition of dipeptide linkages, particularly because other gly-lys and leu-lys structures found in these peptides are not cleaved. The natural substrates of the magainin family, magainin 2, PGLa, XPF, and CPF are some 20 residues long and have in common an amino terminal glycyl residue and a lysyl residue on the carboxyl side of the peptide bond that is cleaved. There is no similarity in primary structure beyond this.

Analysis of the cleavage products of the natural peptides and synthetic analogs can be explained by a model in which magaininase acts on peptides with a particular secondary structure.[69] It is suggested that substrates contain an amphipathic α-helix of at least 12 residues. The helix on one face contains four nonpolar residues thought important for binding; the other face contains the lysyl residue in a similar orientation in all peptide substrates. In these structures, the position of the cleavable linkage is shared on the hydrophilic face. Changing the spacing on the amino terminal side, introducing hydrophilicity to the hydrophobic face, or by inserting proline, an α-helical breaker, in peptides influences dramatically the enzymatic reaction consistent with the view that the enzyme recognizes organized structural elements in its substrate.

Physical studies have shown that these peptides can adopt an α-helical structure in a phospholipid environment,[70,71,72] but no information is yet available about their structures when bound to enzyme. But other peptides with related helical potential like mellitin,[73] β-endorphin, and mastoparan can form an α-helical structure when they interact with calmodulin.[74]

Lipoylation

The glycine cleavage system, a multienzyme complex, catalyzes the oxidation of glycine forming 5,10-methylenetetrahydofolate, CO_2, NH_3, and NADH. The H-protein of the multienzyme complex contains bound lipoic acid. The lipoic acid bound to an ε-amino group of lysine serves to accept a methylamine moiety derived from glycine and transfers the methyl group to tetrahydrofolate.[75] The sequence about the bound lipoic acid shows significant homology with sequences of acyl transferases of α-keto acid dehydrogenases. This similarity in structures could be due to related functions served by the lipoyl group in the different proteins and/ or be needed for effective lipoylation. Site-directed mutagenesis and lipoylation of the mutant proteins suggests that the conserved glutamate$_{56}$ and glycine$_{70}$ are important for effective lipoylation of lysine$_{59}$. Other conserved residues may not be so critical for lipoylation.[76] The region of lipoylation is suggested to be α-helical with glutamate and lysine on the same side of the helix. It is suggested that an interaction of a lipoyl-transferring enzyme with glutamate$_{59}$ may serve to position the enzyme for lipoylation.

Hydroxylation

Aspartic acid and asparagine residues in specific sequences of blood-clotting proteins are known to be hydroxylated by a 2-oxyglutarate-dependent dioxygenase, and the reactive Asp/Asn residue is in a consen-

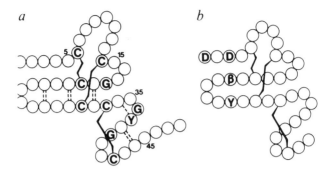

Figure 3.9. (a) Schematic representation of the secondary structural features of (1-48) hEGF. Disulphide bridges are marked with solid lines and hydrogen bonds with dashed lines. EGF-like sequences are found in several proteins and some highly conserved residues are marked; the residue at position 37 is sometimes Phe in the related sequences. The conservation of residues argues strongly for the conservation of the associated structural features. (b) Extrapolation of the features in (a) to the first EGF-like region of Factor IX (residues 47-84) and Factor X (residues 46-83). 'β' represents β-hydroxyaspartate or β-hydroxyasparagine and Phe may replace the marked Tyr. Reprinted with permission from *Proceedings of the National Academy of Science, USA* 86:444 (1989).

sus sequence, Cys-Xaa-Asp/Asn-Xaa-Xaa-Xaa-Xaa-Phe/Tyr-Xaa-Cys. One such sequence is identified as the first epidermal growth factor domain in Factor IX, and a study of the solution structure of the epidermal growth factor by NMR spectroscopy shows this domain consists of a triple-stranded β-sheet conformation.[77] A schematic representation of the structure and the relationship of a proposed structural element from various blood-clotting proteins is shown in Figure 3.9.

The reactive aspartate in the epidermal growth factor domains in these proteins is on the same face as two other aspartates and may constitute a binding site for calcium after hydroxylation. Note that a tyrosyl group is adjacent to the reactive aspartate in the β-sheet. A question unanswered is does the dioxyogenase recognize the β-sheet, the orientation of tyrosyl and aspartate residues with respect to each other, or some combination of these two possible three-dimensional arrangements. Using synthetic peptides of this segment in enzymatic studies, the results support the concept that organized structure is needed for effective modification. Linear peptides don't serve as substrates, but peptides containing disulfide bonds of the epidermal growth factor domain of Factor IX do.[78,79]

INFLUENCES OF NEIGHBORING GROUPS ON MODIFICATION

End Group Effects

Acetylation

Of the vast array of proteins biosynthesized in eukaroytes, more than half contain amino terminal acetyl groups. But modification is not restricted to eukaryotes inasmuch as acetylation is well documented in prokaryotic and viral proteins. The main target for modification is alanine or serine, and reactions may occur during biosynthesis or after chain completion. Consider some factors in the protein substrate that may influence reactions involved in co-translational acetylation of the α-amino group of an eukaryotic protein. The addition of an acetyl group can occur by the action of N α-acetyltransferase and acetyl CoA to a growing peptide chain about 20–50 residues long. The first residue of a nascent eukaryotic peptide chain is methionine. It can be acetylated, it can retain its free α-amino group, or it can be cleaved by a methionine amino-peptidase. An acetyl group may or may not be added to the new amino terminal group. Various studies show that the nature of the penultimate residue adjacent to the methionyl residue is important in the reactions,[80,81,82] but other factors in the amino terminal region up to 40 amino acids contribute.[83,84] If the penultimate residue (X) is bulky, acylation or cleavage of methionine will be hindered. When X is small (radius of gyration less than 1.29 Å), acylation of methionine is favored, but cleavage can also occur. The study of Huang et al.[81] illustrates the importance of the penultimate residue. The plant protein, thaumatin, which is expressed in yeast, was altered in the penultimate position by site-directed mutagenesis of a synthetic gene. Four categories of proteins are expected depending on the interplay of hydrolysis and acetylation. These groups are shown in Figure 3.11. Category 1 results from cleavage of the methionyl-X peptide bond. Category 2 results from cleavage and acetylation (note that X is small here). Category 3 results from lack of cleavage and acetylation (note that X is large). Category 4 results from acetylation of uncleaved methionyl-X proteins.

Another mechanism may explain the four categories of proteins seen in acetylation reactions. After co-translational acetylation of the methionyl group, cleavage of the acetyl methionyl peptide bond may occur by a specific hydrolase.[85] The penultimate residue released could then be acetylated by an acetyltransferase. Such a mechanism has been demonstrated for the processing of *Dictyostelium discoideum* actin[86] and may apply to other eukaryotic proteins.[82,84]

Chart I: Effect of the Penultimate Amino Acid on Cotranslational Modifications of Plant Thaumatin Expressed in Yeast as Nonsecretory Mutants

Category 1

X ———— :

X=

Proline
Valine
Cysteine

Category 2

Ac-X ———— :

X=

Glycine
Alanine
Serine
Threonine

Category 3

M-X ———— :

X=

Isoleucine
Leucine
Methionine
Phenylalanine
Tyrosine
Tryptophan
Lysine
Histidine
Arginine
Glutamine

Category 4

Ac-M-X ———— :

X=

Aspartic Acid
Glutamic Acid
Asparagine

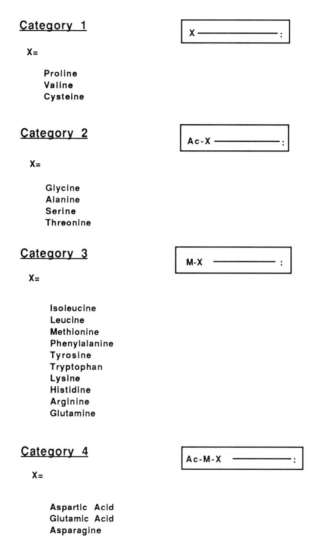

Figure 3.10. Categories of proteins produced by cotranslational modifications of eukaryotic proteins as a function of the penultimate residue. Reprinted with permission from *Biochemistry* 27:7979 (1988). Copyright 1988, American Chemical Society.

Synthetic peptides and purified N-acetyl transferase and methionyl N-acetyl transferase have been used to examine which features in substrates influence acetylation. The penultimate residue, with the exception of proline, had little effect on reactions with N-acetyl transferase,[87,88] but had more effect on the enzyme acting upon the methionyl residue. Amino acids can, however, be acetylated by N-acetyl transferase in peptides that are not often acetylated in proteins. Peptides can be useful models of protein substrates and appropriate for this study, but questions should be raised regarding the quality of these models.

Myristoylation

An interesting parallel exists between myristoylation and acetylation reactions. One, both reactions may occur co-translationally on the α-amino group, and two, glycyl residues can serve as acceptors of the acyl group. Other residues can serve as acceptors in acetylation reactions as pointed out above. But when glycine is modified, are their features in the protein substrate that regulate myristoylation reactions? And are any of these structural features important for determining whether a particular sequence is myristoylated or acetylated? The amino terminal sequence of various myristoyl-proteins is shown in Table 3.4.[89]

This region of the protein substrates is important for the enzymatic reaction as illustrated in site-directed mutagenesis studies[90] and in the use of synthetic peptides as substrates. No striking homology exists, however, in the primary structures of these proteins. Yet certain features, for example, hydrophobicity, size, or charge, may be accommodated by various residues at key sites. An indication of these structural features is illustrated by results obtained with yeast N-myristoyl transferase (Figure 3.11). Serine at position 5 influences myristoylation of peptides and proteins. N-myristoylation of rat Goα, containing serine 5, occurs in *Saccharomyces cerevesia*, but human Gsα, not containing serine 5, does not. Octapeptides of G protein substrates, however, are not

Table 3.4. Amino terminal sequences of known N-Myristoyl proteins

Sequence	Protein
Gly-Asn-Ala-Ala-Ala-Ala-Lys-Lys	cAMP-dependent protein kinase
Gly-Ser-Ser-Lys-Ser-Lys-Pro-Lys	pp60[src]
Gly-Asn-Glu-Ala-Ser-Tyr-Pro-Leu	β subunit of calcineurin
Gly-Gln-Thr-Val-Thr-Thr-Pro-Leu	Murine leukemia virus gag p[15]
Gly-Ala-Gln-Leu-Ser-Thr-Leu-Gly	Cytochrome b$_5$ reductase
Gly-Cys-Thr-Leu-Ser-Ala	G proteins

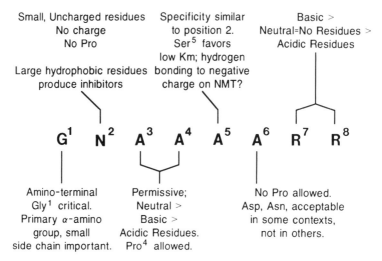

Figure 3.11. Yeast NMT peptide substrate specificity. This figure summarizes the substrate specificity data of yeast NMT. The similarity in the residues allowed at positions 2 and 5, together with the negative effects of acidic residues at positions 2, 5, and 8 on peptide binding, suggests a helical structure for peptide ligands when bound to NMT. However, crystallographic studies will be necessary for the definitive assignment of NMT substrate secondary structure. Reproduced, with permission, from the *Annual Review of Biochemistry* 57:69 (1988). Copyright 1988 by Annual Reviews, Inc.

substrates; acidic residues at position 7 and 8 seem to be negative determinants.[91]

The regulatory protein subunit of cyclic AMP-dependent protein kinase, the first protein demonstrated to be myristoylated, has the amino terminal sequence Gly-Asn-Ala-Ala-Ala-Ala-Lys-Lys. A synthetic analogue of this sequence, with Arg replacing Lys, can be effectively myristoylated. But if Asn is substituted by its charged relative, Asp, no reaction can occur.

Compare these results with those concerning the amino terminal Gly sequences of acetylated proteins. Note that many of these proteins have a preponderance of charged amino acids in the first few residues after glycine. In one cytochrome C, the sequence is identical to that of the regulatory subunit of cyclic AMP-dependent protein kinase, with exceptions at position 2 (Asp for Asn) and position 6 (Gly for Ala). Further studies are needed to determine whether these structural differences in the protein substrate or other factors, for example, enzyme compartmentation, activators or inhibitors, can influence the type of end modification.

Methylation

Enzymatic methylation of the α-amino groups is known in certain proteins involved in macromolecular assembly, for example, myofibrils, nucleosomes, flagella, pilins, and ribosomes. Common sequences found in the first few residues are thought to play a major role in determining the formation of Me-Ala, Me-Met, Me-Phe, Me$_2$-Pro, and Me$_3$-Ala, in proteins.[92,93] Four classes of proteins were identified by sequence homologies. For example, proteins found to have dimethyl proline or trimethyl alanine at the n-terminus have a Pro-Lys sequence at positions 2 and 3, whereas several proteins containing monomethyl methionine have a Gln-Pro sequence at positions 3 and 4. These sequences are believed to be important recognition points for specific methylases.

Arginylation

Transfer of amino acid to the α-amino acid of a completed polypeptide chain can occur by enzymes using amino acyl-t-RNA as substrate. Arginylation, the transfer of arginine from its t-RNA, is an important reaction, because it introduces into the protein a residue targeting the modified protein for intracellular proteolysis. Only particular cytosolic proteins are modified by arginylation. But why? First, it is necessary that the protein contain an aspartyl or a glutamyl residue at the amino terminal position for reaction.[94] Second, it seems that the hydrophobic character of the protein favors modification, as indicated by the strong binding of arginylated proteins on phenyl sepharose and their tendency to aggregate.[95] [^{3}H] arginyl-t-RNA and liver extract were used to find that the substrate ornithine decarboxylase was labeled with a specific activity almost 9,000 times that of total cytosolic proteins,[96] and it was shown that ornithine decarboxylase bound more tightly to octyl and phenyl sepharose than other cytosolic proteins studied.[97]

Amidation

Peptide hormones containing a C-terminal amide are derived from the proteolytic processing of the peptide chain, followed by hydroxylation, and cleavage of a modified C-terminal glycyl residue. Selective proteolysis occurs on the carboxyl side of a prohormone containing consecutive basic residues followed by the action of a carboxypeptidase specific for basic residues, to removes these groups. If glycine is present, enzymatic hydroxylation and cleavage may occur with the transfer of the amino group of the glycyl residue to the C-terminal position,[98] as discussed earlier in Chapter 2. The reactions occur only with an end glycyl residue and are influenced by the amino acid in the penultimate position as are reactions of acetylation and myristoylation. Interestingly, an acidic amino acid in the penultimate position to the "reactive" glycyl residue retards both amidation[99] and myristoylation. Studies with synthetic peptides show that an uncharged residue next to the carboxy terminal glycine is optimal for the amidation reaction.

Prenylation and C-terminal modifications

The incorporation of farnesyl or geranylgeranyl isoprenoids into proteins occurs at the C-terminal region of proteins. A consensus sequence for prenylation is C-A-A-X, where A represents aliphatic residues and X, undefined, is the C-terminal amino acid.[100,101,102] The γ subunits of G-proteins and proteins such as Ras contain this sequence motif and are substrates for prenylation. If a glycine residue is found distal to the cysteinyl residue, no prenylation occurs. The $G_i \alpha 1$ subunit containing a glycyl residue next to the cysteinyl residue in the sequence C-G-L-F is not prenylated, but mutagenesis of the carboxy terminus of $G_i \alpha 1$ from CGLF to CVLS allows for prenylation.[103]

The prenylated protein with a proteolytically-derived C-terminal cysteine is a substrate for methylation by a membrane bound carboxymethyltranferase. But no effective methylation of synthetic peptides occurs in the absence of a modified cysteine side chain. The prenyl nuclei acts to lower the K_m of the substrate. Unnatural saturated hydrocarbon substituents like decyl, tridecyl, and pentadecyl are less effective.[104] Methylation of peptides of G-proteins can be effectively mimicked by using S-farnesylated thioacetic acid as substrate. Thus, recognition of the substrate does not appear to involve the amino terminal portion of the protein substrate. But what seems important is the isoprenyl side chain and the position and reduced state of the sulfur atom of the cysteinyl residue. If the sulfur atom is oxidized to the sulfoxide form or is placed one atom closer or farther away found in the cysteinyl residue, no modification takes place.[105]

CHEMICAL INFLUENCES OF SIDE CHAINS ON MODIFICATION

Glycation

Nonenzymatic reactions of carbohydrate with protein (glycation) have been well characterized; for example, much attention has been given to the reaction of proteins with glucose in diabetes mellitus. Although the processes involved are distinctly different from enzyme-catalyzed reac-

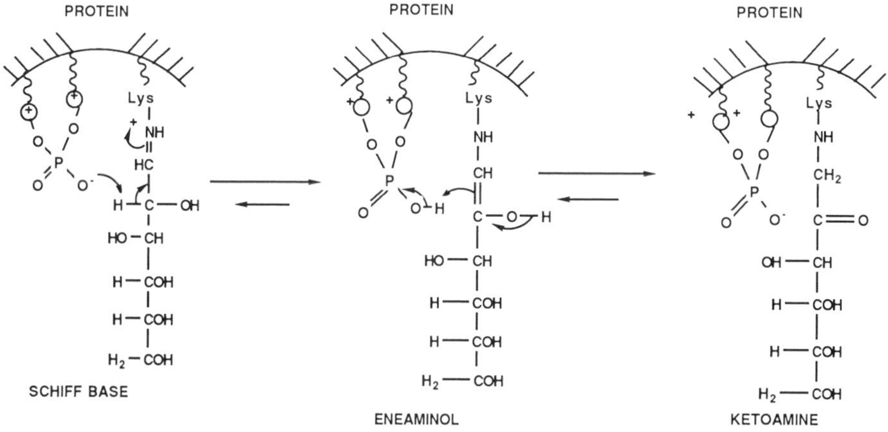

Figure 3.12. Reaction pathway for glycation of protein, showing the catalytic effect of phosphate bound in a basic micro-environment in the structure of a protein. Reprinted with permission from *Journal of Biological Chemistry* 262:7207 (1987). Copyright 1987, The American Society for Biochemistry & Molecular Biology.

tions, an analysis of chemical features of the process illustrates some features of a protein substrate influencing reactivity. The reaction with glucose initially involves the formation of a Schiff base with an α or ε-amino group, an Amadori rearrangement, and subsequent formation of a ketamine. Not all amino groups are equally reactive; for example, the ε-amino group of Lys_1 in ribonuclease is about 10 times more reactive than the α-amino group in forming ketoamine derivatives.[106] The proximity of charged groups, for example, a Lys-Lys sequence and the binding of specific ions, can favor selectivity.[107] Hence, the characteristics of the microenvironment of the protein can have an important role in directing glycation. A scheme for the reaction with glucose accelerated by phosphate binding to a protein site is illustrated in Figure 3.12. Binding of Ca^{+2} influences the conformation of calmodulin, and a study of glycation of calmodulin shows that Ca^{+2} influences the reaction. The process can occur with all seven lysyl groups, but no more than two to three lysyl residues react in a single molecule, and proximity of charge may influence reactivity.[108]

Acetylation

Enzymatic acetylation reactions of ε-amino groups of lysyl residues in histones demonstrate how the environment of modifiable groups in a protein substrate can influence the reaction. The acetylation of histone

H4 can occur in the cytoplasm as part of a scheme for its deposition on DNA in nucleosomes, or it can be modified directly in the nucleosomes as a part of processes involved in altering chromatin structure. The interactions of lysyl residues of histones in these two states could be expected to be different. In nucleosomes, lysyl residues in the amino terminal portion of histone H4 are believed to interact with the negatively charged phosphate groups in DNA,[109] but no such interactions occur in the cytoplasm. Using a partly purified acetyl transferase from *Tetrahymena thermophila*, Lys_{11} is the exclusive site of the first reaction in the free histone, but Lys_7 is the first site modified in the nucleosome.[110] Diacetylated products $Lys_{11,4}$ and $Lys_{7,4}$ were identified for free and bound histone, respectively. That the same reactions occurring *in vitro* occurred *in vivo* indicated that these observations were not artefactual. A study of acetylation reactions in various cell types indicates that acetylation of four different lysyl residues in histone H4 occurs nonrandomly and that different patterns were obtained for HeLa cells in interphase and metaphase.[111]

A comparison of chemical modification of histones with acetic anhydride in free and bound states also points to differences in reactivity of the lysyl residues.[112] Using conditions undisruptive to the structure of the nucleosome, the lysyl residues in the histones involved in enzymatic acetylation are poorly reactive.[113] But, acetylation occurs in high ionic strength solutions, a condition expected to change the interactions between histone and DNA. The lysyl residues in histones in the free state are reactive and show pK_a values of about 9.5, which indicates a slight change in chemical property from the ε-amino groups of exposed lysyl residue that have pK_a values from 9.9 to 10.5.[112]

Incubation of calf thymus histones with acetyl CoA leads to the acetylation of the ε-amino group of a few lysyl residues in the histones. This reaction was not found with other proteins and the efficiency of modification depended on the "native" state showing the significance of tertiary structure in the reactions.[114] A high reaction rate was found with polylysine. Because of the cluster of positive residues in the polymer, pK_a values of certain lysyl residues may be perturbed to lower values that could enhance their reactivity. The enzymatic and chemical studies emphasize the need to consider the natural environment of the "protein substrate" in evaluating modification reactions.

Methylation

In addition to O-esterification, enzymatic methylation reactions also occur on N and S atoms. Sites of modification include (1) the ε-amino group of lysine and the guanidino group of arginine, (2) the α-amino group of phenylalanine, alanine, methionine, and the α-imino group of proline, and (3) the thiol of cysteine and the thioether of methionine.

Several proteins, cytochrome c, myosin heavy chains, and histones have sequences containing a lysyl residue (a methylation site) adjacent to another basic amino acid, lysine or arginine. A Lys-Lys sequence was suggested early as a recognition site for a methylase in cytochrome C.[115] Modification occurs specifically at one site, Lys_{72}, in the sequence Asn-Pro-Lys-Lys-Tyr. But there are other Lys-Lys sequences, for example, at positions 7 and 8, which aren't modified in the native protein. Yet in a CNBr fragment containing this sequence, methylation can occur, a fact suggesting that the structure of the native protein is important for specificity.[116] The x-ray structure of cytochrome C shows that Lys_{72} is near a hydrophobic patch and that Lys_{73} points to the other side of the molecule.[117] Chemical modification with 4-chloro-3,5-dinitrobenzoate also shows specificity. With this reagent, Lys_{72} is the most reactive lysine in the molecule, and Lys_{73} is unreactive just as it was in enzymatic methylation.[118] Binding of the reagent to the hydrophobic patch could contribute to specificity. A Lys-Lys sequence is not essential for all methylation reactions. One lysyl residue can be methylated in calmodulin at Lys_{115} in the sequence Gly-Glu-Lys-Leu.[119] Using calmodulin from *Dicytyostelium discoideum*, it has been found that the same lysyl residue modified in methylation reactions is the site for enzymatic ubiquiti-nation.[120] An analysis of the x-ray structure of the product shows that the modified lysyl residue in calmodulin is nearby a hydrophobic cleft.[121] A hydrophobic region nearby a lysyl residue in protein substrates may be recognized in common by specific enzymes and chemicals.

A single arginyl residue, Arg_{107} in myelin basic protein in the sequence Gly-Arg-Gly-Leu, can undergo enzymatic methylation. Three prolyl residues in the sequence 99-101 are suggested to produce a sharp bend in the molecule, thus bringing together the methylated arginine close to a hydrophobic amino acid, phenylalanine, in an adjacent chain.[122] But the structure of the substrate is not known. It is also not known how lipids influence this structure in the myelin sheath. Enzyme recognition requires something more than an exposed arginyl group, because the peptide Lys-Gly-Arg-Gly-Leu of human myelin basic protein, res. 105-109, is not a substrate for methylation.[123]

Deimination

Peptidylarginine deaminase converts arginyl residues in proteins to citrulline, but only certain residues react. Reactions also can occur with free arginine or arginine derivatives, but the K_m values for small substrates such as benzoyl arginine ethyl ester are much higher than for the reaction with the protein substrate, soybean trypsin inhibitor.[124,125]

Under certain circumstances only a single residue, Arg_{63}, in the the reactive-inhibitor site reacts.[126] Arg_{65}, which also is exposed and adjacent to Arg_{63}, does not react. X-ray crystallography shows that Arg_{63} is in an exposed loop,[127] demonstrating the importance of accessibility for the enzymatic reaction. Mere exposure of a guanidine side chain is insufficient to explain effective catalysis. This is evident from differences in reactivity. A clue to reactivity may be provided by the fact that arginine specific reagents such as phenylglyoxal or 2,3–butanedione inactivate soybean trypsin inhibitor.[126] Perhaps the protein environment around Arg_{63} may change the basicity or chemical properties of Arg_{63} and influence both chemical and enzymic processes.

CHEMICAL CHANGES IN THE PROTEIN INFLUENCE SUBSEQUENT POST-TRANSLATIONAL MODIFICATION REACTIONS

Carboxylmethylation

Two types of carboxymethylation reactions have been demonstrated in eukaroytes. In both cases, the reactions won't take place unless the protein substrate is chemically changed from its original translated form. One modification reaction takes place on the side chain carboxyl group; it is similar then to carboxymethylation reactions occurring in procaryotes but differs in that only aspartic acid is modified.[57] The neighboring residue next to the "aspartate" to be modified influences the formation of the reactive state.

Analyses of one type of a carboxymethylation reaction in proteins reveal that two derivatives of aspartate are formed: (1) a methyl ester of an L-isoaspartic acid residue, and (2) a β-methyl ester of a D-aspartic acid residue. But what chemical changes in the protein substrate might occur so that these products could be formed? To explain the formation of an isoaspartate derivative, it has been suggested, as illustrated in Figure 3.13,[57] that an aspartic acid or an asparagine residue in a protein forms a succinimide ring and that an isoaspartate or aspartate residue is generated by hydrolysis.

Last, a carboxymethyltransferase modifies specifically the α-carboxyl of the isoaspartate residue. An analysis of protein structural features suggests that orientation of the nitrogen atom of the peptide functional group with respect to the side chain carboxyl or carboxamide is important for formation of the imide intermediate. Phi and psi dihedral angles of -120° and 120° were suggested as optimal,[128] and flexible regions of the protein seemed to form the correct orientation when heated,[129] a condition known to promote imide formation.

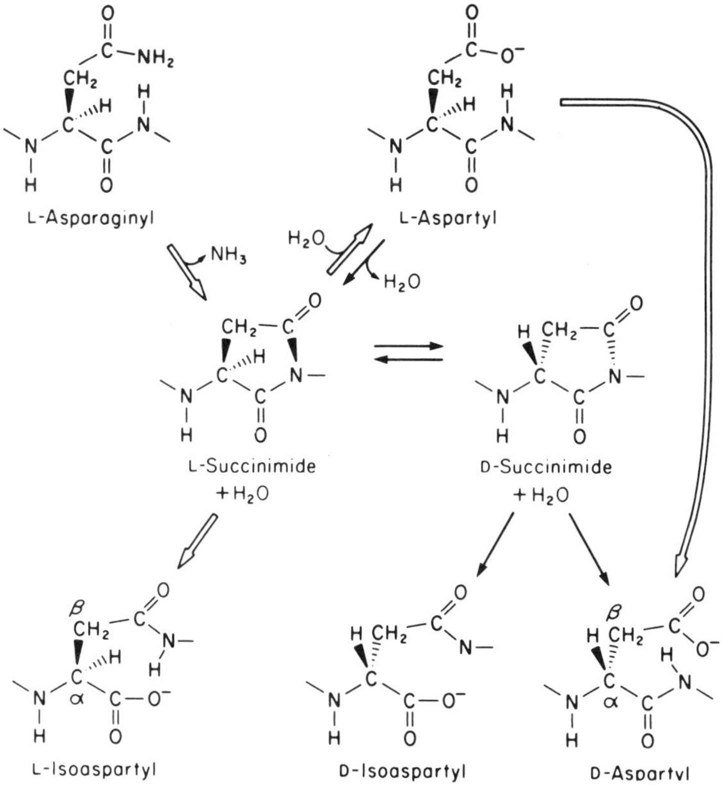

Figure 3.13. Chemical routes to the formation of D-aspartyl and L-isoaspartyl residues from naturally occurring L-aspartyl and L-asparaginyl residues in proteins. The heavy arrows indicate steps supported by experimental data in model systems; the light arrows indicate other feasible reactions. Reproduced, with permission, from the *Annual Review of Biochemistry* 54:479 (1985). Copyright 1985 by Annual Reviews, Inc.

When glycine is in the (n+1) position with respect to an asparaginyl residue, facile imide formation is possible because of the small size of H-atoms on the α-carbon atom. Presumably, bulky R groups do not allow for imide formation and thus only Asn-Gly sequences in proteins are cleaved by alkaline NH_2OH. But chemical features are also important. An adjacent seryl residue may H-bond with the carbonyl group of the carboxyamide and/or the peptide nitrogen and facilitate its deprotonation enhancing chemical reactivity.[130] Protonation of the carboxyl group of aspartic acid provides a better leaving group and stimulates imide formation.[131] Moreover, as suggested for an Asp-Tyr sequence in glucagon,[132] the neighboring residue may affect reactions. In this instance, a tyrosyl residue with a low pK_a may help deprotonate the peptide nitrogen and facilitate the reaction. An Asp-Tyr and an Asp-Gln sequence are modification sites in calmodulin. Flexibility of the structure is considered important but chemical features may also be involved.[133,134]

To generate the second product, a D-methylated isomer, racemization must obviously take place. But whether an imide is involved is uncertain.[57] An imide might facilitate proton abstraction from the α-carbon atom and could explain racemization at particular asparagine or aspartic acid sequences. Interestingly, D-aspartate and the L-isoaspartate are recognized by one carboxymethyltransferase, a finding suggesting similarities in their structures, as depicted in Chapter 2.[135]

Hence, stereochemistry of "aspartate/asparagine" sequences in protein structures contribute significantly to specific carboxymethylation reactions.

A second type of eukaryotic carboxymethylation occurs at the α-carboxyl group of proteins and peptides. With the Ras proteins, the site of modification is the α-carboxyl group of cysteine. But no cysteinyl residue is present in the original translated protein. Again, some alterations are required in the protein structures. As stated earlier, prenylation of the cysteinyl residue initiates a sequence of further enzymatic reactions. Three residues on the carboxy side of the modified cysteinyl group apparently are then removed by proteolytic processing to bare the reacting α-carboxyl functional group.[136,137] Methylation has been described for the α-subunit of cGMP phosphodiesterase in bovine rod outer segments and for human nuclear lamin A. A cysteinyl residue found in the same position as in the Ras proteins suggests that processing occurs before methylation.[136]

SEQUENCES DISTINCT FROM THE MODIFIABLE SITE CAN INFLUENCE POST-TRANSLATIONAL MODIFICATION REACTIONS

Carboxylation

Initial kinetic studies with synthetic peptides showed that carboxylation reactions could occur on small substrates, but the reasons for ineffective catalysis, high K_m values, were not understood. Various blood coagulation factors—for example, prothrombin, protein C, protein A, Factor VII, Factor IX, and Factor X, which can be multiply carboxylated in the amino terminal region, contain a propeptide sequence of 17–24 amino acids between a prepeptide signal sequence and the start of the mature protein. The propeptide is strongly conserved in these proteins, and kinetic studies suggest that this region influences binding of "carboxylatable" sequences. When a 20-residue peptide of the propeptide region of Factor X was used, it was found that this peptide would stimulate carboxylation of the blocked tripeptide, Boc-Glu-Glu-Leu-OMe, by lowering the K_m value approximately six-fold.[138] A 28-residue peptide, HVFLAPQQARSLLQRVRRANT*FLEEV*RK, containing the propeptide region and the designated carboxylatable sequence, is modified with a K_m value of 3.6 µM, whereas FLEEL is carboxylated with a K_m of almost 3 orders of magnitude greater.[139] A synthetic peptide representing the propeptide region, residues -18 to -1, in prothrombin inhibits competitively carboxylation of the 28-residue peptide with a K_i equal to the K_m value, showing this region dominates binding. Binding could occur in an active site region, but where the propeptide binds relative to the reactive sequence is yet to be established. Two-dimensional NMR spectroscopy of the propeptide region shows that this segment can form an amphipathic helix in helix-forming solvents or at low temperatures in aqueous solutions.[140] This structure could serve to anchor the carboxylase. Studies of the matrix Gla protein show that internal sequences also can influence carboxylation reactions.[141] It has been suggested the propeptide or the internal binding region in this case and the carboxylatable sequences may have arisen from an initial gene duplication event.

Biotinylation

The formation of N ε-(biotinyl)lysine, biocytin, in biotin enzymes occurs by the action of holoenzyme synthetase on a variety of apocarboxylases. The sequence surrounding the modifiable lysine, Ala Met Lys Met, is highly conserved in most carboxlyases in prokaryotes and eukaryotes, a finding suggesting that these enzymes are derived from a common ancestral gene some 3,000 million years ago. Thus, it seems that this region is important for recognition by the holoenzyme synthetase, but it alone is insufficient for the biotinylation process because a small peptide bearing this sequence cannot be biotinylated.[142] The modifiable sequence is located in the C-terminal region of most carboxylases with the exception of animal acetyl-CoA carboxylase. And in the C-terminal region, the reacting lysyl residue is 35 amino acids away from the end. Site-directed mutagenesis studies show that in the biotinyl subunit of transcarboxylase (123 residues) the hydrophobic amino acid in the penultimate position is critical for biotination.[143] Exactly how the C-terminal region participates in enzyme recognition of the modifiable lysine is not yet established. In animal acetyl-CoA carboxylase, the modifiable lysine is in the N-terminal region, but an aspartyl residue is

35 residues away and next to it at position 34 is leucine.[144] Thus, the proteins have some common features important for biotinylation although at different parts of the polypeptide chain. Biotinylation seems to require a stable folded domain in its protein substrate for effective reaction.[145,146]

Phosphorylation

Rhodopsin, a membrane bound protein, is well known to be an integral part of the phototransduction network in the rod cells of the retina. Isomerization of the 11-cis retinal covalently bound to rhodopsin in response to light causes a conformational change in the protein which in turn influences nucleotide interactions of the G-protein, transducin. Dissociation of subunits occur and a cascade of reactions takes place influencing the conductance of the plasma membrane in the visual response.[147] Phosphodiesterase, which is activated by the photoexcited rhodopsin, is deactivated by a process which involves phosphorylation of the activated rhodopsin and binding of arrestin.[148] Multiple phosphorylation occurs in the C-terminal region of rhodopsin, but this phosphorylation can only occur when rhodopsin has been activated[149] even though it seems that the C-terminal region is exposed in unactivated rhodopsin and is available for enzymatic modification. Peptides of the C-terminal region can be phosphorylated by rhodopsin kinase but these peptides have K_m values about three orders of magnitude higher than rhodopsin.[150] Studies with activated rhodopsin and peptides suggest that some portion of rhodopsin besides the C-terminal phosphorylatable site is necessary for interaction with rhodopsin kinase to yield effective phosphorylation.[151] Studies with truncated and activated forms of rhodopsin show these forms stimulate phosphorylation of a C-terminal peptide.[152] The results are interpreted to mean that cytoplasmic loops of rhodopsin are involved in binding and activation of rhodopsin kinase. Loop V-VI, a region distant in the linear sequence from the phosphorylation sites, is critical for the activation of rhodopsin kinase.

CROSS-LINKING REACTIONS

Disulfide Bonds

Folding of a protein and formation of native disulfide bonds is not a random process and frequently involves the formation of intermediates with nonnative disulfide bonds. Protein disulfide isomerase is involved in forming native disulfide bonds from fully reduced proteins and with proteins containing nonnative disulfide linkages.[153] The disulfide bond is naturally an important recognizable element of the enzyme, but because the enzyme influences rates of disulfide formation and interchange when conformational changes are rate limiting, it has been suggested that the enzyme will likely have different affinity for various conformational states of the protein substrate.[154] Protein disulfide isomerase is known to catalyze the steps involved in the formation of the native three-disulfide structure of bovine trypsin inhibitor. It is suggested that the initial one- and two-disulfide intermediates are made and remade 10^4–10^5 times before the protein is folded in its final state.[155]

The importance of the conformational state of the protein is illustrated in the reaction of disulfide isomerase with ribonuclease S, an active species missing the amino terminal portion of the native enzyme, ribonuclease A. In the presence of the isomerase, the ribonuclease is inactivated presumably by reshuffling of the disulfide bonds. This inactivation may be expected if ribonuclease S with a different primary structure from the native protein is in a metastable state and if a state of lower free energy can be derived upon formation of new disulfide bonds. If a peptide of the amino terminal region of RNAse A is added after inactivation, reactivation occurs in the presence of the disulfide isomerase leading presumably to the formation of native-disulfide bonding caused by the interaction of the peptide with ribonuclease S.

The insulin receptor is synthesized as a single polypetide chain, but it exists in its mature form as a glycoprotein composed of α- and β- subunits cross-linked by disulfide bonds. The extracellular portion of the molecule contains cysteine-rich domains, but only a few sulfhydryl groups are involved in disulfide bonds linking the receptor in its tetrameric β-α- α-β form.[156] Exactly how these disulfide bonds are formed is not known. It is known that the receptor lacks ability to bind ligand after translation, and it has been suggested that some disulfide bonds initially formed may change through the action of the disulfide isomerase to produce the active form of the receptor.[157]

Another facet to be considered is the enzymology itself; for example, the protein disulfide isomerase and the β-subunit of prolyl-4-hydroxy-lase are products of the same gene.[158] Moreover, evidence exists that a glycosylation site-binding protein and thyroid hormone-binding protein are the same or closely related to the protein disulfide isomerase.[159] Understanding how binding occurs with these varied substrates is the next challenge ahead.

Halogenation and Cross-Linking

Iodination of tyrosyl residues and the coupling of these derivatives are two important post-translational events involved in the formation of the hormones thyroxine and triiodothyronine from the protein, thyroglobulin. Seemingly, the process is wasteful because only a small number of hormone molecules (6–8) are derived from a large dimeric protein of 330K M_r containing 72 tyrosyl residues per subunit. Yet the process is very efficient, particularly at low iodine concentrations, and it has been suggested that the structure of the protein substrate may have evolved to facilitate effective iodination and coupling reactions.[160] The importance of the structure of the substrate is illustrated by the enzymatic reaction's ability to be duplicated by low-level chemical iodination and oxidative coupling with similar specificities; iodination does not require a native thyroglobulin molecule, but the coupling reaction, a transfer of an iodinated phenol nucleus to an iodinated phenol, does.[161,162] A comparison of sites of chemical iodination and iodination catalyzed by thyroid peroxidase at a level of 7 atoms and 25 atoms/mole, a physiological level,[163] shows that the same sites are modified with and without the enzyme. That is, the protein substrate, not the enzyme, dictates the specificity of iodination. But thyroid peroxidase is needed for effective coupling at the hormogenic sites.

Mature bovine thyroglobulin contains 2,750 residues and analysis of the primary structure, based on sequence analysis of its cDNA, shows that the protein contains extensive internal homology.[160] The major coupled site for thyroxine biogenesis is at Tyr_5 when thyroglobulin contains approximately twenty I/mole.[161] Minor sites exist at Tyr_{2555} and Tyr_{2569}. All these sites contain a neighboring acidic amino acid consistent with earlier peptide substrate, studies suggesting that a glutamyl residue next to a tyrosyl residue favored enzymatic iodination by thyroid peroxidase. Tyr_{2748}, the second from the last residue, is the major coupled site for triiodothyronine. Thus, hormogenic sites are at the amino and carboxy terminal regions of the protein. Although the three-dimensional structure of the thyroglobulin molecule is not proven, the transfer of a di-iodophenol nucleus to di-iodinated Tyr_5 appears to occur from di-iodinated Tyr_{2469}, or Tyr_{2522},[164] which suggests that portions of the amino and carboxy terminal regions are in close proximity. In one mechanism, a charge transfer complex is suggested between the modified tyrosyl groups, and the coupling process is shown in Figure 3.14.[165] The residue donating the modified phenol is converted to a dehydroalanyl residue. No biphenyl ether linkage between tyrosyl residues occurs in thyroglobulin without iodination. Chemical features of the modified residues

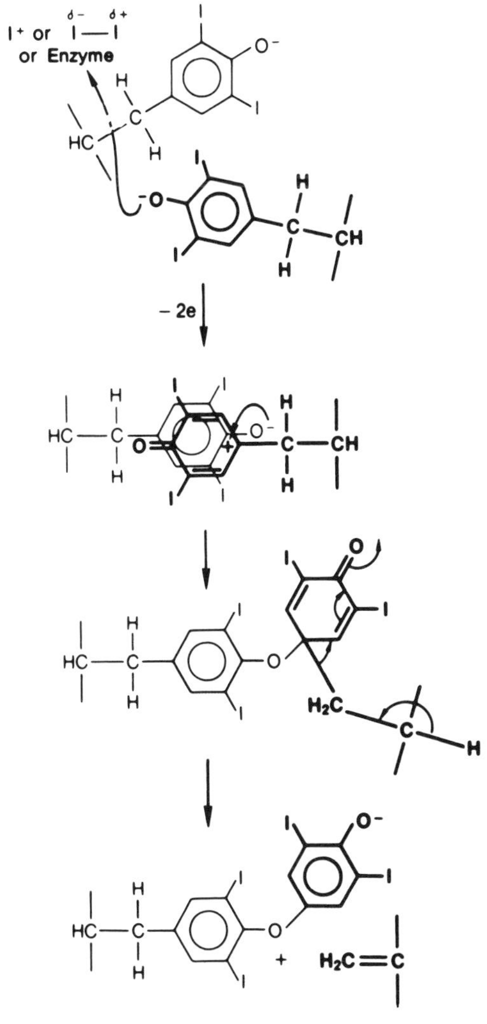

Figure 3.14. Proposed scheme for coupling of iodotyrosine residues in thyroglobulin. Reprinted with permission from *Journal Biological Chemistry* 256:9167 (1981). Copyright 1981, The American Society for Biochemistry & Molecular Biology.

could influence reactions, for example, the di-iodinated tyrosyl group with a pK$_a$ of 6.4 would be a better nucleophile than an unionized tyrosyl group with a pK$_a$ of 10 for the coupling at physiological pH.

N ε–(γ-Glutamic) Lysine Cross-Linking

Specific and naturally occurring complexes of proteins are known to be substrates for intermolecular cross-linking reactions catalyzed by various transglutaminases. In this reaction, an isopeptide linkage is formed from the carboxamide function of a glutaminyl residue and the ε-amino group of a lysyl residue.[166] A few glutaminyl residues in specific proteins can react, and sequence surrounding the glutaminyl residue is important,[167] but it is not the only consideration. This idea is illustrated by a study with the protein substrate, involucrin, which is believed to undergo extensive cross-linking reactions with other membrane proteins as a part of the terminal differentiation of keratinocytes. Using an amine as an acceptor, it was found that one glutaminyl residue near the carboxy terminus reacted. But up to ten other residues could react in CNBr fragments, and sequence analysis of many sites showed striking homology with the reacting glutaminyl residue of the natural protein, thus showing that the structure of the protein substrate limits reaction.[168] In this instance, some reactive sites are hidden, but they may become exposed upon reaction with a protein acceptor and/or binding with other cellular components. Spermine can act as an acceptor in crosss-linking reactions and a major product formed from apolipoprotein B has been identified as N'N''-bis (g-glutamyl) spermine.[169]

In blood clotting, the stabilization of the fibrin structure is provided in part by intermolecular cross-linking, and information about the molecular architecture of the interacting fibrin molecules is made clear by an analysis of the cross-linked γ-γ-chains. The reactions occur between individual fibrin molecules, and the γ-chains are arranged in an equivalent antiparallel array. Sequence analysis showed that coupling occurred in the carboxy terminal regions and two identical linkages were formed as illustrated in Figure 3.15. Because the glutaminyl and the lysyl residues are eight residues apart, these residues could project from the same side of an α-helix and allow interactions between side chains of an oppositely oriented α-helix of a second fibrin molecule.[170] The exact structures involved are yet to be proven.

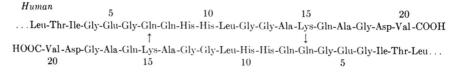

Figure 3.15. Amino acid sequences of carboxy-terminal segments of γ-chains showing locations of reciprocal cross-links between antiparallel neighboring chains of human fibrin. Reprinted with permission from *Adv. in Protein Chemistry* 27:1 (1973). Copyright 1973, Academic Press.

ADP-ribosylation

Poly (ADP-ribosylation) of proteins consists of two main phases: (1) the attachment of an ADP-ribosyl group to the protein and (2) addition of subsequent ADP-ribosyl units to lengthen and branch the chain. Sequences of acceptor sites in polyADP-ribosylation in nuclear proteins show that the major acceptor is the g-carboxyl group of glutamic acid although it seems that the a-carboxyl function is also modified in histone H1.[171] Modification of nucleosomes provides information of topography of protein-DNA complexes. Histone H1 is believed to be in the linker region between nucleosomal particles, and carboxyl groups in the extended amino and carboxy regions are believed to be acceptors. Evidence exists that poly(ADP-ribose) serves as a cross-linking agent for two H1 molecules between nucleosomal particles.[172]

Dityrosine

A linkage between tyrosyl residues in protein substrates can occur both by enzymatic and nonenzymatic means. Hence, juxtaposition of the tyrosyl residues is a key factor in coupling. An oxidative coupling illustrated in Figure 3.16[173] catalyzed by an ovoperoxidase is responsible for the hardening of the membrane in sea urchin eggs.[174,175] Irradiation of calmodulin with UV light also induces dityrosyl formation.[176]

THE PROTEIN CAN CAUSE ITS OWN MODIFICATION

Affinity labeling reagents are known to modify proteins selectively by binding to a site in the protein and then chemically reacting with a specific group in the protein. Sulfonyl fluoride-containing nucleotides are one example of this type of reagent. Free glucose may fit this category as described by its specific interaction and chemical reaction with ribonuclease. Certain modifiers, for example, coenzyme A derivatives, methylated DNA, associated with post-translational modification reactions may have similar effects. In these instances, no separate enzyme may be required for the modification, and the protein will undergo one round of reaction. Modification may occur "autocatalytically". Cases exist in which protein in the absence of any modifiers can cause one post-translational modification reaction and transform itself into a catalyst.

Palmitoylation

In contrast to myristoylation, this reaction occurs post-translationally and on side chains with the sulfhydryl group of cysteine being the main target.[177,178] In rhodopsin, two cysteinyl residues at positions $_{322}$ and $_{323}$ can

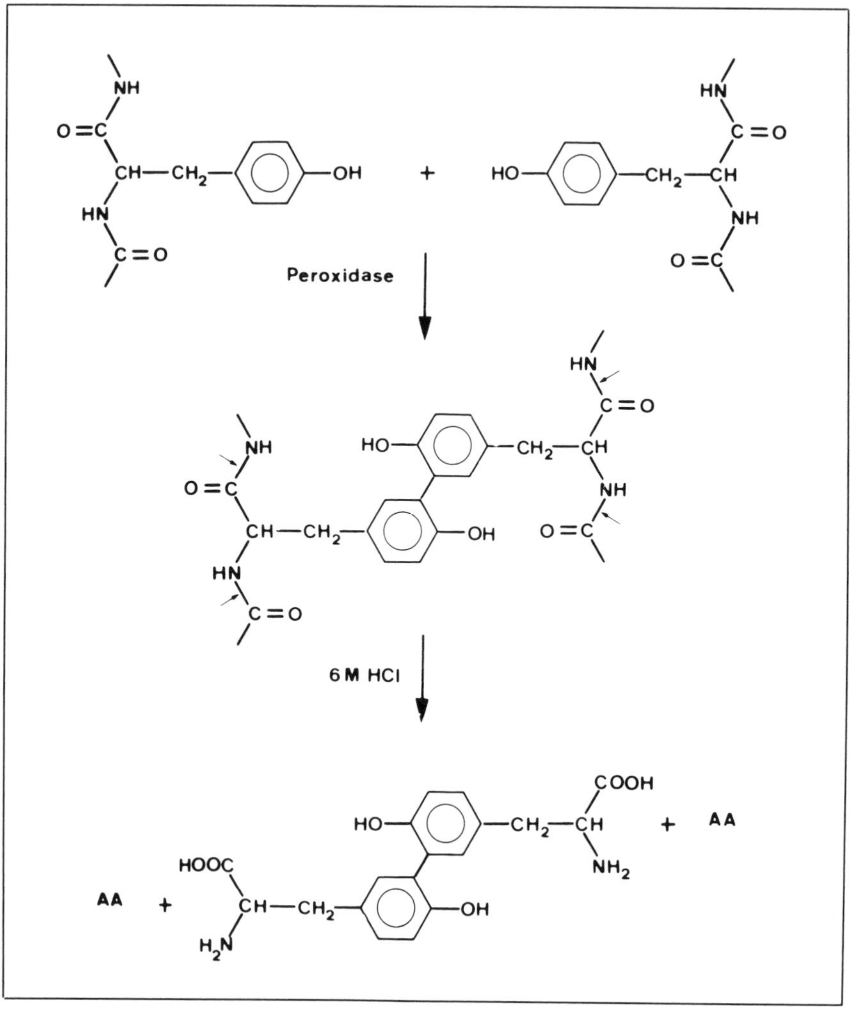

Figure 3.16. Oxidative phenolic coupling of tyrosine residues in polypeptide chains with peroxidase/hydrogen peroxide and formation of dityrosine by acid hydrolysis (indicated by arrows). AA, amino acids. Reprinted with permission from *Methods in Enzymology* 107:377 (1984). Copyright 1984, Academic Press.

be modified,[179] and in the human β_2-adrenergic receptor palmitoylation likely occurs at cysteine$_{341}$.[180] Topographical models suggest that reactions could occur in equivalent positions in the two molecules and that a comparison of sequences between various G-linked receptors shows that a cysteinyl residue is highly conserved. The primary structures of the receptors and rhodopsin at this site have some similarity, but whether there are any common chemical or structural features in this region of the protein substrate influencing reactivity is not known.

A different and interesting case of post-translational modification is provided by the activation of phospholipases A_2 by its substrate. Early it was described that the hydrolysis of monomolecular layers of phospholipids proceeds with an initial lag in the catalytic reaction.[181] It is now understood that during this latency period the substrate ester can acylate a lysyl residue and activate the enzyme more than 100-fold. Octanolylation and palmitoylation can occur at a specific lysyl residue in porcine pancreatic lipase A_2.[182] This process of modification influences subunit structure and transforms the enzyme from the monomer to the fully activated dimer.

With rhodopsin and myelin proteolipid, protein reactions can occur in the absence of added enzyme.[183,184] That these reactions follow saturation kinetics indicates that binding of palmitoyl CoA occurs at a specific site on the protein substrates before modification takes place. With rhodopsin, a cysteinyl residue is suggested to react, and, with myelin proteolipid protein, an oxy group is possibly involved. The sites of reactions with either protein have not been proven, but it has been suggested that *in vivo* and *in vitro* modification sites are identical for the myelin proteolipid protein.[184] Thus, this case of palmitoylation provides a new viewpoint about post-translational modification. The protein substrate contains not only the modifiable group, it binds the reactant, palmitoyl CoA, and has the potential to carry out the reaction itself. Whether the modifiable group has any unusual chemical reactivity is not known. This result does not necessarily mean that in other palmitoylation reactions no separate enzyme is involved. A membrane-bound palmitoyltransferase may be involved in the modification of Ras p21.[185,186]

Methylation

The Ada protein, a transcriptional regulator in *E. coli*, can serve to reverse potential mutagenic products induced by formation of methylphosphotriesters or of O^6-methylguanine in DNA. The Ada protein binds and reacts with the modified DNA-transferring methylated group from methylphosphotriesters and methylated guanine, respectively, to cysteinyl residues in the amino and carboxy terminal regions of Ada. No self or other demethylation reactions of Ada are known, so it seems Ada serves one round of reactions.[187] But methylated Ada at the amino terminal region is known to act as a transcriptional activator of the Ada gene producing more protein for subsequent reactions.[188]

Table 3.5. Active site reigon of *E. coli* O⁶ Me Gua-DNA
methyltransferases compared to those of thymidylate synthases

O⁶ Me G-DNA		
methyltransferase	*E.coli*	—Ala-Ile-Val-Ile-Pro-Cys-His-Arg-Val-Val-Arg
Thymidylate synthase	*E.coli*	—Met-Ala-Leu-Ala-*Pro-Cys-His*-Ala-Phe-Phe-Gln
Thymidylate synthase	T4 phage	—Met-Ala-Leu-Pro-*Pro-Cys-His*-Met-Phe-Tyr-Gln
Thymidylate synthase	*L.casei*	—Met-Ala-Leu-Pro-*Pro-Cys-His*-Thr-Leu-Tyr-Gln
Thymidylate synthase	yeast	—Met-Ala-Leu-Pro-*Pro-Cys-His*-Ile-Phe-Ser-Gln

Reprinted with permission from *Proc. Natl. Acad. of Sciences* 82, 2688, 1985.

A comparison of sequences of Ada of the modified regions with the active site region of thymidylate synthase shows all contain a pro-cys-his triplet and other closely related features (Table 3.5).[189] It was suggested that the histidine might serve as a proton acceptor and activate the thiol function in a mechanism similar to that proposed for papain.[190]

The cysteinyl residue in thymidylate synthase and in other methyl transferases, for example, HhaI methyltransferase, and DNA(cytosine-5)-methyltransferase, also in a pro-cys doublet, is thought to serve as a nucleophile in catalysis, but it is not methylated in the reaction. The prolyl residue may provide a bend in the structure positioning the cysteinyl residue with respect to other residues facilitating a proton transfer. Similarly, the cysteinyl residues in Ada may be activated. A highly reactive sulfhydryl in Ada is furthered indicated by the fact that incubation with methylating agents CH_3I and methylmethane sulfonate activate transcriptional activity of Ada.[191] Further similarity between active site regions of the methylases and Ada is shown by the fact that EcoRII methyl transferase can be methylated at a cysteinyl residue in a pro-cys doublet by a photochemical reaction with S-adenosylmethionine.[192]

Serinolysis and Pyruvoyl Formation

The inactive proform of histidine decarboxylase, characterized from several gram positive bacterial sources, is made as a single polypeptide chain (p), but it undergoes self-induced chemical reactions producing an a-subunit containing an amino terminal pyruvoyl residue, a b-subunit, and ammonia.[193] The created active enzyme uses the pyruvoyl group as an internal coenzyme for decarboxylation of histidine. The auto-activation reactions, occurring between two internal adjacent seryl residues in the p chain, are believed to involve formation of an ester and a b elimination reaction yielding a b chain with a C-terminal serine and the a-subunit with an end dehydroalanyl residue. Hydrolysis is not involved for cleavage of the two chains, but bound serine is; hence, the reaction has been termed serinolysis. Finally, the pyruvoyl group is formed. A scheme for these reactions is presented in Figure 3.17.[194]

Expression of the histidine decarboxylase gene and mutant forms have demonstrated the importance of the diserine sequence for autoactivation.[195] If the second seryl residue is changed to an alanyl residue, no activation occurs because no ester can be formed. A cysteinyl or a threonyl residue in place of the same serine yields a protein that like the wild type protein can form an internal ester, but the reactions to form active enzyme are exceedingly slow; for example, a threonyl mutant made from a *Clostridium perfringens* gene has a half life of 4 days.[196] The elimination reaction may be slow in the mutant enzymes;[197] for example, steric interference is suggested in the threonyl mutant and may account for nonproductive chain cleavage seen in the mutant enzymes. Thus, it is clear that the wild type protein has special properties causing the seryl hydroxyl group to be particularly reactive to promote effective cleavage under physiological conditions.

Glycogenin and Glycosylation

Animal cells store glucose as the polysaccharide glycogen. This highly branched macromolecule, containing a-1,4 and a-1,6 glucosidic linkages, can attain molecular weights up to 60 million. Enzymatic reactions can add on new glucosyl residues to the preformed glycogen molecule, but how does the synthesis of glycogen get started? A big advance came with the discovery that glycogen contains covalently bound protein. This protein named glycogenin[198] is linked through a tyrosyl residue to glycogen.[199,200] It is present in equimolar proportions to glycogen, and it is suggested that this protein and related glycosylated forms play a key role in priming glycogen synthesis.[201,202] Forms of the protein related to glycogenin have been isolated but with bound maltose or maltooligosaccharides, and these species may serve a self-glycosylating function. But how do these sugars become attached? And how do these processes influence glycogen synthesis? A form of glycogenin without bound carbohydrate has been isolated.[203] Apo-glycogenin is not auto-catalytic and is presumably modified in a separate reaction to attach maltose. Then, incubation of UDP-glucose with glycogenin, the self-glycosylating protein containing bound maltose, yields a protein with a maximum of eight saccharide units, which can serve to prime glycogen synthesis.[204,205] Hence, the history of the protein, its state of glycosylation, determines whether it is a substrate, enzyme, or primer.

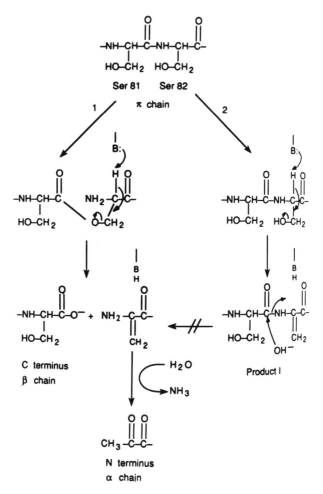

Figure 3.17. Postulated route (S) for the activation of Pro H_{15} DCase and of product 1 formation. Reprinted with permission from *Proceedings of the National Academy of Sciences, USA* 85:8449 (1988).

SUBUNIT STRUCTURE AND COVALENT MODIFICATION

Adenylation

A good example of the definition of the structure of the protein substrate is provided by x-ray crystallography of glutamine synthetase obtained from the bacterium, *S. typhimurium*.[206] Glutamine synthetase consists of 12 identical subunits, each with a M_r of 51,628. The adenylated tyrosyl group, Tyr_{397}, lies at the interface between the subunits and is in close proximity, approximately 6Å, to Trp_{57} in the bottom of a loop in a neighboring subunit. The tyrosyl group is exposed, consistent with spectroscopic evidence and does not show any unusual chemical reactivity. Adenylation of the tyrosyl group is suspected to lead to an interaction of the adenyl group with Trp_{57}, an interaction causing formation of a taut conformer and inactivation of the enzyme. What exactly controls the specificity of adenylation of Tyr_{397} is not known, but the neighboring Trp_{57} in an adjacent subunit may provide direction for adenylation through interaction of the purine portion of ATP during adenylation. Carboxymethylated and urea-denatured glutamine synthetase is not a substrate for adenylation. This finding suggests that a correct tertiary and/or quaternary structure is needed for effective adenylation.[207]

Not all glutamine synthetases from various bacteria undergo adenylation. A tyrosyl residue can be found, however, in an equivalent position in these proteins with those proteins that can be modified. A prolyl residue three residues away from the tyrosine seems necessary for effective adenylation. This proline is missing, for example, in Anabena glutamine synthetase, a poor substrate for modification by adenyltransferase from *E. coli*. Molecular modeling suggests the prolyl residue$_{400}$ does not influence the orientation of the modifiable tyrosine but serves as a recognition point for the transferase.[208]

Phosphorylation

Covalent modification reactions in the pyruvate dehydrogenase complex illustrate an important mechanism that living systems use to regulate metabolic processes.[209] The pyruvate dehydrogenase complex from prokaryotes or eukaryotes is a multienzyme complex consisting of three enzymes: pyruvate decarboxylase-dehydrogenase (E_1), dihydrolipoamide acetyltransferase (E_2), and lipoamide dehydrogenase (E_3). In addition, the eukaryotic complexes contain a small number of tightly bound pyruvate dehydrogenase kinase and phosphatase molecules. The kinase (about 3 molecules/complex) bound to the core acetyltransferase (60 molecules/complex), can cause modification of the bound substrate (about 20 pyruvate dehydrogenase tetramers). First, the kinase is bound tightly to the enzyme complex and cannot be released by chaotropic agents that cause dissociation of the complex; the main binding site is not the substrate subunit of E_1 but actually the core protein, E_2.[210,211] Second, the inactivation rate can be explained by bound enzyme acting on bound E_1 in the multienzyme complex, but rates cannot be explained by bound enzyme of the complex phosphorylating dissociated E_1.[212] Hence, the enzyme may not simply bind to its substrate, phosphorylate it, and dissociate and bind another substrate in the same or a different enzyme complex. Nor does the substrate come off fast enough either. Thus, intracore migration of the substrate to the bound enzyme molecules may explain why the process of modification is so efficient. A similar model may explain dephosphorylation.

ADP-Ribosylation and G-proteins

G-proteins, or GTP-binding proteins, are an important class of proteins known to participate in numerous signal-transduction networks. These G-proteins are protein substrates for various bacterial toxins with ADP-ribosyltransferase activity. In fact, mono-ADP-ribosylation of these proteins which causes a disturbance of the function of these proteins is a causative factor in the diseases diptheria, cholera, and whooping cough. Acceptor sites have been defined in G-proteins, such as cysteine, arginine, asparagine, and diphthamide, a modified histidine, in mono(ADP-ribosylation) reactions for pertussis toxin, choleragen, botulinum toxin, and diphtheria, respectively.

Heterotrimeric and monomeric GTP-binding proteins are major targets for monoADP-ribosylation reactions catalyzed by bacterial toxins containing ADP ribosyltransferase activity. The α-subunit of the heterotrimeric proteins containing αβγ subunits is the site of modification. But what role does subunit structure have in the modification process? Information is lacking regarding the three-dimensional structure of the ADP-ribosylation sites of these proteins, but knowledge of the primary structures and structural information of the GTP-GDP binding regions of c-H-Ras p21;[213,214] elongation factor Tu from *E. coli*,[215,216] a structurally related molecule to the eukaryotic substrate EF-2; and of reconstitution studies with α-subunits and βγ subunit pairs provides insight about the process and specificity of the reaction.

Table 3.6 shows portions of the primary structures of the α-subunits of heterotrimeric GTP-binding proteins related to the action of cholera and pertussis toxin.

In Gsα, the arginyl residue marked by an asterik is modified by cholera toxin. The nearby arginyl residue$_{177}$ is not. Moreover, note that Gi and Go proteins with related sequences are not ADP-ribosylated on either arginyl residue. A comparison of the two substrate sequences in Gs and Td with the nonsubstrates Gi and Go suggests that a threonyl residue in place of a cysteinyl or seryl residue between the two arginyl residues restricts ADP-ribosylation. The change of the sequence from Val-Lys to Asp-Met in Gzα seems critical and lack of reaction is likely due to the presence of an inhibiting acidic function adjacent to the arginine.[219]

Table 3.6. Primary structures of ADP-ribosylation sites of GTP-binding proteins

Protein	Modification Site	Enzyme
Gsα	Q D L L R C R*V L T S G I	Cholera Toxin[217]
Tdα	Q D V L R S R*V K T T G I	
Goα	D I L R T R V K T T G	
Giα	Q D V L R T R V K T T G I	
Gzα	E D I L R S R D M T T G I[218]	
Giα	N N L K D C*G L F	Pertussis Toxin[217]
Goα	N N L R G C*G L Y.	
Tdα	E N L K D C*G L F	
Gsα	M H L R Q Y E L L	
Gz	N N L K Y I G L C[218]	

Primary structure alone, however, is insufficient to explain specificity because both arginyl residues in unidecapeptides of these regions in Go, Td, Gi, and Gs serve as acceptors of ADP-ribose with cholera toxin.[220]

G_2, a loop region of the Ras protein, contains a threonyl residue that interacts with Mg^{+2} bound to GTP. An analogous region containing threonine (shown in bold letters in the table) is proposed for the α-subunit. Note that this region contains an invariant arginyl residue which is necessary for GTP hydrolysis. Its modification by ADP-ribosylation or substitution by lysine, glutamate, or alanine[221] cause inhibition of the GTPase activity. It has been suggested that the arginyl residue might cause a side chain to be in the correct position for a nucleophilic attack on the γ-phosphorous of GTP. Its modification or substitution could distort the required structure.[214] Presumably the structure of this region important for GTP hydrolysis is important for the specificity of ADP-ribosylation.

The cysteinyl residue$_{352}$ of a total 355 residues is the site of modification in Gi for pertussis toxin. In the nonsubstrates Gs and Gz the cysteinyl residue is missing but some sequence homology still exists. X-ray crystallograpy studies of related GTP-GDP- binding proteins show that this region is not involved in nucleotide binding. The carboxy terminal portion, however, is likely involved in an interaction with its receptor, and modification interferes with transfer of information to the nucleotide site.[222,223]

Reconstitution experiments with βγ and α-subunits provided a means to evaluate the importance of subunit structure in ADP-ribosylation reactions catalyzed by pertussis toxin. Approaches similar to this could find applications in the study of other post-translational modification reactions. In Figure 3.18 the effects of varying amounts of βγ on reactions with α are illustrated.[224]

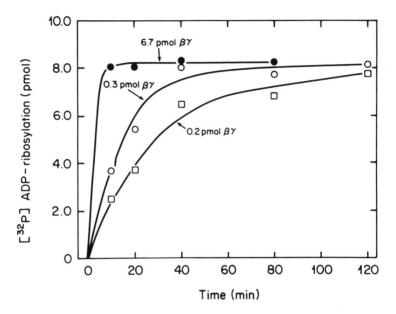

Figure 3.18. Time course of ADP-riboxylation of G_{oα} at increasing levels of βγ. The assay mixtures contained 10 pmol of G_{oa} and either 0.2 (open square), 0.3 (open circle), or 6.7 pmol (closed square) of brain βγ and the reactions were terminated at the indicated times. Reprinted with permission from *Biochemistry* 28:611 (1989). Copyright 1989, American Chemical Society.

First, it should be noted that the rates of the reactions depend on βγ. Without any βγ, little or no reaction occurs, a condition indicating that the α-subunit in the combination with βγ is the effective protein substrate for ADP-ribosylation. Second, even though less than stoichiometric amounts of βγ are used compared with α, the same extent of ADP-ribosylation is achieved. This finding suggests that the heterotrimeric product formed dissociated into a βγ subunit pair and an ADP-ribosylated α-subunit and that the βγ pair released can then serve to affect subsequent reactions.

The fixation of nitrogen catalyzed by nitrogenase is due to a complex of proteins, dinitrogenase, and dinitrogenase reductase. MonoADP-ribosylation of a single arginyl residue in the reductase catalyzed by an ADP ribosyltranferase in the photosynthetic bacterium, *Rhodospirillum rubrum*, causes inactivation of nitrogenase.[225] The site of modification is

Gly-Arg-Gly-Val-Ile-Thr. No modification can occur if the protein is oxygen-denatured or on simple peptides of this sequence. A glycohydrolase exists in the photosynthetic bacterium that reverses the process. In this instance, the whole protein substrate is not required because ADP-ribosylated arginyl derivatives are acted upon by the glycohydrolase.[226] The protein complex between dinitrogen reductase from *Clostridium pasteurianum* and dinitrogenase from *Azotobacter vinelandii* cannot be ADP-ribosylated, and it has been proposed that ADP-ribosylation interferes with complex formation,[227] similar to effects of ADP-ribosylation on complexes between receptor-GTP binding proteins.

SEQUENTIAL PROCESSES AND SYNERGISTIC EFFECTS

ADP-ribosylation

Elongation factor (EF-2) in eukaryotes and in archaebacteria contains a modified histidyl residue that is a target for ADP-ribosylation caused by diphtheria toxin. The sequences of the ADP-ribosylated sites, Asp-Ala-Ile-X-Arg, are identical in rat liver, beef liver, yeast, and in wheat germ.[228] The acceptor-modified histidyl residue, 2-[3–carboxyamido-3–(trimethylammonio)propyl]histidine, commonly called diphthamide (Figure 3.19), is derived post-translationally in a series of reactions.

The 4-carbon side chain of the CG-1 precursor is derived from methionine, and subsequent modifications are shown in the figure. The N-atom, N1, designated by the arrow is the site of ADP-ribosylation.[229] A series of mutants obtained from Chinese hamster ovary cells and S. *cerevisiae*, defective in diphthamide biosynthesis,[230,231] shows that methylation of side chain to make diphthine is needed for diptheria toxin sensitivity. Amidation, although not necessary, influences the rate of ADP-ribosylation, and the slow rate found with EF-2 in archaebacteria may be due to the presence of diphthine. Thus, the modified histidyl residue is of central importance for effective ADP-ribosylation, but the exact role of the surrounding protein structure awaits further investigation.

Phosphorylation

The enzymic interconversion of glycogen phosphorylase is the first known case for regulation of enzyme activity by phosphorylation-dephosphorylation and information about the structures of the substrates, phosphorylase *b* and *a*, serves as a basis for understanding the specific processes involved but for other reactions of phosphorylation-dephos-

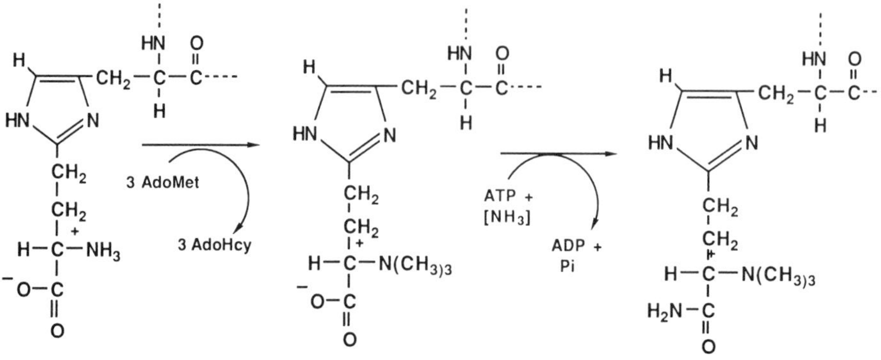

CG-1 PRECURSOR   DIPHTHINE   DIPHTHAMIDE

Figure 3.19. Proposed structures of dipthamide, diphthine and the CG-1 precursor amino acid. Steps in the precursor's conversion and the enzyme activities required are indicated. Reprinted with permission from *Journal Biological Chemistry* 263:3840 (1988). Copyright 1988, The American Society for Biochemistry & Molecular Biology.

phorylation. The phosphorylation of phosphorylase b occurs at residue, $Ser_{14,}$ in each monomeric unit, and phosphorylation of the seryl group in one monomeric unit can influence the subsequent modification of the seryl group in the adjoining subunit of the dimer. Kinetic studies show the second reaction is enhanced by the first phosphorylation event.[232] X-ray analysis of the T state of phosphorylase b suggests that the phosphorylatable amino terminal region, (residues 10-18), Arg_{10}-Lys_{11}-Gln_{12}-$Ile_{13-}Ser_{14}$-Val_{15}-Arg_{16}-Gly_{17}-Leu_{18}, is in an irregular extended conformation making intra-subunit contacts with helices $\alpha3$, $\alpha4$, and $\alpha16$.[233] Phosphorylation occurs only with phosphorylase kinase. Results of studies with synthetic peptides suggest that the basic residues, Lys_{11} and $Arg_{16,}$ along with the intervening residues are important for recognition by the enzyme, phosphorylase kinase. Because Arg_{16} serves as a negative determinant for phosphorylation by cyclic AMP-dependent protein kinase, the orientation of this residue in the protein substrate may be critical for specific enzyme recognition.[234]

A comparison of the x-ray structures of phosphorylase b and a shows that new inter- and intra-subunit contacts are formed as a consequence of phosphorylation.[235] The residues (5–16) disordered in phosphorylase b form a distorted 3_{10} helix, (res. 14–16) extended, after phosphorylation and bind in a cavity between the interface of the two subunits promoting the active R conformation of phosphorylase a. Prominent interactions occur with the phosphoryl group and Arg_{69} and Arg'_{43} of its own and neighboring subunits, respectively. In the process of binding, the C-terminal segment, residues 838-841, stabilizing the inactive T state of phosphorylase b, is displaced by the phosphorylated N-terminal region. The structure formed in phosphorylase a (due to the phosphoryl group) is not possible in phosphorylase b because of unfavorable charge repulsion of positive charges of Lys_9, Arg_{10}, Lys_{11}, Arg_{16}, Arg_{69}, and Arg'_{43}. A schematic ribbon profile of phosphorylase a and b showing the amino terminal region and interactions between the subunits is illustrated in Figure 3.20.

The protein substrate, phosphorylase a, is acted upon specifically by protein phosphatases 1 and 2a. Alkaline phosphatase and other nonspecific phosphatases cannot dephosphorylate the native protein but act effectively on phosphorylated peptides derived from phosphorylase a.[236] The orientation of Arg_{16} along with other positive charges from both subunits may act as positive determinants for specific protein phosphatases but negative determinants for other phosphatases such as alkaline phosphatase.

A cluster of basic residues on the amino terminal side of the phosphorylatable seryl residue of phosphorylase seen in other kinase substrates, for example in a sequence, Arg-Arg-X-Ser, or in a related structure, is common in protein substrates for cyclic AMP-dependent protein kinase, and it has been suggested that structural changes similar to those observed in phosphorylase may be induced in other protein kinase substrates as a consequence of phosphorylation.[235] Circular dichroism measurements with synthetic peptides show that phosphorylation induces a conformational change and that a strong intrapeptide salt linkage is formed between the phosphoryl group and the guanidino group of arginine[237] similar to an interaction observed in phosphorylase.

Many proteins can be mutiply-modified, and results with phosphorylation point out how these reactions can occur in specific instances, but the concepts and approaches are also relevant to other types of modification reactions. What structural changes occur in the protein substrate as the reactions proceed? With multiple reactions, do these reactions occur randomly? Or is there a sequence of events?

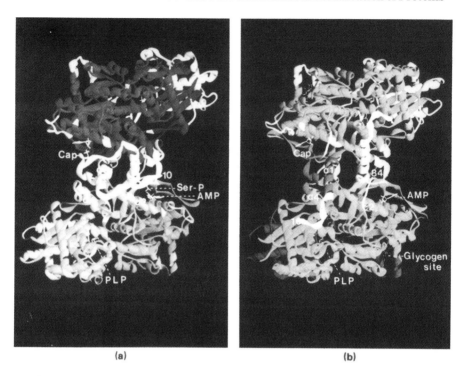

(a)   (b)

Figure 3.20.   T state of GPb (b) dimers. The N-terminal segment of phosphorylase b shown by the region starting with Ser 10 makes intramolecular contacts in (b). Upon phosphorylation of Ser 14, order is induced in the N-terminal region and new contacts of this segment are formed at the subunit interface of phosphorylase a (a). Reprinted with permission from *Journal of Molecular Biology* 218:233 (1991). Copyright 1991, Journal of Molecular Biology.

A neurofilament protein (M_r of 150 kDa) contains the 13 amino acid sequence, Lys-Ser-Pro-Val-Pro-Lys-Ser-Pro-Val-Glu-Glu-Lys-Gly, which is repeated six successive times in the carboxy terminal portion of the protein.[238] This region of the structure seems highly phosphorylatable, and it has been proposed that phosphorylation could lead to an alteration of the projected side arms in neurofilaments.[239] A study of the structure of synthetic peptides of the repeat region suggests that the segment would have a repeating secondary structure of β-turns and 3_{10} or α-helices.[240] The peptide, Ac-Pro_5-Lys_6-Ser_7-Pro_8-Val_9-Glu_{10}-Glu_{11}-Lys_{12}-Gly_{13}-$NHCH_3$, illustrated in Figure 3.21 shows a salt linkage between the ε-amino group of Lys_6 and the carboxyl group of Glu_{10}. A loop peptide (Res. 1–13) stabilized by the salt linkage may facilitate phosphorylation of Ser_2 and Ser_7. After phosphorylation, the structure is altered presum-

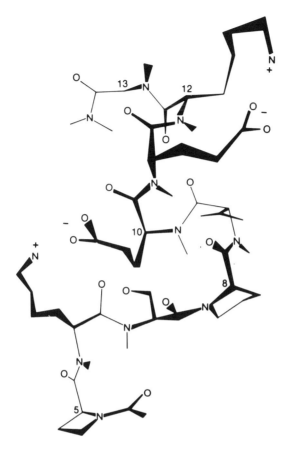

Figure 3.21. Schematic representation of the salt-bridged loop of Ac-Pro-Lys-Ser-Pro-Val-Glu-Glu-Lys-Gly-NHCH$_3$. Reprinted with permission from *Journal of Protein Chemistry* 7:365 (1988). Copyright 1988 Plenum Press.

ably by forming salt bridges between Lys$_6$ and Ser$_2$ or Ser$_7$. Changes in structure in the phosphorylatable sequence by forming and breaking salt linkages could alter the course of phosphorylation in the natural protein substrate.

The phosphorylation of glycogen synthase occurs in the amino and carboxy terminal regions of the protein, and seven different phosphorylation sites have been identified. A cluster of these sites exists in the carboxy terminal portion and phosphorylation causes inactivation of the enzyme. The concept of synergistic phosphorylation was noted when pretreatment of glycogen synthase with casein kinase II, an enzyme notcausing inactivation, potentiated inactivation of the enzyme by glycogen synthase kinase (GSK-3).[241,242] The use of a synthetic peptide of the phosphorylation site with these two kinases revealed the molecular mechanism for this initial observation. The peptide, AcPRPAS(3a)VPPS(3b)PSLS(3c)RHSS(4)PHQS(5)EDEEP-amide, contains five different phosphorylation sites. Note that although there are no exact repeat sequences, as in the neurofilament protein, the phosphorylation sites are spaced evenly, four residues apart throughout the structure. Using this peptide, it was found that casein kinase II phosphorylates site 5, which then allows for subsequent and sequential phosphorylation of the sites 4, 3c, 3b, and 3a, respectively, by GSK-3.[243] The recognized structural elements are not yet proved, but the primary sequence of SXXXS(P) is likely a recognizable element of (GK-3) kinase. Because the sequence is rich in prolyl residues, a b-turn may be involved as suggested for other protein kinase substrates, including reactions occurring at tyrosyl sites.[244]

Hydroxylation

Prolyl hydroxylation can occur at the 3–position although less frequently than modification of the 4-position by a prolyl-3–hydroxylase. Type I and type II collagens contain the sequence Gly-3Hyp-4Hyp-Gly, and studies with various protocollagens suggest that the main sequence for hydroxylation is Gly-Pro-4Hyp-Gly.[245] As conformational changes are suggested to occur after 4-hydroxylation, the new structure may be recognized by the 3–hydroxylase. Multiple hydroxylations of neighboring prolyl residues occur at the 4-position in the plant protein, extensin. Whether a sequential relation exists is not proven, but the relation between 4- and 3–hydroxylation is similar to the synergistic relation documented for different protein phosphorylation reactions.

PROTEIN SUBSTRATE-MODIFIER INTERACTIONS

Because the substrate of post-translational modification reactions are complex macromolecules, various mechanisms exist by which modifiers can influence post-translational modification reactions. A classification can be based on the binding of modifiers to protein substrates. In case (1) a covalently bound modifier, for example, a phosphoryl, acetyl, methyl group introduced in an enzymatic reaction is taken to cause a change in the protein substrate and influence subsequent modification reactions.

In case (2), a modifier, noncovalently bound, for example a nucleotide, metal ion, protein, is suggested to alter the characteristics of the protein and influence modification reactions.

Case 1

Numerous examples exist that demonstrate the biosynthesis of a mature protein is directed by a series of sequential covalent modification reactions, with the protein substrate structure changing throughout the process. In reactions of collagen biosynthesis, reactions occur co- and post-translationally and some reactions occur intra- and extracellularly. Overall the reactions include peptide bond cleavage, hydroxylation, glycosylation, oxidation, cross-linking of oxidized residues, and disulfide bond formation. But how is the sequence of events established for this specific case and for other proteins bearing multiple modifications? One definitive way is to isolate intermediate forms and analyze what modifications have occurred. Pulse-label and chase experiments with ^{14}C and ^{12}C proline in cartilage cells followed by analysis of gel filtration profiles of procollagen was used to establish that hydroxylation precedes disulfide bond formation. Because gel filtration profiles of procollagen molecules bearing radioactive hydroxyproline in the early stages of the biosynthesis were unchanged by the disulfide rupturing agent, mercaptoethanol, it was concluded that these hydroxylated molecules don't contain disulfide bonds.[246] But other approaches, for example, use of mutants with defects in collagen biosynthesis, identification of enzymes in organelles involved in post-translational modification reactions, and the use of specific inhibitors helped establish the pathway of biosynthesis.

A novel separation method based on affinity chromatography has general applications in the study of multiple-modification reactions. With histone and HMG acetylation and phosphorylation, it was not known whether single molecules could be labeled with both phosphate and acetate. Using radioactive thioacetate and inorganic phosphate with duck erythrocytes, incorporation of both isotopes was observed into the nuclear proteins. The isolation of thioacetylated proteins from other proteins was accomplished with mercury-affinity column chromatography. Some thioacetylated histones and HMG proteins also contained phosphate, a finding proving that dual labeling occurred in a single molecule.[247] A similar approach can be taken with inorganic thiophosphate and mercury-affinity chromatography, and it seems variants of this method could apply to other systems.

Activation or inhibition of post-translational modification reactions can be attributed to changes in the properties of the protein substrate due to the presence of a covalently bound modifier. For example, synergistic phosphorylation of glycogen synthase, protein phosphatase inhibitor II[248] and the regulatory subunit (RII) of cyclic dependent protein kinase[249] illustrate the positive reinforcement of one event on subsequent reactions. Ubiquitination of eukaroytic proteins serves as a mechanism for initiating cellular degradation of specific proteins. But the process of ubiquitination depends on the "history" of the protein. One instance shows that the presence of a modifier, an acetyl group on the α-amino group, blocks effective ubiquitination.[250] Factors in the protein substrate leading to effective processing include the nature of the amino acid at the amino terminus (the N-end rule)[251] and a mobile segment containing modifiable lysyl residues. The end amino acids are designated as primary, secondary, and tertiary destabilizing elements. A primary amino acid, for example, arg, leu, may send a signal directly for ubiquitination; a secondary amino acid, for example, asp, glu, can lead to a coupling of a destabilizing amino acid via reactions with various amino acyl-t-RNA protein transferases and then ubiquitination; a tertiary amino acid residue, for example, asn apparently is first deamidated, then acylated by the transferase, and then ubiquitinated.[252] But the presence of a motile segment with a modifiable lysyl residue is also important. A genetic construct of mouse dihydrofolate reductase yielding "destabilizing" amino acids at the n-terminus was not effectively degraded in yeast until a 38 segment containing two lysyl residues, 15 and 17, of β-galactosidase were fused to the protein. Either of the two lysyl residues can be ubiquitinated followed by reactions leading to the formation of a chain of up to 20 ubiquitin molecules.[253] Hence, many reactions may take place in the history of a protein before it is targeted for its final post-translational modification reaction of proteolysis. Other paths of degradation exist including acetylated proteins in ubiquitination reactions.[254] An oxidized methionyl residue may signal subsequent ubiquitination reactions.

Case 2

The reversible interaction of a ligand with a protein substrate can be studied by various physical and chemical techniques. Kinetics can distinguish whether modifiers act on the substrate, the enzyme, or both and defines how these modifiers act. Binding constants may be evaluated directly by such procedures as gel-filtration, equilibrium dialysis, ultrafiltration, quantitative affinity chromatography, and a variety of spectroscopic techniques. The study of the action of ligands on the interconversion reactions of glycogen phosphorylase by many of these techniques along with the x-ray crystallographic results provides an excellent example of how modifiers interacting with the protein substrate can regulate enzymatic reactions.[235] Glycogen phosphorylase b and a exist in two major conformational states, T and R. With phosphorylase b, ligands that appear to stabilize the T state inhibit the phosphorylation. X-ray crystallography suggests a change occurs in orientation of the N-terminal tail when phosphorylase b is changed from the R to the T state.[255] In general, ligands stabilizing the active R state of the phosphorylated form, a, inhibit dephosphorylation and those that promote the inactive T state activate dephosphorylation. AMP causes inhibition, and its binding promotes tighter binding of the phosphoryl group and leads to a shortening of the hydrogen bonds between the oxygens of Ser_{14}-P with Arg_{69} and Arg'_{43} by 0.2Å. Glucose activates the reaction with the phosphate group becoming more motile and exposed. ^{31}P-NMR studies of thiophosphorylated phosphorylase a in the presence of glucose were helpful in this instance, because the chemical shift of the thiophosphoryl group is distinct from other phosphoryl group resonances such as bound pyridoxal phosphate. NMR studies of thiophosphorylated forms of other proteins, for example, multiple-phosphorylated species, may provide insight about structural changes induced by phosphorylation and how ligands affect various post-translational modification reactions.

REFERENCES

1. Patthy, L., and Thez, J., *European J. Biochem.* 105, 387–393 (1980).
2. Klopman, G., *J. Am. Chem. Soc.* 90, 223–234 (1968).
3. Salem, L., *J. Am. Chem. Soc.* 90, 543–552 (1968).
4. Fleming, I., Frontier Orbitals and Organic Chemical Reactions, New York: John Wiley, p. 37.(1987).
5. Pearson, R.G., *J. Am. Chem. Soc.* 85, 3533–3539 (1963).
6. Pearson, R.G., *J. Org. Chem.* 54, 1423–1430 (1989).
7. Fersht, A., *Enzyme Structure and Mechanism*, New York: W.H. Freeman and Co., (1985).
8. Janssen, M.J. in *Sulfur in Organic and Inorganic Chemistry*, (Senning, A., Ed.), New York, Marcel Dekker, 355–377 (1972).
9. Ghosh, S.S., Bock, C.S., Rokita, S.E., and Kaiser, E.T., *Science* 231, 145–148 (1986).
10. Jencks, W.P., *Chem. Rev.* 85, 511–527 (1985).
11. Martin, B., Wu, D-L., Jakes, S., and Graves, D.G., *J. Biol. Chem.* 265, 7108–7111 (1990).
12. Fersht, A.R., and Jencks, W.P., *J. Am. Chem. Soc.* 92, 5442 (1970).
13. Martin, B.L., and Graves, D.J., *J. Biol. Chem.* 261, 14545–14550 (1986).

14. Soman, G., Narayanan, J., Martin, B.L., and Graves, D.J., *Biochem.* 25, 54113–4119 (1986)
15. Oppenheimr, N.J., *J. Biol. Chem.* 253, 4907–4910 (1978).
16. Schmidt, D.E.,Jr., and Westheimer, F.H., *Biochemistry* 10, 1249–1253 (1971).
17. Rose, G.D., Gierasch, L.M., and Smith, J.A., *Adv. Prot. Chem.* 37, 1–109 (1985).
18. Marshall, R.D., *Ann. Rev. Biochem.* 41, 673–702 (1972).
19. Schick, D.K., and Lennarz, W.J., in *The Biochemistry of Glycoproteins*, New York: Plenum Press, p. 35 (1981).
20. Beeley, J.G., *Biochem. Biophys. Res. Commun.* 76, 1051–1055 (1977).
21. Mononen, I., Karjalainen, E., *Biochim. Biophys. Acta.* 788, 364–367 (1984).
22. Chou, P.Y., and Fasman, G.D., *Biochem.* 13, 222–245 (1974).
23. Aubert, J-P., Biserte, G., and Loucheux-Lefebvre, M-H., *Arch. Biochem. Biophys.* 175, 410–418 (1976).
24. Marshall, R.D., *Biochem. Soc. London Transactions* 40, 17–25 (1974).
25. Bause, E., and Legler, G., *Biochem. J.* 195, 639–644 (1981).
26. Bause, E., Hettkamp, H., and Legler, G., *Biochem. J.* 203, 761–768 (1982).
27. Bause, E., *Biochem. J.* 209, 331–335 (1983).
28. Rathod, P.K., Tashjian, A.M.,Jr., and Abeles, R.H., *J. Biol. Chem.* 261, 6461–6469 (1986).
29. Hortin, G., Stern, A.M., Miller, B., Abeles, R.H., and Boime, I., *J. Biol. Chem.* 258, 4047–4050 (1983).
30. Aubert, J-P., Helbecque, N., and Loucheux-Lefebvre, M-H., *Arch. Biochem. Biophys.* 208, 20–29 (1981).
31. Ronin, C., and Aubert, J-P., *Biochem. Biophys. Res. Commun.* 105, 909–915 (1982).
32. Aubert, J.P., Chiroutre, M., Kerckaert, J.P., Helbecque, N., and Loucheux-Lefebre, M.H., *Biochem. Biophys. Res. Commun.* 104, 1550–1559 (1982).
33. Abbadi, A., Boussard, G., Marraud, M., Pichon-Pesme, V., and Aubry, A., In *Second Forum on Peptides* (Aubry, A., Marraud, M., and Vitoux, B., Eds.) Colloque INSERM, 174, 375–378, John Libbey Eurotext Ltd.(1989).
34. Imperali, B., and Shannon, K.L., *Biochemistry* 30, 4374–4380 (1991).
35. Imperali, B., Shannon, K.L., Unno, M., and Rickert, K.W., *J. Am. Chem. Soc.* 114, 7944–7945 (1992).
36. Miletich, J.P., and Broze, G.J., Jr., *J. Biol. Chem.* 265, 11397–11404 (1990).
37. Aubert, J.P., Biserte, G., and Loucheux-Lefebvre, M-H., *Arch. Biochem. Biophys.* 175, 410–418 (1976).
38. Young, J.D., Tsuchiya, D., Sandin, D.E., and Holroyde, M.J., *Biochem.* 18, 4444–4448 (1979).
39. Kivirikko, K.I., and Myllyla, R., in *The Enzymology of Post-translational Modification of Proteins*, (Freedman, R.B., and Hawkins, H.C., Ed.) New York: Academic Press, 54–104 (1980).
40. Ananthanarayanan, V.S., Attah-Poku, S.K., Mukkamala, P.L., and Rehse, P.H., *Proc. Int. Symp. Biomol. Struct. Interactions, Suppl. J. Biosci.* 8, 209–221 (1985).

41. Rapaka, R.S., Renugopalakrishan, V., Urry, D.W., and Bhatnagar, R.S., *Biochem.* 17, 2892–2898 (1978).
42. Bansal, M., and Ananthanarayanan, V.S., *Biopolymers* 27, 299–312 (1988).
43. Atreya, P.L., and Ananthanarayanan, V.S., *J. Biol. Chem.* 266, 2852–2858 (1991).
44. Small, D., Chou, P., and Fasman, G.D., *Biochem. Biophys. Res. Commun.* 79, 341–345 (1978).
45. Pinna, L.A., Meggio, F., and Marchiori, F., in *Peptides and Protein Phosphorylation*, (Kemp, B.E., Ed.) Boca Raton, Florida: CRC Press, p.145–169 (1990).
46. Meggio, F., Perich, J.W., Meyer, H.E., Hoffmann-Posorske, E., Lennon, D.P.W., Johns, R.B., and Pinna, L.A., *Eur. J. Biochem.* 186, 459–464 (1989).
47. Graves, D.J., and Graves, P.R., unpublished results.
48. Wilmot, C.M., and Thornton, J.M., *J. Mol. Biol.* 203, 221-232 (1988).
49. Granot, J., Mildvan, A.S., Bramson, H.N., and Kaiser, E.T., *Biochemistry* 20, 602–610 (1981).
50. Rosevear, P.R., Fry, D.C., Mildvan, A.S., Doughty, M., O'Brian, C., and Kaiser, E.T., *Biochemistry* 23, 3161–3173 (1984).
51. Bramson, H.N., Thomas, N.E., Miller, W.T., Fry, D.C., Mildvan, A.S., and Kaiser, E.T., *Biochemistry* 26, 4466–4470 (1987).
52. Leszczynski, J.F., and Rose, G.D., *Science* 234, 849–855 (1986).
53. Bek, E., and Berry, R., *Biochemistry* 29, 178–183 (1990).
54. Perlman, D., and Halvorson, H.O., *J. Mol. Biol.* 167, 391–409 (1983).
55. Folz, R.J., Nothwehr, S.F., and Gordon, J.I., *J. Biol. Chem.* 261, 14752–14759 (1986).
56. Nothwehr, S.F., and Gordon, J.I., *J. Biol. Chem.* 264, 3979–3987 (1990).
57. Clarke, S., *Ann. Rev. Biochem.* 54, 479–505 (1985).
58. Stewart, R.C., and Dahlquist, F.W., *Chem. Rev.* 87, 997–1025 (1987).
59. Terwilliger, T.C., Wang, J.Y., and Koshland, D.E., Jr., *J. Biol. Chem.* 261, 10814–10820 (1986).
60. Nowlin, D.M., Bolinger, J., and Hazelbauer, G.L., *J. Biol. Chem.* 262, 6039–6045 (1987).
61. Terwilliger, T.C., Wang, J.Y., and Koshland, D.E., Jr., *Proc. Natl. Acad. Sci.* 83, 6707–6710 (1986).
62. Mowbray, S.L., Foster, D.L., and Koshland, D.E., Jr., *J. Biol. Chem.* 260, 11711–11718 (1985).
63. Kehry, M.R., Bond, M.W., Hunkapiller, M.W., and Dahlquist, F.W., *Proc. Natl. Acad. Sci.* 80, 3599–3603 (1983).
64. Rice, M.S., and Dahlquist, F.W., *J. Biol. Chem.* 266, 9746–9753 (1991).
65. Tanaka, M., Sato, K., and Uchida, T., *J. Biol. Chem.* 256, 11397–11400 (1981).
66. Cooper, J.B., Chen, J.A., van Holst, G-J., and Varner, J.E., *Trends in Biochem. Sci.* 12, 24–27 (1981).
67. van Holst, G-J., and Varner, J.E., *Plant Physiol.* 74, 247–251 (1984).
68. Zasloff, M., *Proc. Natl. Acad. Sci.* 84, 5449–5453 (1982).

69. Resnick, N.M., Maloy, W.M., Guy, H.R., and Zasloff, M., *Cell* 66, 541–554 (1991).
70. Marion, D., Zasloff, M., and Bax, A., *FEBS. Lett.* 227, 21–25 (1988).
71. Matsuzaki, K., Harada, M., Handa, T., Funaskoshi, S., Fuji, N., Yajima, H., and Miyajima, K., *Biochem. Biophys. Acta.* 981, 130–134 (1989).
72. Williams, R.W., Starman, R., Taylor, K.M.P., Gable, K., Beeler, T., Zasloff, M., and Covell, D., *Biochemistry* 29, 4490–4495 (1990).
73. Maulet, Y., and Cox, J.A., *Biochemistry* 22, 5680–5685 (1983).
74. O'Neil, K.T., and DeGrado, W.F., *Trends Biochem. Sci.* 15, 59–64 (1990).
75. Fujiwara, K., Okamurai-Ikeda, K., and Motokawa, Y., *J. Biol. Chem.* 265, 17463–17467 (1990).
76. Fujiwara, K., Okamura-Ikeda, K., and Motokawa, Y., *FEBS Lett.* 293, 115–118 (1991).
77. Cooke, R.M., Wilkinson, A.J., Baron, M., Pastore, A., Tappin, M.J., Campbell, I.D., Gregory, H., and Sheard, B., *Nature* 327, 339–341 (1987).
78. Stenflo, J., Holme, E., Lindstedt, S., Chandramouli, N., Huang, L.H.T., Tam, J.P., and Merrifield, R.B., *Proc. Natl. Acad. Sci.* 86, 444–4447 (1989).
79. Gronke, R.S., VanDusen, W.J., Garsky, V.M., Jacobs, J.W., Sardana, M.K., Stern, A.M., and Friedman, P.A., *Proc. Natl. Acad. Sci.* 86, 3609–3613 (1989).
80. Sherman, F., Stewart, J.W., and Tsunasawa, S., *BioEssays* 3, 27–31 (1986).
81. Huang, S., Elliot, R.C., Liu, P.S., Koduri, R.K., Weickmann, J.L., Lee, J.H., Blair, L.C., Ghosh-Dastidar, P., Bradshaw, R.A., Bryan, K.M., Einarson, B., Kendall, R.L., Kolcacz, K.H., and Saito, K., *Biochem.* 26, 8242–8245 (1987).
82. Arfin, S.M., and Bradshaw, R.A., *Biochem.* 27, 7979–7984 (1988).
83. Augen, J., and Wold, F., *Trends in Biochem. Sci.* 11, 494–497 (1986).
84. Persson, B., Flinta, C., vonHeijne, G., and Jornvall, H., *Eur. J. Biochem.* 152, 523–527 (1985).
85. Radharkrishna, G., and Wold, F., *J. Biol. Chem.* 261, 9572–9575 (1986).
86. Redman, K.L., and Rubenstein, P.A., *Methods in Enzymol* 106, 179–192 (1984).
87. Lee, F-J.S., Lin, L-W., and Smith, J.A., *J. Biol. Chem.* 265, 11576–11580 (1990).
88. Lee, F-J.S., Lin, L-W., and Smith, J.A., *J. Biol. Chem.* 265, 3603–3605 (1990).
89. Towler, D.A., Gordon, J.I., Adams, S.P., and Glaser, L., *Ann. Rev. Biochem.* 57, 69–99 (1988).
90. Kaplan, J.M., Mardon, G., Bishop, J.M., and Varmus, H.E., *Mol. Cell Biol.* 8, 2435–2441 (1988).
91. Duronio, R.J., Rudnick, D.A., Adams, S.P., Towler, D.A., and Gordon, J.A., *J. Biol. Chem.* 266, 10498–1054 (1991).
92. Stock, A., Clarke, S., Clarke, C., and Stock, J., *FEBS Lett.* 220, 8–14 (1987).
93. Stock, A., in *Adv. in Exp. Med. and Biol.* 231, 387–399 (1987).
94. Deutch, C.E., *Methods in Enzymol.* 106, 198–205 (1984).
95. Bohley, P., Kopitz, J., and Adam, G., *Biol. Chem. Hoppe-Seyler* 369, 307–310 (1988).

96. Kopitz, J., Rist, B., and Bohley, P., *Biochem. J.* 267, 343–348 (1990).
97. Kopitz, J., Diplomarbeit, Universitat Tubingen, 1988.
98. Bradbury, A.F., and Smyth, D.G., *Biosciences Reports* 7, 907–915 (1987).
99. Bradbury, A.F., and Smyth, D.G., *Biochem. Biophys. Res. Commun.* 112, 372–377 (1983).
100. Hancock, J.F., Magee, A.M., Childs, J.E., and Marshall, C.J., *Cell* 57, 1167–1177 (1989).
101. Casey, P.J., Solski, P.A., Der, C.J., and Buss, J.E., *Proc. Natl. Acad. Sci.* 86, 8323–8327 (1989).
102. Schaefer, W.R., Kim, R., Sterne, R., Thorner, J., Kim, S-H., and Rine, J., *Science* 245, 379–385 (1989).
103. Jones, T.L.Z., and Spiegel, A.M., *J. Biol. Chem.* 265, 19389–19392 (1990).
104. Stephenson, R.C., and Clarke, S., *J. Biol. Chem.* 265, 16248–16254 (1990).
105. Tan, E.W., Perez-Sala, D., Canada, F.G., and Rando, R.R., *J. Biol. Chem.* 266, 10719–10722 (1991).
106. Watkins, N.G., Thorpe, S.R., and Baynes, J.W., *J. Biol. Chem.* 262, 7207–7212 (1985).
107. Watkins, N.G., Neglia-Fisher, C.I., Dyer, D.G., Thorpe, S.R., and Baynes, J.W., *J. Biol. Chem.* 262, 7207–7212 (1987).
108. Kowluru, R.A., Heidron, D.B., Edmondson, S.P., Bitensky, M.W., Kowlru, A., Downer, N.W., Whaley, T.W., and Trewhilla, J., *Biochem.* 28, 2220–2228 (1989).
109. Allfrey, V.G., Di Paola, E.A., and Sterner, R., *Methods in Enzymol.* 107, 224–240 (1984).
110. Chicoine, L.G., Richman, R., Cook, R.G., Gorovsky, M.A., and Allis, C.D., *J. Cell Biology* 105, 127–135 (1987).
111. Turner, B.M., and Fellows, G., *European J. Biochem.* 179, 131–139 (1989).
112. Malchy, B.L., *Biochem.* 16, 3922–3927 (1977).
113. Malchy, B.L., and Kaplan, B.L., *J. Mol. Biol.* 82, 537–545 (1974).
114. Paik, W.K., Pearson, D., Lee, H.W., and Kim, S., *Biochim. Biophys. Acta* 213, 513–522 (1970).
115. Delange, R.J., Glazer, A.N., and Smith, E.L., *J. Biol. Chem.* 244, 1385–1388 (1969)
116. Paik, W.K., and DiMaria, P., *Methods in Enzymol.* 106, 274–287 (1984).
117. Swanson, R., Trus, B.L., Mandel, N., Mandel, G., Kallia, O.B., and Dickerson, R.E, *J. Biol. Chem.* 252, 759–774 (1977).
118. Brautigan, D.L, Miller, S.F., and Margoliash, E., *J. Biol. Chem.* 253, 130–139 (1978).
119. Siegel, F.L., in *Advances in Exp. Med. and Biol.* 231, 341–351 (1987).
120. Gregori, L., Marriott, D., West, C.M., and Chau, V., *J. Biol. Chem.* 260, 5232–5235 (1985).
121. Babu, Y.S., Bugg, C.E., and Cook, W.J., *J. Mol. Biol.* 204, 191–204 (1988).
122. Brostoff, S., and Eylar, E.H., *Proc. Natl. Acad. Sci.* 68, 765–769 (1971).
123. Schafer, D.J., *J. Chem. Soc.*, Perkin I, Part 3, 2295 (1974).
124. Takahara, H., Oikawa, Y., and Sugawara, K., *J. Biochem.* 94, 1945–1953 (1983).

125. Takahara, H., Okamato, H., and Sugawara, K., *J. Biochem.* 99, 1417–1424 (1986).
126. Takahara, H., Okamato, H., and Sugawara, K., *J. Biol. Chem.* 260, 8378–8383 (1985).
127. Sweet, R.M., Wright, H.T., Janin, J., Choitha, C.H., and Blow, D.M., *Biochemistry* 13, 111–117 (1974).
128. Clarke, S., *Int. J. Peptide Prot. Res.* 30, 808–821 (1987).
129. Wearne, S.J., and Creighton, T.E., *Proteins* 5, 8–12 (1989).
130. Kossiakoff, A.A., *Science* 240, 191–194 (1988).
131. Luo, S., Liao, C-X., McClelland, J., and Graves, D.J., *Int. J. of Peptide and Protein Chem.* 29, 728–733 (1987).
132. Ota, I.M., Ding, L., and Clarke,S., *J. Biol. Chem.* 262, 8522–8531 (1987).
133. Ota, I.M., and Clarke, S., *J. Biol. Chem.* 264, 54–60 (1989).
134. Ota, I.M., and Clarke, S., *Biochem.* 28, 4020–4027 (1989).
135. Murray, E.D.Jr., and Clarke, S., *J. Biol. Chem.* 259, 10722–10732 (1984).
136. Clarke, S., Vogel, J.P., Deschenes, R.J., and Stock, J., *Proc. Natl. Acad. Sci.* 85, 4643–4647 (1988).
137. Gutierrez, L., Magee, A.I., Marshall, C.J., and Hancock, J.F., *EMBO J.* 8, 1093–1098 (1989).
138. Knobloch, J.E., and Suttie, J.W., *J. Biol. Chem.* 262, 15334–15337 (1987).
139. Ulrich, M.M.W., Furie, B., Jacobs, M.R., Vermeer, C., and Furie, B.C., *J. Biol. Chem.* 263, 9697–9702 (1988).
140. Sanford, D.G., Kanagy, C., Sudmeier, J.L., Furie, B.C., Furie, B., and Bachovchin, W.W., *Biochemistry* 30, 9835–9841 (1991).
141. Price, P.A., Fraser, J.D., and Metz-Virca, G., *Proc. Natl. Acad. Sci.* 84, 8335–8339 (1987).
142. Goss, N.H., and Wood, H.G., *Methods in Enzymol.* 107, 261–278 (1984).
143. Murtif, V.L., and Samols, D., *J. Biol. Chem.* 262, 11813–11815 (1987).
144. Hong, D-H., Moon, T-W., Lopez-Casillas, F., Andrews, P.C., and Kim, K-H, *European J. Biochem.* 189, 239–244 (1989).
145. Cronan, J.E.Jr., *J. Biol. Chem.* 265, 10327–10333 (1990)
146. Reed, K., and Cronan, J.E.Jr., *J. Biol. Chem.* 266, 11425–11428 (1991).
147. Stryer, L., *Cold Spring Harbor Symp. Quant. LIII*, 283–294 (1988).
148. Wilden, U., Hall, S.W., and Kuhn, H, *Proc. Natl. Acad. Sci.* 83, 1174–1178 (1986).
149. Palczewski, K, and Hargrave, P.A., in *Mechanism of Phototransduction*, New York, Springer-Verlag, in press.
150. Palczewski, K., McDowell, J.H., and Hargrave, P.A., *Biochemistry* 27, 2306–2313 (1988).
151. Fowles, C., Sharma, R., and Akhtar, M., *FEBS Lett.* 238, 56–60 (1988).
152. Palczewski, K., Buczylko, J., Kaplan, M.W., Polans, A.S., and Crabb, J.W., *J. Biol. Chem.* 266, 12949–12955 (1991).
153. Freedman, R.B., and Hillson, D.A., in *The Enzymology and Post-translational Modification of Proteins*, (Freedman, R.B., and Hawkins, H., Ed.), New York: Academic Press, 158–212 (1980).
154. Creighton, T.E., Hillson, D.A., and Freedman, R.B., *J. Mol. Biol.* 142, 43–62 (1980).

155. Creighton, T.E., *Biochem. J. Reviews* 270, 131–145 (1990).
156. Finn, F.M., Ridge, K.D., and Hofman, K, *Proc. Natl. Acad. Sci.* 87, 419–423 (1990).
157. Olson, T.S., and Lane, D.L., *FASEB J* 3, 1618–1624 (1989).
158. Pihlajaniemi, T., Helaakoski, T., Tasanen, K., Myllyla, R., Huhtala, M-J., Koivu,J., and Kivirikko, K.I., *The EMBO J.* 6, 643–649 (1987).
159. Habib, M.G., Noiva, R., Kaplan, H., and Lennarz, W.J., *Cell* 54, 1053–1060 (1980).
160. Mercken, L., Simons, M-J., Swillens, S., Massaer, M., and Vassart, G., *Nature* 316, 647–651 (1985).
161. Rolland, M., Montfort, M-F., and Lissitzky, S., *Biochim. Biophys. Acta* 303, 338–347 (1973).
162. Turner, C.D., and Rawitch, A.B., in *The Enzymology of Post-translational Modification of Proteins* 2, (Freeman, R.B., and Hawkins, H., Ed.), New York: Academic Press, 95–121 (1985).
163. Xiao, S., Pollock, H.G., Taurog, A., and Rawitch, A.B., Personal Communication (1991).
164. Palumbo, G., *J. Biol. Chem.* 262, 17182–17188 (1987).
165. Gavaret, J-M., Cahnmann, H.J., and Nunez, J., *J. Biol. Chem.* 256, 9167–9173 (1981).
166. Folk, J.E., and Finlyason, J.S., *Adv. Prot. Chem.* 31, 1–120 (1977).
167. Gorman, J.J., and Folk, J.E., *J. Biol. Chem.* 259, 9007–9010 (1984).
168. Simon, M., and Green, H., *J. Biol. Chem.* 263, 18093–18098 (1988).
169. Cocuzzi, E., Piacentini, M., Beninati, S., and Chung, S.I., *Biochem. J.* 265, 707–713 (1990).
170. Doolittle, R.F., *Adv. Prot. Chem.* 27, 1–109 (1973).
171. Ueda, K., and Hayaishi, O., *Methods in Enzymol.* 106, 450–461 (1984).
172. Smulson, M.E., *Methods in Enzymol.* 106, 512–522 (1984).
173. Amado, R., Aeschbach, R., and Neukson, H., *Methods in Enzymol.* 107, 377–388 (1984).
174. Foerder, C.A., and Shapiro, B.M., *Proc. Natl. Acad. Sci.* 74, 4214–4218 (1977).
175. Deits, T., Farrance, M., Kay, E.S., Medill, L., Turner, E.E., Weidman, P.J., and Shapiro, B.M., *J. Biol. Chem.* 259, 13525–13533 (1984).
176. Malencik, D.A., and Anderson, S.R., *Biochemistry* 26, 695–704 (1987).
177. Sefton, B.M., *Mol. Biol. of Intracell. Protein Sorting and Organelle Assembly*, 215–219 (1988).
178. Grand, R.J.A., *Biochem. J.* 258, 625–638 (1989).
179. Ovchinnikov, Y.A., Abdulav, N.G., Bogachuk, A.S., *FEBS. Lett.* 230, 1–5 (1988).
180. O'Dowd, B.F., Hnatowich, M., Caron, M.G., Lefkowitz, R.J., and Bouvier, M., *J. Biol. Chem.* 264, 7564–7569 (1989).
181. Verger, R., Mieras, M.C.E., and DeHaas, G.H., *J. Biol. Chem.* 248, 4023–4034 (1973).
182. Tomaselli, A.G., Hui, J., Fisher, J., Zurcher, N.H., Reardon, I.M., Oriaku, E., Kezdy, F.J., and Heinrikson, R.L., *J. Biol. Chem.* 264, 10041–10047 (1989).

144   Co- and Post-Translational Modification of Proteins

183. O'Brien, P.J., St. Jules, R.S., Reddy, T.S., Bazan, N.G., and Zatz, M., *J. Biol. Chem.* 262, 5210–5215 (1987).
184. Bizzozero, O.A., McGarry, J.F., and Lees, M.B., *J. Biol Chem.* 262, 13550–13557 (1987).
185. Magee, A.I., *J. Cell Science* 97, 581–584 (1990).
186. Grand, R.J.A., and Owen, D., *Biochem. J.* 279, 609–631 (1991).
187. Lindahl, T., Segwick, B., Sekiguchi, M., and Nakeabeppu, Y., *Ann. Rev. Biochem.* 57, 133–157 (1988).
188. Teo, I., Sedgwick, B., Kilpatrick, M.W., McCarthy, T.V., and Lindahl, T., *Cell* 45, 315–324 (1986).
189. Demple, B., Sedgwick, B., Robins, P., Totty, N., Waterfield, M.D., and Lindahl, T., *Proc. Natl. Acad. Sci.* 82, 2688–2692 (1985).
190. Drenth, J., Jansonius, J.N,.Koekoek, R., Swen, H.M., and Wolthers, B.G., *Nature* 218,929–932 (1968).
191. Takahashi, K., Kawazoe, Y., Sakumi, K., Nakabeppu, Y., and Sekiguchi, M., *J. Biol. Chem.* 263, 13490–13492 (1988).
192. Som, S., and Friedman, S., *J. Biol. Chem.* 266, 2937–2945 (1991).
193. Recesi, P.A., and Snell, E.E., *Biochemistry* 12, 365–371 (1973).
194. van Poelje, P.D., and Snell, E.E., *Proc. Natl. Acad. Sci.* 85, 8449–8453 (1988).
195. van Poelje, P.D., and Snell, E.E., *Ann. Rev. Biochem.* 59, 29–53 (1990).
196. van Poelje, P.D., Kamath, A.V., and Snell, E.E., *Biochemistry* 29,10413–10418 (1990).
197. Vanderslice, P., Copeland, W.C., and Robertus, J.D., *J. Biol. Chem.* 263, 10583–10585 (1988).
198. Kennedy, L.D., Kirkman, B.R., Lomako, J., Rodriguez, J.R., and Whelan, W.J., in: Membranes and Muscle (Berman, M.C., Gevers, M.,and Opie, L., eds.) pp. 65–84 (1985).
199. Rodriguez, I.R., and Whelan, W.J., *Biochem. Biophys. Res. Commun.* 132, 829–835 (1985).
200. Smythe, C., Caudwell, F.B., Fergurson, M., and Cohen, P., *EMBO J.* 7, 2681–2685 (1988).
201. Lomako, J., Lomako, W.M., and Whelan, W.J., *FASEB J* 2, 3097–3101 (1988).
202. Lomako, J., Lomako, W.M., and Whelan, W.J., *FEBS Lett.* 264, 13–16 (1990).
203. Lomako, J., Lomako, W.M., and Whelan, W.J., *Carbohydrate Research* 227, 331–338 (1992).(1992).
204. Lomako, J., Lomako, W.M., and Whelan, W.J., *FASEB J* 4, A711 (1990).
205. Lomako, J., Lomako, W.M., and Whelan, W.J., *FEBS Lett.* 279, 223–228 (1991).
206. Almassy, R.J., Janson, C.A., Hamlin, R., Xuong, N-H, and Eisenberg, D., *Nature* 323, 304–309 (1986).
207. Martensen, T.M., *Curr. Topics in Cell Reg.* 27, 171–181 (1985).
208. Cader, B.M., Hemmingsen, J., and Villafranca, J.J., in *Techniques in Protein Chemistry II*, (J.J.Villafranca, Ed.), San Diego: Academic Press, p.313–323 (1991).

209. Reed, L.J., and Yeaman, S.J., in *The Enzymes* 18, (Boyer, P.D.and Krebs, E.G., eds.), Orlando, Fl.:Academic Press, p. 77–95 (1987).
210. Linn, T.C., Pelley, J.W., Pettit, F.H., Hucho, F., Randall, D.D., and Reed, L.J., *Arch. Biochem. Biophys.* 148, 327–342 (1972).
211. Linn, T.C., Pettit, F.H., and Reed, L.J., *Proc. Natl. Acad. Sci.* 64, 227–234 (1969).
212. Brandt, D.R., and Roche, T.E., *Biochemistry* 22, 2966–2971 (1983).
213. deVos, A.M., Tong, L., Milburn, M.V., Matias, P.M., Jancarik, J., Noguchi, S., Nishimura, S., Miura, P.M., Ohtsuka, E., and Kim, S-H., *Science* 239, 888–893 (1988)
214. Bourne, H.R., Sanders, D.A., and McCormick, F.,*Nature* 349, 117–127 (1991).
215. Jurnak, F., *Science* 230, 32–35 (1985)
216. la Cour, T.F.M., Nyborg, J., Thirup, S., and Clark, B.F.C., *The EMBO J.* 4, 2385–2388 (1985).
217. Gilman, A.G., *Ann. Rev. Biochem.* 56, 615–49 (1987).
218. Fong, H.K.W., Yoshimoto, K.K., Eversole-Cire, P., and Simon, M.I., *Proc. Natl. Acad Sci.* 85, 3066–3070 (1988).
219. Kharadia, S., and Graves, D.J., *J. Biol. Chem.* 262,17379–17383 (1987).
220. Kharadia, S., and Graves, D.J., unpublished results.
221. Freissmuth, M., and Gilman, A.G.,*J. Biol. Chem.* 264, 21907–21914 (1989).
222. Moss, J., and Vaughan, M., *Adv. in Enzymol.* 61, 303–379 (1988).
223. Bourne, H.R., *Nature* 321, 814–815 (1986).
224. Casey, P.J., Graziano, M.P., and Gilman, A.G., *Biochemistry* 28, 611–615 (1989).
225. Lowery, R.G., and Ludden, P.W., *J. Biol. Chem.* 263, 16714–16719 (1988).
226. Pope, M.K., Saari, L.L., and Ludden, P.W., *Anal. Biochem.* 160, 68–77 (1987).
227. Murrell, S.A., Lowery, R.G., and Ludden, P.W., *Biochem. J.* 251, 609–612 (1988).
228. Brown, B.A., and Bodley, J.W., *FEBS. Lett.* 103, 253–255 (1979).
229. Van Ness, B.G., Howard, J.B., and Bodlely, J.W.,*J. Biol. Chem.* 255, 10710–10715 (1980).
230. Chen, J-Y.C., and Bodley, J.W., *J. Biol. Chem.* 263, 11692–11695 (1988).
231. Moehring, J.M., and Moehring, T.J., *J. Biol. Chem.* 263, 3840–3844 (1988).
232. Harris, W., and Graves, D.J., *Arch. Biochem. Biophys.* 276, 102 (1990).
233. Barford, D., Hu, S.-H., and Johnson, L.N.,*J. Mol. Biol.* 218, 233–260 (1991).
234. Carlson, G.M., Bechtel, P., and Graves, D.J., *Adv. in Enzymol.* 50, 41–113 (1979).
235. Sprang, S.R., Acharya, K.R., Goldsmith, E.J., Stuart, D.I., Varvill, K., Fletterick, R.J., Madsen, N.B., and Johnson, L.N., *Nature* 336, 215–221 (1988).
236. Graves, D.J., Fischer, E.H., and Krebs, E.G., J. Biol. Chem. 235, 805-809 (1960).
237. Hider, R.C., Ragnarsson, U., and Zetterqvist, O.,*Biochem. J.* 229, 485–489 (1985).

238. Myers, M.W., Lazzarini, R.A., Lee, V.M.Y., Schlaeper, W.W., and Nelson, D.L., The *EMBO J.* 6, 1617–1625 (1987).

239. Lee, M-Y, Otvos, L.Jr., Carden, M.J., Hollosi, M., Dietzschold, B., and Lazzarini, R.A., *Proc. Natl. Acad. Sci.* 85, 1998–2002 (1988).

240. Otvos, L.Jr., Hollosi, M., Perczel, A., Dietzschold, B., and Fasman, G.D., *J. of Protein Chem.* 7, 365–375 (1988).

241. Picton, C., Woodgett, J., Hemmings, B.A., and Cohen, P., *FEBS Lett.* 150, 191–195 (1982).

242. DePaoli-Roach, A.A., Ahamed, Z., Camici, M., Lawrence, J.C.Jr., and Roach, P., *J. Biol. Chem.* 258, 10702–10709 (1983).

243. Fiol, C.J., Mahrenholz, A.M., Wang, Y., Roske, R.W., and Roach, P.J., *J. Biol. Chem.* 262, 14042–14048 (1987).

244. Tinker, D.A., Krebs, E.G., Feltham, I.C., Attah-Poku, S.K., and Ananthanarayanan, V.S., *J. Biol. Chem.* 263, 5024–5025 (1988).

245. Tryggvason, K., Risteli, J., and Kivirikko, K.I., *Biochem. Biophys. Res. Commun.* 76, 275–281 (1977).

246. Prockop, D.J., Berg, R.A., Kivirikko, K.I., and Uitto, J., in (Ramachandran, G.N., and Reddi, A.H., Ed.), *Biochemistry of Collagen*, New York: Plenum Press, p.163–273.(1972).

247. Sterner, R., and Allfrey, V.G., *J. Biol. Chem.* 258, 12135–12138 (1983).

248. DePaoli-Roach, A.A., *J. Biol. Chem.* 259, 12144–12152 (1984).

249. Hemmings, B.A., Aitken, A., Cohen, P., Rymond, M., and Hofmann, F., *Eur. J. Biochem.* 127, 473–481 (1982).

250. Hershko, A., Heller, H., Eytan, E., Kaklij, G., and Rose, I.A., *Proc. Natl. Acad. Sci.* 81, 7021–7025 (1984).

251. Bachmair, A., Finley, D., and Varshavsky, A., *Science* 234, 179–185 (1986).

252. Bachmair, A., and Varashavsky, A., *Cell* 56, 1019–1032 (1989).

253. Chau, V., Tobias, J.W., Bachmair, A., Marriott, D., Ecker, D.J., Gonda, D.K, and Varshavsky, A., *Science* 243, 1576–1583 (1989).

254. Mayer, A., Siegel, N.R., Schwartz, A.L., and Ciedhanover, A. *Science* 244, 1480–1483 (1989).

255. Johnson, L., personal communication.

4

The Modified Protein-Isolation and Analysis of Structure

Many observations suggest that a post-translational modification reaction has taken place in response to a biological signal. For example, incubation of cells in culture, in the presence of a hormone or growth factor with radiolabeled inorganic phosphate [^{32}Pi], adenosine [^{3}H], or inorganic sulfate [^{35}S] followed by polyacrylamide SDS gel electrophoresis of the proteins in cell extracts can provide evidence that some modification reaction has taken place. Deprivation of Swiss 3T3 cells by inhibitors of HMG-CoA reductase blocks cell replication and causes a change in cell morphology. With added ^{14}C-mevalonic acid, radioactive protein can be identified.[1] But what are the reactions? The incorporation of [^{32}Pi] suggests phosphorylation, but other possibilities exist, for example, ADP-ribosylation or other reactions of derivatives containing phosphate. Also, the labeling obtained from the use of [^{3}H] adenosine, [^{35}S] sulfate, or [^{14}C] mevalonic acid doesn't define what chemical events account for the reactions. In animals, administration of coumarol, a vitamin K antagonist, causes both the formation of an abnormal prothrombin molecule and defective blood coagulation. During aging, the crystallin in the lenses of our eyes changes. To understand better the biological and chemical aspects of these processes, it is necessary to isolate the proteins involved and to characterize their structures.

Isolation of Modified Proteins

Sophisticated methods are available for the purification of proteins, and these techniques can be used effectively to separate modified proteins from their unmodified forms. Two general methods apply. First, the modified protein can be separated from its unmodified form based on the properties of the added group. Lectin affinity chromatography is useful for separating various glycosylated proteins.[2] Table 4.1 shows some common lectins that have been used to separate glycoproteins containing different carbohydrates.[3]

Aminophenyl boronate cellulose has been used in the purification of ADP-ribosylated proteins,[4] because this cellulose derivative can bind carbohydrate containing proteins with cis-diols. Recently, a novel approach has been developed for the purification of ADP-ribosylated proteins using biotinylated β-nicotinamide adenine dinucleotide.[5] Biotinylated NAD is a substrate for enzymatic ADP-ribosylation and the resulting modified protein can be separated from unmodified forms by affinity chromatography with avidin-agarose; this method was used to purify a nitric oxide stimulated ADP-ribosylated protein to homogeneity from rat brain. Antibodies obtained to various functional groups, for example, complex carbohydrates and phosphorylated tyrosyl residues, are particularly effective for protein purification. Monoclonal antibodies have been isolated and characterized that specifically precipitate proteins containing phosphorylated tyrosyl residues,[6,7,8] and antibodies made against poly(ADP-ribose) have been used to identify nuclear proteins from rat thymocytes and hepatocytes modified by poly(ADP-ribosylation) reactions.[9]

Second, the modification reaction may change the charge, solubility, or associative state of the protein to allow its separation from the unmodified protein. A mono-S-cation exchanger bearing a sulfonic acid group can separate phosphorylated basic high mobility group nuclear proteins (HMG) from unphosphorylated HMG in a single step.[10] Glycogen phosphorylase contains two seryl residues, one per monomer, which are phosphorylated by phosphorylase kinase. Anionic HPLC separates not only the fully phosphorylated species from the dephosphorylated form, but also the hybrid molecule, in which only one of the monomers is phosphorylated.[11] Isoelectric focusing is particularly applicable to the separation of post-translational modification products in which the charges are different. S-thiolated proteins containing mixed disulfides with glutathione can be separated by this technique.[12] For example, thin-

Table 4.1. Lectins commonly used for glycoprotein purification

Lectin	Specificity	Comments
Concanavalin A (ConA)	Structures containiing a-linked mannose such as the N-acetylglucosaminyl core disaccharide in Asn-linked glycoproteins	Broad use due to ubiquitious α-Mannose containing structures, pH optimum 5.6. As with other mannose-binding lectins, Ca^{2+} and Mn^{2+} are part of the binding site. Buffers may be Ca^{2+}/Mn^{2+} free. Avoid chelators.
Lentil lectin	Similar to ConA, narrower specificity in that substitution to the core saccharide may enhance or diminish binding.	Sometimes employed sequentially with ConA.
Peanut agglutinin	Recognizes Gal(β1-3)-GalNAc	Common core structure to many O-linked membrane glycoproteins; however, often requires sialidase treatment to expose binding disaccharide
Jacalin	Recognizes Gal(β1-3) GalNAc	Unlike peanut lectin, this lectin will recognize substituted disaccharide, eliminating the need for sialidase treatment. Specific for O-linked oligosaccharides.
Wheat germ agglutinin	Recognizes N-acetylglucos-amine dimer or trimer structures/chitobiose. Also lower affinity inter-actions with sialic acid, mannose-containing structures	Widely used for membrane glyco-protein purification, which frequently contain Asn-linked oligosaccharides.
Ricinus communis agglutinin I (RCA I)	Reacts with terminal (nonreducing) Gal, less strongly with GalNAc.	May be useful with both N- and O-linked structures. Since these residues are often not terminal, but penultimate, treatment with exoglycosidases (neuraminidase, fucosidase) may uncover binding properties.
Soybean agglutinin	Recognizes terminal α- or β-GalNAc.	Complementary to RCA I lectin

Reprinted with permission from *Methods in Enzymology* 182, 529. Copyright 1991, Academic Press, Inc.

gel isoelectric focusing (IEF) used in a preparative mode separates glutathione disulfide containing proteins obtained from liver cytosol with pI's of 6.4 and 6.1.[13] A g-carboxylated glutamyl containing protein,

Protein C, can be separated readily from partly carboxylated Protein C based on the fact that the carboxylated form has high affinity for Ca^{+2}. Calcium binding induces a conformational change in the carboxylated protein that influences its binding to chromatographic supports. For example, sufficiently modified recombinant Protein C in the presence of Ca^{+2} binds to an affinity column containing monoclonal antibody. Insufficiently carboxylated Protein C passes through the column. EDTA then can be used to elute the carboxylated form. Alternatively, anion exchange chromatography can be used.[14]

Application of these techniques and a variety of others can help develop better procedures for the isolation of modified proteins. Before detailed chemical, physical, and biological studies are undertaken, it is essential to establish that the sample is pure. General techniques of gel electrophoresis (nondenaturing, isoelectric focusing, and SDS-PAGE) are extremely useful and provide information about purity, charge isomers, and quaternary structure. The isolated modified protein could be a complex of identical or different polypeptide chains. Or, it could be one single chain with different extents of modification on each molecule. Where different polypeptide chains are known, further fractionation is necessary before amino acid analysis and sequencing can be initiated. A variety of methods are available for chain separation, for example, gel-permeation chromatography, HPLC, gel electrophoresis, and antibody affinity chromatography. The four subunits of phosphorylase kinase can be separated readily by reverse-phase HPLC.[15]

AMINO ACID ANALYSIS OF MODIFIED PROTEINS

Two basic questions need to be answered: What is the amino acid composition of the protein? What modifications have taken place? Because of the complexity of the chemical properties of the amino acids and their derivatives, no one procedure can be used for all proteins. Hydrolysis of the protein is usually done by acids, bases, or enzymes. The classical procedure of acid hydrolysis (6N HCl for 20–24 hours at 110° in vacuo) and its variants[16] yields those amino acids and derivatives that are stable in strong acid. Tryptophan, cystine or cysteine, asparagine, and glutamine are destroyed, and other amino acids, such as methionine, serine, and threonine are partly broken down. The addition of various antioxidants can minimize some of these losses, but these reagents have no effect on reactions such as the hydrolysis of the amide-containing residues. Chemical derivatives of side chains and alpha-amino and alpha-carboxyl groups in proteins, for example, various acylated forms, esters, and glycosides, are particularly sensitive to the acid conditions

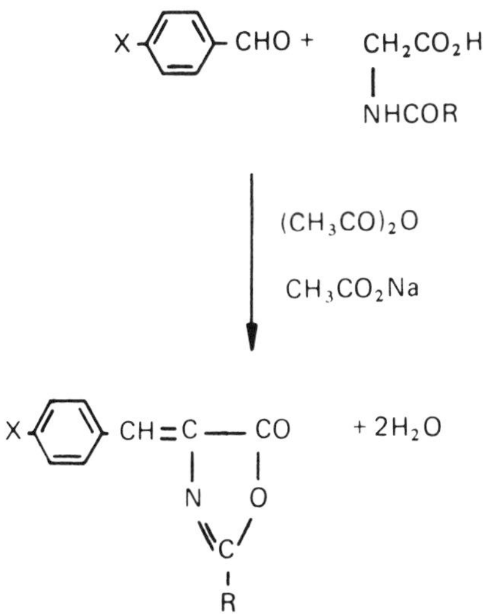

$$X\text{—}\langle\bigcirc\rangle\text{—}CHO \; + \; \underset{\underset{NHCOR}{|}}{CH_2CO_2H}$$

$$\downarrow \; (CH_3CO)_2O$$
$$CH_3CO_2Na$$

$$X\text{—}\langle\bigcirc\rangle\text{—}CH=\underset{\underset{R}{|}\underset{\diagdown C \diagup}{}}{\underset{\underset{N}{|}}{C}}\text{—}\underset{\underset{O}{|}}{CO} \quad + 2H_2O$$

Figure 4.1. Erlenmeyer's azlactone reaction. Initial cyclization of the acylamino acid is followed by a Perkin type condensation with an aromatic aldehyde to generate the azlactone. (Reprinted with permission from *Biochemical Journal* 253:839 (1988). Copyright 1988, The Biochemical Society and Portland Press.)

used for complete hydrolysis. Nevertheless, because acid hydrolysis is effective and some derivatives are stable in acid, the procedure will continue to find use. Base hydrolysis can be employed for tryptophan and derivatives like O-phosphotyrosine and O-sulfotyrosine. When 5N KOH at 155°C is used for 30 min., 80 percent of the phosphorylated tyrosyl residues can be recovered from modified proteins.[17] A procedure for the analysis of O-phosphotyrosyl residues in proteins by base hydrolysis has been described.[18]

O-phosphoseryl residues are sensitive both to acid and base used to hydrolyze proteins. These residues can be detected if the protein is treated first in base in the presence of ethanethiol before hydrolysis. By this treatment, phosphoseryl residues in proteins are converted to the stable S-ethyl cysteinyl derivative. Hydrolysis with HCl followed by derivitization with phenylisothiocyanate yields the phenythiocarbamylderivative (PTC) of S-ethylcysteine.[19] The modified residue is easily detected by HPLC. To determine the residues phosphorylated in a phosphoprotein labeled with the isotope [P^{32}], a common procedure is to subject the protein to partial acid hydrolysis (5.7 M HCl for 2 hrs. at 110°) followed by high-voltage electrophoresis or TLC and radioautography.[20,21,22] Electrophoresis in two dimensions separates serine, threonine, and tyrosine phosphate, but because acid hydrolysis of the protein causes the breakdown of these esters, the procedure is useful only for identifi-cation, and not for quantification. Partial acid hydrolysis also is useful for identifying myristoylated proteins.[23] The amino-termi-nal myristoyl linkage to glycine, unlike acetylglycine, hydrolyzes slowly in 0.05M HCl at 100°. The released myristoylglycine can be reacted with p-nitrobenzaldehyde, yielding an azlactone derivative (Figure 4.1), which can be identified by reverse-phase HPLC.[24]

A mild method for hydrolysis that could effectively cleave the peptide bonds and not the modified group is highly desirable. Enzymatic proce-dures, for example, use of proteases coupled to insoluble matrices, have been used in determining amino acid composition.[25] The major problem here is to find a method that will quantitatively cleave all the peptide linkages in the modified protein. Because the reaction with various proteases could well depend on the nature of the modifying group attached to the amino-acid side chain, this method, though useful, may not release all modified amino acids from the protein. For example, trypsin catalyzes the hydrolysis of peptide bonds in which the C-terminal portion comes from an arginyl or lysyl residue, but it will not cleave peptide linkages containing modified groups such as trimethyllysine or ADP-ribosyl arginine. Bulky carbohydrates in glycoproteins can also inhibit digestion. Pronase catalyzed hydrolysis yields glycopeptides from N-linked glycoproteins; and, in favorable cases a single asparagine residue N-linked to an oligosaccharide might be obtained. But this is usually not the case. Enzymes may also be used to remove the modifying group itself, for example, glycosidases[26] may act to remove bound carbo-hydrate or to trim off parts of its complex structure. An N-glycosidase, peptide-N-(N-acetyl-B-glucosylaminyl) asparagine amidase, can cleave both proteins and peptides, releasing the oligosaccharide chain.[27] In this case, the asparagine residue is converted into an aspartic acid residue in the protein. Extra aspartic acid produced in proteolytic digests of enzymically deglycosylated protein would suggest modification at aspar-agine.

Modified and unmodified amino acids can be separated and quantified by a variety of techniques. Reaction with phenyliso-thiocyanate pro-duces PTH-derivatives that can be separated on a reverse-phase HPLC system.[28] Quantification is possible because of the absorbance of the PTH-derivatives in the UV-region of the spectrum. In some cases, electrophoresis or TLC chromatography may be sufficient to separate the products. When applicable, group-specific reagents or isotopic count-ing can be used to determine amounts of specific amino acid derivatives. In Tables 4-2 and 4-3, a summary of methods used for the analysis of modified amino acids in proteins is illustrated. In many cases, only the instabilities can be described.

The majority of the post-translationally modified amino acids in proteins shown in Tables 4.2 and 4.3 are not stable to acid conditions normally used to hydrolyze proteins. Although conventional amino acid analysis (acid hydrolysis and chromatography of the released amino

Table 4.2. Stability and analysis of modified amino acids in proteins

Amino Acid	Modification	Stability	Procedure
Arginine	N^G-monomethyl N^G, N^G-dimethyl N^G, N^{1G}-dimethyl	stable in acid, unstable in base	acid hydrolysis, ion exchange chromatography[29]
	N^G-ADP ribosyl	unstable in acid and base $t_{1/2}$ is 18' and 6' at 66° in 0.25 M HCl and 0.25M NaOH, respectively	linkage cleaved by NH_2OH (see text)[30,31]
	N-phospho	unstable in acid and base 100% conversion to Pi in 0.1M HCl at 60° for 10'	enzymatic hydrolysis, paper chrom.[32]
Aspartic acid	β-methyl ester α-methyl ester	unstable in acid and base	enzymatic hydrolysis, ion exchange chrom.[33,34]
	β-phosphate	unstable in acid and base	$NaBH_4$ treatment, acid hydrolysis and det. of homoserine[35]
	erythro-β-hydroxy	stable in acid	acid hydrolysis and ion exchange chrom.[36]
	0-(ADP ribosyl)n	unstable in strong acid and base, $t_{1/2}$is 10-20' in 0.1M NaOH at 25°C.	no general method available[37]
	β-carboxy	stable in base	hydrolysis in KOH, ion exchange chrom.[38]
Asparagine	N-glycosyl	unstable in acid and base	deglycosylation with trifluoromethane sulfonic acid, acid hydrolysis, and det. of glucosamine[39,40]
	N-methyl	unstable in acid and base	acid hydrolysis, ion exchange chrom. and det. of methylamine[41]
	erythro-β-hydroxy	unstable in acid and base	acid hydrolysis and and ion exchange chrom. det. of β-hydroxy aspartic acid[42]
Cysteine	disulfide (cystine) mixed sulfides	unstable in acid and base	reaction with 2-nitro-5-thiosulfo-benzoate in the presence of excess Na_2SO_3[43,44]

Table 4.2. (continued)

Amino Acid	Modification	Stability	Procedure
	S-phospho	unstable in acid and base $t_{1/2}$ at 37°, pH 3.0 is 0.4 hrs.; relatively stable at pH< 2.0 at 37°	hydrolysis of thiophosphate by $I_2(t_{1/2} \ll 1$ min)[45,46]
	S-glycosyl	unstable in acid	hydrolysis by 0.5 N HCl for 8 hrs at 105°, det. of sugar by gas liquid chrom.[47]
	S-ADP-ribosyl	unstable in acid	linkage cleaved by $Hg(OAc)_2$[48] (see text)
	S-acyl (fatty acid)	unstable in acid and base	thioacyl group cleaved and acyl function reduced to an alcohol by $NaBH_4$[49]
	S-palmityl	unstable in acid and base	methanolic KOH yields methyl palmitate[50]
	S β-(2-Histidyl)	stable in acid	6N HCl hydrolysis, ion exchange chrom.[51]
	S-(sn-1-glyceryl)-	stable in acid	4N HCl for 15 hrs at 100°C., ion exchange chrom., (S-glyceryl cysteine)[52]
	S-[sn-1-(2,3-diacyl)-glyceryl]	stable in acid	performic acid, acid hydrolysis, high voltage electrophoresis, (S-glyceryl cysteine sulfone)[53]
	S-alkyl S-hydroxylated farnesyl S-prenyl	Stable to base	treat with CH_3I,[54,55] reduction, hydrolysis, (S-methyl cysteine) base hydrolysis or proteolytic digestion, release of prenyl group by Raney Ni.[56,57]
	S-retinyl	unstable to acid	released by hydroxylamine or thiol[58,59]
Selenocysteine		Unstable to acid	alkylation with iodo-acetic acid, acid hydrolysis, and ion exchange chrom. or dinitrophenylation, acid hydrolysis, and TLC[60]
Glutamic acid	0-(ADP-ribosyl)n	unstable in acid and base	linkage cleaved by NH_2OH, pH 7.0, or 0.1M NaOH at 25°[61]
	4-carboxy	unstable in acid, stable in base	2M KOH and ion exchange chrom[62,63]
	0-methyl	unstable in acid and base	0.2 M NaOH 30' at 30°C to release methanol[64]
	0-phospho	unstable in acid and base	$NaBH_4$ will yield α-amino-δ-hydroxy-valeric acid[35]

Table 4.2. (continued)

Amino Acid	Modification	Stability	Procedure
	N-putrescine, spermidine	unstable in acid and base	enzymatic hydrolysis ion exchange chrom.[65]
Glutamine	N-methyl	unstable in acid and base	acid hydrolysis, det. of methylamine; pronase treatment, paper electrophoresis, det. N-methyl glutamine[66]
Histidine	$\pi(\tau)$-phospho	unstable in acid	3 M KOH autoclave at 15 PSI for 3 hrs, paper chrom. Anion exchange chrom. or reverse phase HPLC[67,68]
	2-[(3-Carbamoyl-3-trimethylammonio)-Propyl] (diphthamide)	linkage to his. stable in acid, conversion to COOH form (dipthine)	acid hydrolysis, ion exchange chrom. (^{3}H labeled mat.)[69]
	2-[(3-carbamoyl-3-tri methylammonio)-propyl]-t -(ADP-ribosyl)	somewhat unstable in acid and base	$t_{1/2}$ in 2 NHCl at 110° is 4 hrs, $t_{1/2}$ in 1N NaOH at 110° is 4 hrs; phosphodiesterase and trypsin digestion[69,70]
	π-methyl	stable in acid	6N HCl, 24 hrs 110°,[71,72] ion exchange chrom. or HPLC
	4–iodo	unstable in acid[73]	
Lysine	ε-N-methyl ε-dimethyl ε-trimethyl	stable in acid	6N HCl, 24 hrs 110°, ion exchange chrom.[72]
	ε-N-phospho	unstable in acid	3M KOH autoclave at 15 psi for 3 hrs, chrom.[67]
	ε-N-acetyl	unstable in acid and base	proteolytic digestion, ion exchange chrom. hydrolysis of putative ε-N-acetyl lysine to lysine[74]
	ε-N-(4–amino-2-hydroxy-butyl)-(hypusine)	stable in acid	acid hydrolysis, ion exchange chrom.[75]
	ε-N-biotin	unstable in acid	enzymatic digestion avidin chrom.[76]
	δ-hydroxy	stable in acid	acid hydrolysis, ion[77] exchange chrom.
	δ-hydroxy-ε-N-trimethyl	stable in acid	acid hydrolysis, chromatography[78]
	δ-hydroxy-glycosyl	stable in base	hydroxylysine glycosides released from glycopeptide or proteins by treating in 2N NaOH for 24 hrs at 108°, ion exchange chrom.[79]

Table 4.2. (continued)

Amino Acid	Modification	Stability	Procedure
	ε-(γ-glutamyl)	unstable in acid and base	enzymatic digestion, ion exchange chrom.[80]
	Isopeptide Ubiquitination (polypeptide of 8565 M_r)	unstable in strong acid or base, stable in 0.1N NaOH or 1M NH_2OH at pH 9.0	trypsin digestion, Edman digestion of a branched peptide[81]
	allylysine (α-amino adipic acid δ-semialdehyde)	unstable-exact prop. not reported	peptide treated with $NaBH_4$, acid hydrolysis, identification of ε-hydroxy norleucine[82,83]
Methionine	sulfoxide (nonenzymatic)	unstable in acid	CNBr treatment, reduction, acid hydrolysis, ion exchange chrom. and det. of methionine[84]
Selenomethionine		stable to acid	acid hydrolysis with mercaptoethane-sulfonic acid and ion exchange chrom.[85]
Phenylalanine	β-hydroxy glycoside	unstable in acid	deglycosylation by HF, enzymatic hydrolysis, ion exchange chrom.[86]
	dihydroxy (Dopa)	easily oxidized	acid hydrolysis in the presence of phenol, ion exchange chrom.[87]
	o-bromo	stable in acid	acid hydrolysis, ion exchange chrom.[88]
Proline	3-OH 4-OH	stable in acid	acid hydrolysis, ion exchange chrom .[89]
	4-OH-glycosyl	stable in 0.5N KOH at 4°	enzyme digestion, limited hydrolysis with 0.44N $Ba(OH)_2$[90,91]
Serine	O-phospho	unstable in acid and base	partial acid hydrolysis, 6N HCl, 110° up to 3 hrs or 2N HCl, 100° for 8 hrs[92]
	β-(ADP-riboxyl-phospho)	unstable in acid and base, stable to NH_2OH	$t^{1/2}$ is 1 hr in 0.1 N NaOH at 37°[93]
	O-β-glycosyl O-β-(N-acetyl galactoaminyl) O-β-mannosyl O-β-galactosyl O-β-xylosyl	linkage labile in 0.09 N LiOH at 100° for 1 hr	enzymatic procedures to produce glycopeptides, release of carbohydrate by β-elimination followed by $NaBH_4$ reduction[94,92]
	O-β-(N-acetyl-α-glucosamine-1-phosphoryl)	2N HCl, 100° for 18 hrs causes 76% hydrolysis of phosphodiester linkage	alkaline hydrolysis, det of N-acetyl glucosamine-1-phosphate; partial acid hydrolysis for phosphoserine[95]

Table 4.2. (continued)

Amino Acid	Modification	Stability	Procedure
	O-β(deoxycytosine-5-phosphoryl) of	unstable in acid and base	proteolytic digestion release of nucleotide ColE1 DNA by phosphodiesterase[96]
Threonine	O-phospho	unstable in acid and base	partial acid hydrolysis in 6N HCl, 110° for 3 hrs, electro phoresis [92]
	O-glycosyl	unstable in base	enzymatic procedure to produce glycopeptides, release of carbohydrate and reduction by alkaline $NaBH_4$[94]
	O-fucosyl	unstable in acid	enzymatic procedures, mass spectrometry[97]
Tyrosine	O-phospho	stable in base labile in acid $t^{1/2}$ is 5 hrs in 1 N HCl at 100°C	5N KOH at 155° for 30', electrophoresis [92]
	O-sulfo	stable in base, 95% hydrolysis after 5' at 100° with 1 N HCl	hydrolysis with 0.2M $Ba(OH)_2$ at 110° for 24 hrs, electrophoresis[98]
	O-nucleotidyl	unstable in acid and base	enzymatic digestion, release of nucleotidyl portion by phosphodiesterase[99]
	O-nucleotidyl-5'-phospho (*E.coli* DNA)	unstable in acid and base	partial acid hydrolysis, ident. of O-phosphotyr.[100]
	-halo -bromo, iodo, chloro 3-,and 3,5-	stable in acid	acid hydrolysis in 6N HCl, 105° for 20 hrs in 0.1% thiogly colic acid, gel filtration[101]
	(β-hydroxy) glycosyl	unstable in acid	alkaline $NaBH_4$ cleavage and reduction, hydrolysis in 6N HCl, ion exchange chrom.[86]
	dityrosine	stable in acid	acid hydrolysis, reaction with dabsyl chloride, reverse phase HPLC[102]
	isodityrosine	stable in acid	acid hydrolysis, TLC [103]

acids) is limited, it is valid for some of the modifications and represents an important step for the analysis and characterization of a modified protein. What might be learned? First, information can be gathered about the amount of the normal 20 amino acids in the protein, which provides some insight into the characteristics of the protein, for example, its hydrophobicity and acid-base character. If no amino-acid derivative is found after acid hydrolysis of a modified protein, the results could indicate the attachment of the modifying group through ester, glycoside, or acyl linkages. An alkyl linkage, however, is generally more stable. Hence, these studies may provide a direction for further experimentation proving the structure of the added group.

Table 4.3. Modification reactions of the α-amino and α-carboxyl groups

End Group	Modification	Stability	Procedure
α-Amino	-methyl (mono, di, tri)	stable in acid	5.7N HCl 110°, 24 hrs ion exchange chrom.[104]
	N-monomethyl methionine, alanine	stable in acid	identified in first cycle of Edman degradation[105]
	N-dimethyl proline	stable in acid	N-terminal blocked peptide isolated and hydrolyzed in acid, electrophoresis[106]
	acyl derivatives N-acetyl glycine, alanine, serine, methionine and aspartic acid (dominant)	unstable in acid and base	isolation of N-terminal peptide, release of N-terminal acetylated amino acid by acylpeptide hydrolase, identification by reverse phase HPLC or physical methods[107,108]
α-amino	N-myristoyl	unstable in acid and base	isolation of N-terminal peptide, acid hydrolysis, det of f.a. by gas chrom-mass spectrometry[109,110] treatment of N-terminal peptide with dipeptidyl carboxypeptidase, GC/MS[111] hydrolysis by heating for 2 hrs at 85° in 2N methanolic-HCl partial acid hydrolysis, derivitization with p-nitrobenzaldehyde and HPLC [24]
	N-glucuronamide	unstable in acid and base	NaBH$_4$ reduction, pronase digestion, acid hydrolysis of glucuronyl deriv. amino acid analysis by ion exchange chrom.[112]
	pyroglutamate (5-oxyproline)	unstable in acid and base	pyroglutamate-peptide linkage cleaved by pyrrolidone carboxyl peptidase[113]
	formyl	unstable in acid and base	pronase digestion, electrophoresis, hydrazinolysis of n-formyl amino acid[114]
	pyruvoyl	unstable in acid and base	NaBH$_4$ reduction, acid determination of lactate by chromatography or with lactate dehydrogenase.[115]
	ketobutyryl	unstable in acid and base	tritiated NaBH$_4$ reduction, acid hydrolysis, paper chrom. and determination of a hydroxy butyric acid[116]

Table 4.3. (continued)

End_Group	Modification	Stability	Procedure
a-carboxyl-	amido	unstable in acid and base	treatment with protease dansylation, extraction of acetate, dansyl aa-amide in ethyl-chromatography[117,118,119]
	ethanolamine-	unstable in acid and base	glycopeptide produced by protease digestion[120]
	O-methyl	unstable in acid and base	release of methanol by incubation in 1M NaOH at 37° for 24 hrs .[121]

Second, knowledge of the chemical properties of linkages of known groups attached to a protein provides a base for evaluating structures. For example, ADP-ribose may be attached to proteins via different linkages. These derivatives are unstable to acid hydrolysis, but other chemicals may be used to find out how the ADP-ribose group is bound. A protocol has been developed to identify the nature of linkages to proteins labeled by reaction with NAD and ADP-ribosyltransferases.[122] The isolated proteins obtained from a gel slice are incubated in water, 0.2 M HCl, 2 mM Hg(OAc)$_2$, or 2 M NH$_2$OH. The acid causes release of ADP-ribosides linked by acetal linkages to serine or threonine; the mercurial acetate specifically releases those ribosides linked by a thiol of a cysteine residue, and hydroxylamine releases ADP-ribose bound to a carboxyl group of glutamate or aspartate or ADP-ribose bound by an N-riboside to a guanidinyl function of arginine. These latter two are readily distinguished by differences in their sensitivity to hydroxylamine. Stabilities to chemicals and pH of these and other ADP-ribose derivatives joined to various side chains have been reported recently.[123] After treatment the proteins are re-electrophoresed and examined to define what type of linkages are involved. This is particulary important in the study of ADP-ribosylation because not all modification products are due to direct enzymatic modification. Free ADP-ribose is not an intermediate in enzymatic ADP-ribosylation, but it can be derived from NAD and chemically react with amino groups giving stable ketoamine adducts.[123] Hence, the incorporation of radioactive label from ^{32}P-NAD does not necessarily mean an enzymatic ADP-ribosylation has occurred and illustrates why chemical characterization is so important in the study of post-translational modification reactions. For the proteins containing bound isoprenoids are referred to as prenylated proteins. A farnesyl or geranylgeranyl moiety is linked as an allylic thioether to a cysteinyl side chain. The chemical characteristics of the linkage make it susceptible to reaction with nucleophiles and a procedure has been developed that releases the prenyl group by a reaction with 2-napthol under alkaline conditions.[124] The products are characterized as napthopyrans (Figure 4.2). The napthopyran derivatives of the farnesylated and geranylgeranylated proteins separate on reverse-phase HPLC and are detected easily by their strong absorbance at 246 nm. This method can be used to show whether proteins are prenylated and provides information about types of modifications.

Third, in some modifications shown in Table 4.2 and 4.3, portions of the peptide chain may be removed by proteotylic-processing mechanisms—for example, by removal of acetyl methionine from the amino terminus of newly synthesized actin,[125] release of a peptide segment from

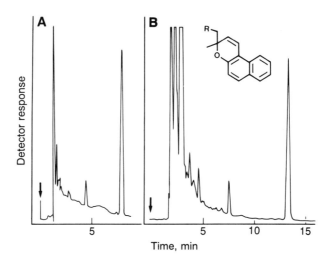

Figure 4.2. (A) HPLC of naphthol cleavage products from 9.5 mg of CHO protein. The column was eluted with acetonitrile at a flow of 1.5 ml/min. Detection was at 360 nm with a full-scale deflection of 0.01 A unit. The arrow indicates sample injection. In this chromatogram the farnesylcysteine derivative is eluted at 4.3 min and the geranylgeranylcysteine derivative at 7.6 min. (B) HPLC of naphthol cleavage products from 2.4 mg of CHO protein. The column was eluted with a gradient of 95% acetonitrile/water to acetonitrile over 15 min. Detection was at 246 nm with a full-scale deflection of 0.04 A unit. The arrow indicates sample injection. In this chromatogram the farnesylcysteine derivative is eluted at 7.7 min and the geranylgeranylcysteine derivative at 13.2 min. (Inset) Structure of methylnaphthopyran. For the farnesyl derivative R is geranyl. In the geranylgeranyl derivative R is farnesyl. For the dimethylnaphthopyran R is CH_3. Reprinted with permission from *Proceedings of the National Academy of Sciences, USA* 88:9668 (1991).

the carboxy terminus prior to the attachment of a glycosyl phosphatidyl anchor,[120] or the release of signal peptides in the transport of proteins through membranes.[126] Besides trimming mechanisms, amino acids residues may be added in post or co-translational modification reactions. Amino acyl-tRNA: protein transferases can transfer amino acids to the amino terminus of protein. The enzyme L-leucyl-tRNA: protein transferase transfers leucine, phenylalanine, and methionine to specific proteins containing an amino terminal arginine or lysine.[127] A tubulin-tyrosine ligase has been identified that catalyzes the addition of a tyrosyl residue to the carboxy terminus of tubulin.[128] Amino acid analysis might then produce results different from those expected based on the knowledge of a cDNA structure. Obviously, the determination of the amino acid sequence would be more conclusive in proving additions or deletions, but the results obtained from amino acid analysis could influence the direction of further studies.

IDENTIFICATION OF MODIFIED AMINO ACIDS BY PHYSICAL METHODS

A peptide containing a modified group can be obtained by enzymatic or by chemical fragmentation of the protein, and the structure of the modified group can be analyzed by physical methods.

A useful technique in searching for post-translational modification products is HPLC.[129] A digest of the protein is obtained and applied to a reverse-phase column. The retention time of peptides on reverse-phase HPLC can, to a reasonable extent, be predicted on the basis of the amino acid composition of the peptides; the more apolar amino acids present, the longer the retention time. For a peptide of less than 40 amino acids, the predicted retention time (T) determined by the retention constant of each amino acid (D_j) and by the molar ratio of amino acids in the peptide (n_j) is given by the equation $T = A \ln(1 + \Sigma D_j n_j) + C$. A and C are constants and depend on the column, but the important point is that a predicted linear relationship exists. If a peptide contains a modified amino acid, the retention time quite often varies from the predicted value, and a change in binding can indicate what modification has taken place. Using an elution buffer of pH 2.0 phosphorylation will reduce the retention time, whereas introducing a myristyl group will increase retention dramatically. Obviously, the method does not prove structure, but it does provide a potential rationale for screening peptides for different types of modification and is helpful for a tentative identification, particularly when no radioisotope or chromophoric substance is present.

Nuclear magnetic resonance spectroscopy can be used in certain instances to identify the added groups, and their mode of attachment (see section in chapter 5 on ^{31}P-NMR spectroscopy). The oxidation of methionine to its sulfoxide in proteins is not simple to detect because of its acid instability. But, ^{1}H-NMR of modified protein can detect directly the sulfoxide of methionine. Bovine pancreatic trypsin inhibitor shows a singlet for the methyl group of methionine at 2.42 ppm. After oxidation to the sulfoxide, this signal is lost and two new singlets appear, 0.6 and 0.46 ppm downfield.[130]

Another method of proving structures is mass spectrometry.[131] In fact, mass spectrometry is preferred in many instances, because it can provide structural information about chemically sensitive molecules, it can be used to analyze mixtures of peptides, it has excellent sensitivity, and it may be applied for the characterization of the structures of not only small molecules but large molecules as well. The presence of γ-carboxyglutamic acid in prothrombin was first proved by the analysis of peptide fragments by mass spectrometry. The determination of blocking groups at the amino-terminus of proteins, for example, acetyl, pyroglutamyl, and myristyl functions have been identified by mass spectral analysis. And it was found by mass spectrometry that urinary plasminogen activator contains a new carbohydrate protein linkage, a fucosyl residue bound to an oxygen atom of a threonyl side chain.[97] It seems new modifications are being found almost routinely.

Two general procedures may be applied. By the first method, the peptide is chemically derivatized and introduced into the ion source of the mass spectrometer. The derivatization permits volatility of the sample and sets it up for defined fragmentation by electron impact or by chemical ionization useful in establishing sequences. In the second case, "soft ionization" methods are utilized. No prior derivatization is required in many cases, but derivatization is still done, for example, as in the analysis of certain carbohydrate chains released from protein for sequence studies. Fast-atom-bombardment mass spectrometry[132] (FAB-MS) belongs to the latter class of soft ionization methods. A sample, in this case a peptide or a modified amino acid, is dissolved in an acidic viscous liquid matrix, for example, either glycerol or thioglycerol containing 5 percent acetic acid or other acids such as HCl. Exposure of a probe containing the sample to high-energy atoms or ions produced by a gun in the mass spectrometer causes the sample to ionize and sputter off—the ions are then collected and analyzed in the detector region of the mass spectrometer. Ions $(M + H)^+$ or $(M - H)^-$ and their fragments can be selected and analyzed. One mode may be more useful than the other, depending on the nature of the modified site; for example, little $(M + H)^+$ may be identified for a sulfated tyrosyl peptide, due to desulfation, but in the $(M - H)^-$ mode, the sulfated tyrosyl peptide may be readily identified.[133] New adaptations of the procedure, for example, collision activated mass spectrometry of the $(M + H)^+$ ions in a FAB/tandem mass spectrometer, provide high sensitivity and specificity over that achieved by FAB alone—a fact that is particularly important in sequence analysis.[134] An important advantage of the latter methodology is that it overcomes both incomplete fragmentation of the parent ion and interference from materials in the matrix. As pure peptides are not needed, the analysis can be performed on partly-purified fractions derived from protease digestion of a modified protein. Peptides in the M_r range of 2000 are readily analyzable. From 0.1–1.0 nmole is usually required, but newer methods allow for analysis in the picomole range.

Another procedure is plasma desorption and analysis of molecular ions in a time-of-flight mass spectrometer. By this method, sample ions coated onto a thin film of nitrocellulose are bombarded with fission fragments derived from the spontaneous fission of 252 Cf. Sample ions are accelerated, allowed to drift through a flight tube, and detected at the end. The time-of-flight relative to the reference, a complementary fragment produced by 252 Cf decay of the sample, measured by a proximate detector, provides a measure of the mass to charge ratio of the sample. The relationship between time of flight and mass/charge is given by the equation $T = C_1 + m/z + C_2$ where C_1 and C_2 are constants. One advantage is that the spectra are dominated by molecular ions and therefore provide a clear cut determination of the molecular weight of the modified peptide. If the molecular weight of the unmodified peptide is known, it is a straightforward procedure determining what group has been added.

A procedure known as FAB-mapping has been developed to analyze peptides derived from proteins by enzymatic or chemical digestion.[135] If the sequence of the protein is known, molecular ions of the fragments can be predicted based on the cleavage method used. In the case of posttranslational modification, the signal will deviate from the predicted m/z ratio of the underivatized peptide. A phosphoryl group will give an additional 80 mass units; an acetyl group, 42 mass units; an ADP-ribosyl function, 641 mass units; and a methyl group, 14 mass units. The procedure has been used successively in the analysis of a multiple-phosphorylated peptide segment of buffalo casein.[136] FAB-MS has been used to examine the *in vivo* phosphorylation state of glycogen synthase.[137] The relative abundance of the molecular ions provides information about stoichiometries of phosphorylation—that is, whether a peptide segment is mono or diphosphorylated—and the results obtained by this procedure correspond well with phosphate content obtained earlier by colorimetric methods. Methods of evaluating the types and sites of modification are being developed for the analysis of a digest of a whole protein. Tryptic digest of calmodulin and elastase digests of enzymatically methylated calmodulin have been analyzed by FAB-mass spectrometry to prove sites and structures of the modification products.[138] Reviews of the applications of mass spectrometry to the study of post-translational modification reactions are helpful.[139,140]

CHEMICAL STRUCTURE OF THE MODIFIED GROUP

How do we determine the structure of complex groups covalently attached to protein, for example, the carbohydrate in a glycoprotein, the fatty acids in a specific protein, or the phosphatidylinositol glycans of certain membrane proteins? Chemical modification and characterization of derivatives by a variety of methods, such as gas chromatography, mass spectrometry, and NMR spectrometry, are important methods utilized to establish structure. Carbohydrates can be linked directly to the protein through N- and O-glycosidic linkages. Although there are many carbohydrates found in living systems, only a few are conjugated to protein.[141] Common monosaccharides found in proteins include the hexoses: galactose, mannose, and glucose; hexosamines: N-acetyl galactosamine and N-acetyl glucosamine; a deoxyhexose, L-fucose; pentoses: xylose and arabinose; sialic acid, and acetyl neuraminic acid. To release carbohydrates as monosaccharides, acid hydrolysis is the preferred method because glycosidic bonds are labile in acid but stable in base. Exceptions are O-glycosidic linkages to serine and threonine, which are completely cleaved by 0.05–0.1M NaOH in 24 hours at room temperature. Nevertheless, acid hydrolysis poses some problems due to the sensitivity of certain carbohydrates to acid. Good recovery of glucosamine and galactosamine can be achieved in 4 M HCl at 100°C for 16 hours, but sialic acid will be completely destroyed. Mannose and galactose have intermediate stabilities (a 23% loss can occur upon heating for 5 hours in 2 M HCl).[141] Methanolysis in dry HCl is commonly used to identify sugars, because a higher recovery is obtained with this system.

To identify the released sugars, a variety of methods can be used, for example, ion exchange chromatography of aminosugars on an amino acid analyzer, and gas-liquid chromatography for neutral sugars—but the main task is to determine how these different sugars are linked to each other and how the carbohydrate is bound to the protein.[141] The problem is easily addressed in proteins like ovalbumin or ribonuclease b, both of which have a single carbohydrate per mole, but becomes quite complex in proteins with a high percentage of carbohydrate, particularly because not all protein molecules of a single type may have the same carbohydrate bound to it. The micro-heterogeneity could result from differences in completion of the oligosaccharide chains during biosynthesis or to some later degradative process. To determine the structure of the carbohydrate chains, the most common approach is probably first to hydrolyze the glycoprotein by extensive proteolysis with a protease like pronase. Glycopeptides can be isolated by many of the methods used to separate other biomolecules, but procedures such as lectin affinity chromatography[142] and monoclonal antibodies to specific carbohydrate antigens can be used effectively.

Analysis can be performed by enzymatic, chemical, and physical methods. Electrochemical methods provide a sensitive means of detecting organic materials. Aliphatic amines and alcoholic functions, constituents of glycosylated moieties in glycoproteins, however, are not usually detected by usual amperometric methods because of their low electroactivity. But at gold and platinum electrodes oxidative reactions can occur and using a pulsed amperometric procedure, carbohydrates may be detected.[143] Picomole quantities of oligosaccharides and glycopeptides derived from glycoproteins, separated by HPLC, may be analyzed successfully by these methods.[144,145] Table 4.4 shows retention times of complex carbohydrates of structures found in N-linked glycoproteins. Because the sugar OH groups have different pKa values, the carbohydrates can be separated by using anionic HPLC under alkaline conditions. It seems this method will have many applications in the study of glycoproteins.

A variety of exoglycosidases can be used to remove monosaccharides from the nonreducing end of a bound oligosaccharide to give information about their sequence in the glycopeptide. For example, the use of α-mannosidase and β-N-acetylglucosoaminidase on a glycopeptide derived from soybean agglutinin was helpful in deducing the sequence, (mannose)$_5$-GlcNAc-(mannose)$_4$-GlcNAc.[141] Chemical procedures are also commonly used.[141] Permethylation of free hydroxyl groups by reagents such as methyl iodide and dimethyl sulfinyl carbanion, followed by acid hydrolysis and analysis of modified oligosaccharides, can provide information about how the units are connected. Periodate cleaves vicinal hydroxyl groups in carbohydrates. After periodate treatment, reduction with NaBH$_4$ and acid hydrolysis is performed. Analysis of the polyalcohol formed and of any released formic acid provides useful information about the structure of the saccharides and of its linkages.

Physical methods, particularly mass spectrometry and NMR spectroscopy, are increasingly being used in the structural analysis of oligosaccharides. First, the query does the protein contain covalently bound carbohydrate is easily answered by mass spectrometric methods. A sample of protein (10–50μg) is subjected to acetolysis, that is, to heating with a mixture of acetic acid, acetic anhydride, and H$_2$SO$_4$. The released and derivitized oligosaccharides can be extracted in CHCl$_3$ and analyzed by FAB-MS.[146] A glycoprotein containing oligosaccharides of mannose will yield peracetylated hexose oligomers; a complex-type carbohydrate may give repeats of derivatives of N-acetyl-lactosamine.

Table 4.4. Structures and retention times of neutral oligosaccharides by HPAE-PAD.

Compound	Oligosaccharide structure	Retention time, min.
1	Fuc(α1–3)GlcNAc(β1–2)Man	3.0
2	Gal(β1–4)GlcNAc(β1–6)Man	10.0
3	Gal(β1–4)GlcNAc(β1–2)Man / Fuc(α1–2)	4.7
4	Gal(β1–4)GlcNAc(β1–3)Gal(β1–4)Glc	18.1
5	Gal(β1–3)GlcNAc(β1–3)Gal(β1–4)Glc	26.5

6	Gal(β1–4)GlcNAc(β1–2)
	    \\
	      Man–OH
	    /
	Gal(β1–4)GlcNAc(β1–4)
		7.8

7	Gal(β1–4)GlcNAc(β1–2)
	    \\
	      Man
	    /
	Gal(β1–4)GlcNAc(β1–4)
		23.2

8	Gal(β1–4)GlcNAc(β1–2)
	    \\
	      Man
	    /
	Gal(β1–4)GlcNAc(β1–6)
		24.2

9	Gal(β1–4)Glc(β1–2)
	    \\
	      Man
	    /
	Gal(β1–4)Glc(β1–6)
		32.7

10	Gal(β1–4)GlcNAc(β1–9)
	    \\
	      Man(α1–2)Man
	    /
	Gal(β1–4)GlcNAc(β1–3)
		28.7

11	Gal(β1–4)GlcNAc(β1–2)Man(α1–6)
	    \\
	      Man
	    /
	Gal(β1–4)GlcNAc(β1–2)Man(α1–3)
		33.0

12	Fuc(α1–3)GlcNAc(β1–2)Man(α1–6)
	    \\
	      Man
	    /
	Fuc(α1–3)GlcNAc(β1–2)Man(α1–3)
		5.6

Table 4.4. continued

Compound	Oligosaccharide structure	Retention time, min.
13	Fuc(α1–3) \ Gal(β1–4)GlcNAc(β1–2)Man(α1–6) \ Man / Gal(β1–4)GlcNAc(β1–2)Man(α1–3) / Fuc(α1–3)	9.2
14	Gal(β1–4)GlcNAc(β1–2)Man(α1–6) \ Man / Gal(β1–4)GlcNAc(β1–2)Man(α1–3) / Gal(β1–4)GlcNAc(β1–2)	35.4
15	Gal(β1–4)GlcNAc(β1–6) \ Gal(β1–4)GlcNAc(β1–2)Man(α1–6) \ Man / Gal(β1–4)GlcNAc(β1–2)Man(α1–3)	37.3
16	Gal(β1–4)GlcNAc(β1–6) \ Gal(β1–4)GlcNAc(β1–2)Man(α1–6) \ Man / Gal(β1–4)GlcNAc(β1–2)Man(α1–3) / Gal(β1–4)GlcNAc(β1–4)	39.4

Reprinted with permission from *Proceedings of the National Academy of Sciences, USA* 85:3289 (1988).

Sialic acid can be readily identified in a bound form. Procedures can then be formulated for specific cleavage and for analysis of O-and N-linked glycans. Alkali and NaBH$_4$ cleave and derivitize the reducing end of an O-linked saccharide, and hydrazine cleaves specifically N-linked saccharides.[147] The oligosaccharides in glycopeptides in the M$_r$ range of 200-10,000 are amenable to structural analysis by FAB mass spectrometry. Peracetyl and permethyl derivatives yield necessary fragment ions to determine the size of the carbohydrate, the branching patterns, and the sequence of the monomeric units.[148] Also, the presence of modifying groups such as phosphate, methyl, acyl, and sulfate can be identified readily. Extensive information about the structure of large polylactosamino glycans of cell surface glycoproteins has been obtained by mass spectral analysis. FAB mass spectra of permethylated glycans containing N-acetyl hexosamine residues are easily interpretable because of the cleavage occurring at these sugars. The fragmentation pattern[149] for a diasialylated biantennary structure derived from a cell surface glycoprotein N-linked to an asparagine residue is shown in Figure 4.3.

The daughter ion compared with the sequence ion containing an N-acetyl amino sugar can give information about branched linkages. One such ion that may be obtained is shown in Figure 4.4.

This daughter ion, 32 mass units lower than the sequence ion, results from the loss of methanol from position 3 of the permethylated derivative. If, however, an additional sugar is bound to the 3 position, for example, fucose in a 1,3 linkage, a signal 206 units can be seen. Hence, branching patterns may be defined.

In the analysis of the oligosaccharide component by NMR spectroscopy, it is common to first release the bound carbohydrate chains by enzymatic or chemical means. Endoglycosidases can be used to hydrolyze glycosidic bonds, N- and O-linked to amino acidic side chains. Treatment with 0.1 M NaOH causes effective β-elimination of saccharides O-linked to seryl or threonyl side chains in inner portions of peptides. Higher amounts of base (1M) are needed for cleavage of other linkages. If hydrolysis is carried out in the presence of $NaBH_4$ the reducing end is converted to an alcohol, and further degradation of the carbohydrate chain is prevented.[150] ^{1}H-NMR spectroscopy in one dimension gives important information about the oligosaccharides and can prove the structure if it is the same as a structure previously determined. But the analysis usually demands more experimentation, particularly because the proton resonances occur in a narrow region of 3–5 ppm. Although complete structural analysis is virtually impossible by single-dimensional NMR, it is now possible by the application of 2-D NMR techniques, such as proton homonuclear 2-D correlation spectroscopy.[151]

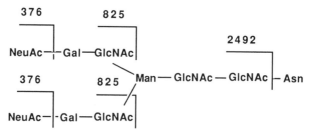

Figure 4.3. Mass spectral fragmentation pattern of a disialyated carbohydrate.

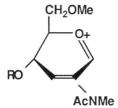

CH₂OMe

RO

AcNMe

Figure 4.4. Daughter ion derived from a N-acetylamino sugar.

According to this method, two types of measurements are performed. First, because the protons in the monosaccharides are linearly arranged around the ring, information about the identities and stereochemistries of the saccharides can be obtained by analyzing their J -connectivities and coupling constants. Correlation spectroscopy (COSY) and spin echo correlated spectroscopy (SECSY) can be used to measure connectivities through chemical bonds. The anomeric proton resonates at 4–5 ppm, and this region of the spectrum is virtually free of other signals. From this reference point, the connectivity of the anomeric proton to a proton(s) at an adjacent carbon atom can be defined. This is followed by determining connectivities of protons around the structure which describes the structure. For example, an aldohexuronic acid has five nonexchangeable protons; and its pattern of coupling, H_1—H_2—H_3—H_4—H_5, can distinguish it from other saccharides. Table 4.5, which shows the J-connectivity patterns for various saccharides, indicates how some of the types of sugars can be identified by this type of analysis.

Evaluation of the vicinal coupling constants provides information about stereochemistry. Together, the information derived from J-connectivities, coupling constants, chemical shifts, and the integrated one-dimensional spectrum define which saccharides are present. How these units are connected (the primary structure) can be identified by 2-D nuclear Overhauser spectroscopy (2-D NOESY). Here, dipole couplings through space are evaluated. The dipole coupling between an anomeric proton (H1) of a saccharide with nonanomeric protons of a second saccharide defines the sequence and the linkage site. Figure 4.5 illustrates the NMR spectrum of an idealized oligosaccharide.

At the top is the reference one-dimensional NMR spectrum. Note that the resonances of the anomeric protons are downfield and well-separated from the other protons. The middle portion (2D-SECSY) shows the trail of connectivities of protons for different saccharides, which determines the type of sugar. The bottom portion (2D-NOSEY) connects anomeric protons with other protons across the glycosidic linkage and within a given unit. If the saccharide contains a ketose lacking an anomeric proton, for example, a NeuAc residue, the determination of linkage then

Table 4.5. Oligosaccharide residue spin systems and their characteristics

Number of protons	J-connectivity pattern	Symbol	Residue constitution	Example residues
5	o-o-o-o-o	AHMRV	Aldohexuronic acid	Glucuronic acid, iduronic acid
6	o-o-o-o⊲	AHMR(XZ)	Aldopentose	Arabinose, ribose, xylose, lyxose
7	⊲>-o⊲	A(HJ)MR(XZ)	2-Deoxyaldopentose	2-Deoxyribose
7	o-o-o-o-o⊲	AHMRV(XZ)	Aldohexose	Glc, Gal, Man, GlcNAc, GalNAc, ManNNc
8	⊲>-o-o⊲	A(HJ)MRV(XZ)	2-Deoxyaldohexose	2-Deoxyglucose
8	o-o-o-o-o-o●	AHMRVX₁	6-Deoxyaldohexose	Fuc, Rha, Qui
9	⊲>-o-o-o●	A(HJ)MRVX₁	2,6-Deoxyaldohexose	Digitoxose
9	⊲>-o-o-o-o⊲	(AB)HMRSV(XZ)	3-Deoxy-2-nonulosonic acid	NeuAc, NeuGy
2 + 5	⟨ o-o-o⊲	AB-AHM(XZ)	2-Hexulose	Fructose, tagatose, psicose, sorbose

Reprinted with permission from *Methods in Enzymology* 138:38 (1987). Copyright 1987 Academic Press, Inc.

requires further study. NMR spectroscopy is a powerful tool that will be increasingly used to determine both primary and secondary structures of complex carbohydrates.

Fatty acids can be bound directly to proteins as thio esters, oxy esters, or amides on the amino terminus. Acid hydrolysis of the protein releases the free fatty acid. Extraction in ether or chloroform, followed by esterification by diazomethane yields a derivative easily characterized by gas chromatography or by GC/MS spectrometry. Isolation of a peptide of the amino terminus of the catalytic subunit of bovine cardiac muscle cyclic AMP-dependent protein kinase, and analysis by FAB mass spectrometry have proved the presence of covalently bound myristic acid. Lack of sequencing by Edman chemistry had suggested that the fatty acid was at the amino terminus but chemical ionization-mass spectrometry has proved that the linkage was to the n-terminal glycine residue.[109] Palmitic acid is linked to the thiol function of a cysteinyl residue in proteins such as p21^{ras}, vesicular stomatitis virus G glycoprotein, alphavirus E2 glycoprotein, mammalian transferrin receptor, rhodopsin, apolipoprotein A, and ankyrin.[152] An alcoholic function of a threonyl residue in rhodopsin may also be modified by a nonenzymatic process.[153] Lamin B, a structural protein in the nuclear envelope in Swiss 3T3 cells, contains a covalently bound lipid derived from mevalonic acid. An isoprenoid structure is thought to be linked to a cysteinyl residue.[57]

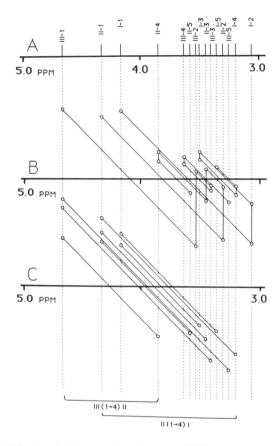

Figure 4.5. Idealized 2-D correlated proton NMR analysis of a trisaccharide: (A) 1-D spectrum for comparison; (B) 2-D SECSY spectrum; and (C) 2-D NOESY spectrum (only anomeric dipole couplings are shown). Multiplicities in resonance lines of the 1-D spectrum and in the contours of the 2-D spectra have been eliminated for clarity. The trisaccharide is assumed to have an NMR-transparent aglycone, such as the trideuteromethyl group. Thus, resonance 1-1 "sees" only the two intraresidue dipole couplings in the 2-D NOESY spectrum. Reprinted with permission from *Methods in Enzymology* 138:38 (1987). Copyright 1987, Academic Press.

Incubation of Lamin B in 8M guanidine HCl at 100°C in the presence of activated Raney Ni causes the release of apolar material, which can be extracted with pentane. Before and after reduction, the gas-liquid chromatographic profiles are identical to that observed for Raney Ni treatment of S-farnesylcysteine. Moreover, that the mass spectra of Lamin B-released material are identical with those derived from S-farnesylcysteine gives convincing evidence of the structure of the bound isoprenoid.

Glycosyl-Phosphatidyl Anchors

The family of proteins containing the complex structure of carbohydrate, ethanolamine, and phosphatidylinositol structure is attracting much attention because of the interaction of these proteins with membranes, and because of the changes in these interactions that may take place in biological signaling. The modified group serves to anchor the proteins; and certain biological events can cause hydrolysis of specific linkages, thus releasing protein or a portion of the modified group. A typical structure of the glycosyl-phosphatidyl anchor is shown in Figure 4.6.

It consists of three major parts: a phosphatidylinositol lipid containing fatty acids esterified to the 1,2 positions of glycerol; a middle glycan portion attached at the reducing end with glucosamine to inositol and attached at the nonreducing end to ethanolamine through a phosphodiester linkage with mannose-6-phosphate; and ethanolamine bound to the carboxyl terminus of the protein.[120] Exactly how was the structure established? Consider the example of glycosyl-sn-1,2-dimyristyl phosphatidylinositol covalently linked to a variant surface glycoprotein of the parasitic protozoan, *Trypsanosoma brucei*. Glycoproteins from specific clones of the protozoan have been isolated. Compositional analysis of the protein from the clone, MIT at 1.6 group 1 shows that the glycosyl-phosphatidyl(GPI)-membrane-anchor contains ethanolamine (1), mannose-6-phosphate (1), mannose (2), galactose (2-5), glucosamine (1), inositol (1), phosphate (2), glycerol (1), and fatty acid, 14:0 (1) per mole of GPI.[154]

The Phosphatidylinositol Portion

Treatment of the variant surface glycoprotein with phosphatidylinositol-specific phospholipase C causes the release of both the protein from the membrane and diacylglycerol, thereby suggesting phosphatidyl inositol is present and important in anchoring the protein to the membrane. Acid hydrolysis and determination of fatty acid as methyl esters by gas-liquid chromatography have proved the presence of myristic acid.[155] A compositional analysis of glycopeptides derived from the membrane bound protein (mfVSG) and from the soluble-phospholipase-treated protein(sVSG) have shown that glycerol and myristic acid are present only in mfVSG. This result proves that myristic acid is associated with the anchor and is likely esterified to glycerol. That acetolysis of mfVSG in anhydrous acetic acid and acetic anhydride yields 1,2-dimyristyl-

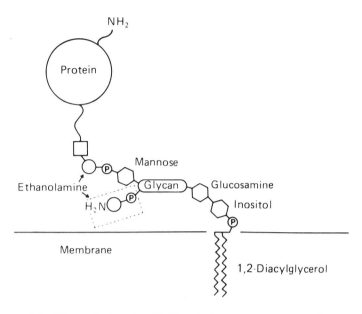

Figure 4.6. Glycosyl-phosphatidylinositol structure proposed to anchor proteins to the membrane. The arrangement of the components in this pictorial representation of the anchoring structure is based on work with VSG, although composition and specific degradation studies of the anchors in other proteins are consistent with this model. The structure shown in the box is proposed to be present in Thy-1, human erythrocyte AChE and possibly other proteins (excepting VSG). The structure has three regions. (1) A phosphatidyl-inositol molecule whose 1,2-diacylglycerol moiety is embedded in the bilayer and is responsible for anchoring; removal of this diacylglycerol by PI-PLC results in release of the protein from the membrane. (2) A glycan of varied structure and composition. The linkage between the glycan and the membrane phosphatidylinositol molecule is via a glycosidic linkage with a glucosamine that has a free amino group. (3) An ethanolamine is amide-linked via its amino group to the α-carboxyl of the C-terminal amino acid. The nonreducing end of the glycan contains a mannose 6-phosphate, which is phosphodiester-linked to the hydroxyl of this ethanolamine residue. The purpose of this figure is to indicate the likely arrangement of components in the anchoring structure. The size and orientation of those components and of the protein and lipid domains are therefore not necessarily portrayed accurately. For example, the C-terminal cysteine in Thy-1 is disulphide-linked to Cys-9 in the mature protein and AChE, APase and 5'-nucleotide probably exist as dimers in the membrane. Reprinted with permission from *Biochemical Journal* 244:1 (1987). Copyright 1987, The Biochemical Society and Portland Press.

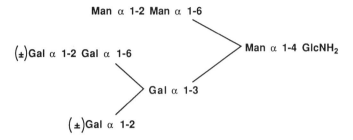

Figure 4.7. Structure and microheterogeneity of a glycan portion of a glycan anchor.

3-acetyl glycerol confirms the linkage and proves the structure.[156] Phospholipase A_2 removes one-half of the bound myristic acid, which indicates that the stereochemical arrangement of the glyceryl moiety is sn-1,2-dimyristyl-3-phosphate. Periodate oxidation, sodium borodeuteride reduction, and acid hydrolysis of a demyristylated glycan-phosphatidylinositol fraction yields 1,4–dideuterothreitol. A 1,6 substituted inositol yields this derivative. Thus, the 1 position is bound to the phosphate; and the 6 position, to the glycan.[157] Myo-inositol in the D-form is the likely isomer present.

The Glycan Portion

Treatment of glycopeptide from sVSG with nitrous acid causes deamination of glucosamine and subsequent breakdown of its glycosidic linkage. The reaction with nitrous acid is very useful because of its high degree of specificity. The formation of Myo-inositol-1-phosphate indicates that glucosamine is in a glycosidic linkage with it. The product 1,4–dideuterio threitol resulting from periodate cleavage and sodium borodeuteride reduction along with the results from 2-D NMR (NOE-measurements), shows that the linkage is α1,6. Compositional analysis, methylation results, FAB-mass spectrometry, and NMR spectroscopy studies have been instrumental in establishing the entire structure.[158] For example, chemical methylation makes mannose in a branching position evident. A mannose derivative containing methyl groups at the 2 and 4 positions shows that the other OH groups must be tied up and place the mannose at a branch with linkages to other saccharides at the 1, 3, and 6 positions. A compositional analysis of the glycan portion indicates that fractions contain varying amounts of galactose (2–5 residues/mole), with the microheterogeneity appearing at the antennae-branch, shown in Figure 4.7 by the plus-minuses at the ends of the proven glycan structure.

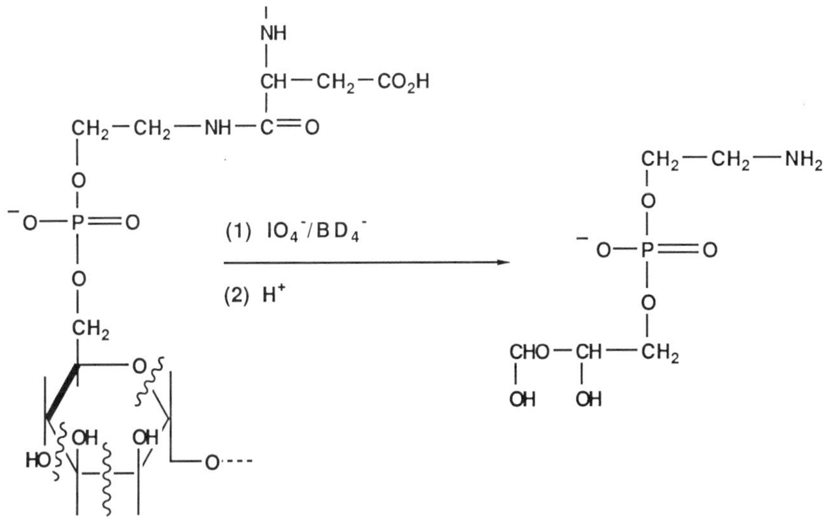

Figure 4.8. Chemical cleavage of the ethanolamine portion.

The Ethanolamine Portion

How is the ethanolamine linked to the saccharide and to the protein? Partial acid hydrolysis of a glycopeptide derived from mfVSG produces ethanolamine phosphate and mannose-6-phosphate, suggesting that the two units are linked by a phosphodiester bridge. a-mannosidase releases no mannose from the glycopeptide, but after cleavage of phosphate with HF, a-mannosidase does release mannose, suggesting the linkage of phosphoethanolamine with a terminal mannose residue. A proof of linkage is obtained by periodate treatment, sodium borodeuteride reduction, and acid hydrolysis followed by mass spectral analysis of a TMS-derivative. Figure 4.8 illustrates the cleavages occurring to produce the phosphoethanolamine-glycerol derivative.

The Linkage to the Protein

Pronase digestion produces a fragment, that upon acid hydrolysis and amino acid analysis yields aspartic acid, ethanolamine, and glucosamine. No reaction can be obtained for the amino group of ethanolamine with the amine reacting reagent, dansyl chloride. But, after acid hydrolysis, a reaction can be obtained, which suggests that the amino portion of ethanolamine is in a linkage with the α or β carboxyl group of aspartic acid. Here, Edman chemistry is extremely useful in deducing the exact linkage. One cycle of the Edman reaction releases aspartic acid and a free amino group of ethanolamine, which reacts with dansyl chloride.[159] Cleavage of a β-linked aspartyl peptide linkage would not be expected to occur.[160] A bound α-carboxyl group indicates that aspartic acid is the C-terminal amino acid in the modified protein.

THE PRIMARY STRUCTURE OF THE MODIFIED SITE

Edman Chemistry

To understand how modification reactions have taken place, it is necessary to define the chemical sites of the reaction. What are the best procedures to use? What are the unique problems involved in sequencing a particular modification site, for example, a phosphorylated tyrosyl, threonyl, histidyl, or seryl residue; a γ-carboxylated glutamyl residue; a carboxymethylated aspartyl residue; or a glycosylated amino acid side chain in a peptide?

In protein chemistry, the most important reaction for determining primary structure is the Edman reaction. An automated procedure can provide extensive information about the amino acid sequence of a protein or a peptide, but it is unlikely that the complete structure of the modified molecule could be deduced by using this one method. The procedure is effective in determining some sites of modification, but there are problems with specific modification sites and the procedures used in automatic sequencing. The most notable problem relates to the presence of specific modifying groups on the α-amino function of the peptide chain. Groups such as the formyl, acetyl, myristyl, and pyroglutamate completely block the chemical reaction with phenylisothiocyanate. Therefore, no information can be obtained about the modified amino acid, its neighbors, or other sites of modification in the protein by direct sequencing of proteins with these N-blocked derivatives. But not all substituted α-amino derivatives stop the reaction. For example, an α-N-methylated amino acid residue is perfectly reactive with phenylisothiocyanate because of the basicity (nucleophilicity) of the amino-nitrogen atom. The Edman reaction consists of three phases, as illustrated in Figure 4.9.

These phases, (1) coupling, (2) cleavage, and (3) isomerization constitute one cycle; identification of the amino acid derivative depends upon how well these reactions have occurred, whether any of these conditions cause breakdown of the modified amino acid, and whether the solvents used in sequencing can dissolve effectively the modified

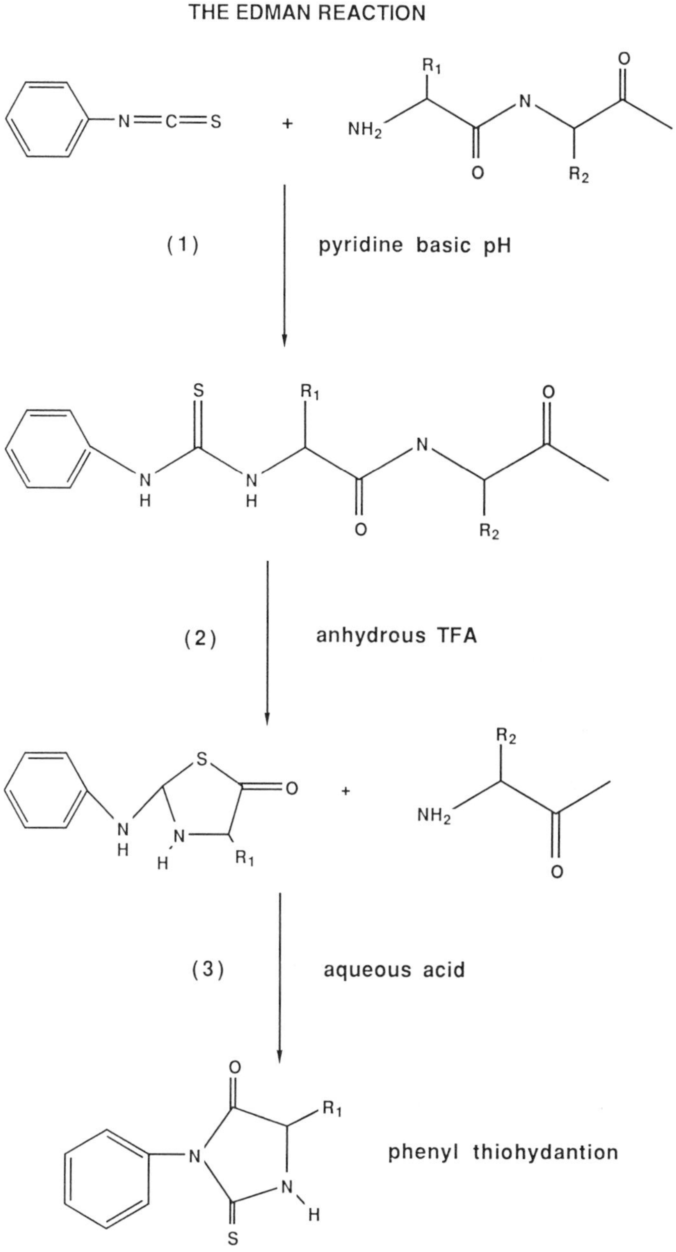

THE EDMAN REACTION

Figure 4.9. Chemistry of the Edman degradation.

phenylthiohydantoin. To illustrate what problems can occur in sequencing modified sites, a few cases will be considered.

Phosphorylation Sites

Serine and threonine. Phosphorylated esters of threonine and serine in peptides are reasonably stable in solutions used in sequencing, but are extremely labile when the phosphorylated amino acid goes through an Edman cycle. It is not known at which stages of the Edman cycle the breakdown occurs or what chemical reactions have taken place. Little or no PTH derivative of the phosphorylated amino acids can be identified but inorganic phosphate is formed. Inorganic phosphate, however, is essentially insoluble in the solvents used to extract the PTH-amino acid derivatives. Hence, automatic sequencing of phosphorylated seryl or threonyl peptides does not effectively define the position(s) at which phosphorylation has occurred. Alternatives taken are (1) to convert the peptide before sequencing to a form that will yield a PTH-derivative indicative of a phosphorylated amino acid, or (2) to analyze for inorganic phosphate and phosphorylated peptide at different cycles of the Edman reaction. In the first case, a phosphorylated peptide is treated with base to induce a β-elimination reaction producing inorganic phosphate and a modified amino acid. The chemistry for the reaction of a phosphorylated seryl peptide in base is shown in Figure 4.10.

The next step is to add a chemical across the double bond that will yield a derivative that can be successfully sequenced. Many nucleophilic reagents have been used, for example, sulfite, methylamine, mercaptoethanol, pyridoxamine, and ethanethiol. Ethanethiol yields S-ethylcysteine, and sequencing of the modified peptide yields its PTH-derivative at the original site of phosphorylation.[161] The method is extremely useful, but drawbacks are that (1) the procedure does not work for phosphorylated threonyl sequences, because β-elimination for a phosphorylated threonyl peptide requires more stringent conditions than for a phosphorylated seryl peptide and because these conditions can cause degradation of the peptide; and (2) if an O-glycosylated seryl residue is also present, it will undergo the same chemistry, β-elimination, and derivatization as a phosphorylated seryl peptide will. Thus, other experimentation would be needed to distinguish these two modifications. The second alternative for sequencing is to interrupt the automatic sequencing of the original phosphopeptide, remove a portion of the sample, for example, on pre-cut discs, extract it with 50 percent formic acid, and perform HPLC on the eluant. By using peptides labeled with the radioisotope, [^{32}P], and by measuring the radioactivity in the HPLC fractions, the proportion of inorganic phosphate and phosphopeptide could be determined at different cycles.[162] Before reaching a phosphoryl-

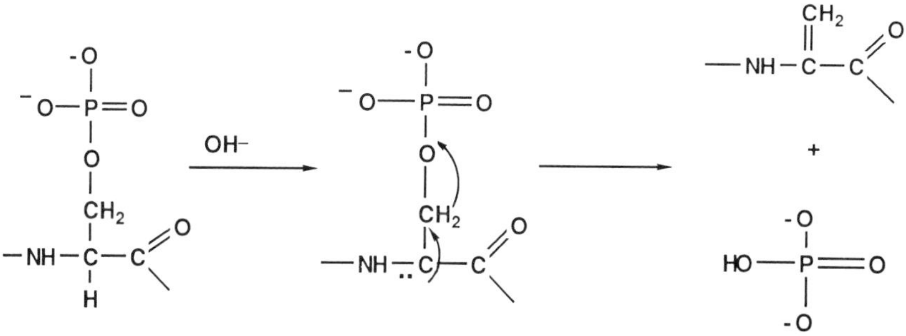

Figure 4.10. β-elimination of phosphate from a phosphoseryl containing peptide.

ated site, the percentage of phosphopeptide and inorganic phosphate would be high and low, respectively. But after the cycle with the phosphorylated residue, the proportion of inorganic phosphate rises strikingly. The procedure can be used for peptides with multiple-modi-fied seryl sites and applies to phosphothreonyl sequences also. But certain points need to be considered. First, to develop a protocol for selection of the samples, the sequence of the unmodified peptide must be known. Second, a phosphoseryl residue causes a significant drop in the yield of subsequent amino-acid residues released in the Edman reaction. When multiple-modified sites exist, the lowering of yield could make the sequencing of limited quantities of peptides difficult. If the chemistry were known for the sensitivity of phosphorylated sites in the Edman cycle, improvements still could be made in phosphopeptide sequencing.

Tyrosine. A phosphorylated tyrosyl residue is stable to the conditions used in Edman sequencing, but automatic procedures using a gas-phase sequencer may not reveal a PTH-derivative at the site of phosphorylation. The major problem, in this case, is not due to degradative reactions, but is related to the solubility of the phosphotyrosyl derivative in solvents used for extraction and identification of PTH-amino acid derivatives. When the peptide contains several tyrosyl residues,[163] insufficient solublization of the ATZ-phosphotyrosine may result in carryover may occur at subsequent cycles and obfuscate assignment of the site(s) of phosphorylation. Successful results, however, have been obtained with a [32P]-phosphotyrosylpeptide derived from modified-lysozyme in the spinning cup sequencer,[164] and a study of synthetic phosphopeptides in a gas-phase sequenator shows that PTH-phosphotyrosine can be detected with a yield of 30–40 percent.[165] Another concern is that the sequence, itself, can influence the yield of the reaction. A phosphopeptide derived from glycogen synthase with the sequence arginyl-phosphotyrosine is not effectively cleaved by the Edman reaction.[166]

Histidine. A phosphorylated histidyl residue can be formed in proteins by the action of a specific protein kinase.[167] The linkage, however, is extremely labile to acid and precludes the use of Edman chemistry to define sites of modification.

Carboxylation

Glutamic acid. Addition of a carboxyl group on the γ-carbon atom produces a substituted malonic acid. Malonic acid and its derivatives are well known to decarboxylate under acid conditions; the product of decarboxylation of the substituted glutamic acid is glutamic acid. Sequencing proteins with γ-carboxyglutamic acid does not normally yield a new PTH-derivative at the sites of carboxylation. Small amounts (low yields) were obtained for PTH-glutamate presumably arising from an acid-induced decarboxylation reaction of a γ-carboxyglutamyl residue. The PTH-derivative of γ-carboxyglutamate is present too, but it is not readily extracted from the sequencer unless a polar solvent like 2-butanol is used.[168] Hence, a γ-carboxyglutamate residue in a protein could be missed if the standard automated Edman procedure were used. A low yield for PTH-glutamate in a sequence analysis might prompt further studies, for example, one concerning a change in extraction conditions for the PTH-derivatives. Or, methylation of the carboxyl groups may be done to yield dimethyl γ-carboxyglutamate. After derivitization in the Edman cycle, this ester is soluble in butyl chloride and can be readily identified by HPLC. This procedure has been used to identify γ-carboxyglutamate in vitamin K dependent proteins on polyvinylidene difluoride membranes.[169]

Aspartic acid. The β-carboxy derivative, also a substituted malonic acid, is even more unstable in acid solutions than the glutamyl derivative[170] is and its sequencing by automated Edman chemistry has not been reported.

Glycosylation

The attachment of oligosaccharides to protein through its various linkages restricts automated sequencing. The sheer bulk of the oligosaccharide could influence the effectiveness of the cyclization steps and its polar character would influence the solubilization of any PTH-derivative in the organic solvents used. One approach taken then is thus to remove by chemical or enzymic methods the complex saccharide from the protein, leaving a stub attached. Chemical deglycosylation with trifluoromethane sulfonic acid (0° for 2.5 hrs.) removes peripherally bound sugars—but only some of the N-acetyl galactosamine O-linked to seryl and threonyl residues, and little or none of the N-acetylglucosamine N-linked to asparaginyl residues.[39] Structural analysis of a carcinoembryonic antigen, a highly N-linked glycoprotein, has been reported by gas-phase microsequencing on glycopeptides derived from the protein after chemical deglycosylation.[171] An O-linked site in human interleukin-2 also has been described by automated sequencing.[172]

A sensitive procedure that detects glycoproteins in the sub-nanomole range can be used effectively to screen peptides obtained from an enzymatic digest to determine which peptides contain bound carbohydrate.[173] This successful procedure can facilitate the analysis and spare precious materials. The glycopeptide or protein bound to a polyvinylpyrrilidone membrane is reacted with various biotinylated lectins. This is followed by the addition of avidin and horseradish peroxidase. If glycosylated material is present, the cross-linked peroxidase will give a brown spot in the presence of the substrates, diaminobenzidine and H_2O_2. The detection limit is about 50-100 picomoles. By using this method with carboxypeptidase P, glycosylated peptide was found and the sequence analysis defined a single site of attachment with the sequence, Y-X-N-T, where X represents a null in the sequence and is the suggested site of attachment. Mass spectrometry showed the isolated glycopeptide had a mass characteristic of a bound carbohydrate chain containing two N-acetylglucoasamine and five mannose residues.

Enzymatic reactions have been described that introduce ADP-ribose groups into proteins from NAD. Both polymeric and mono meric adducts occur, and different amino acid residues can serve as acceptors. Because of the polar nature of the modifying group, it is not surprising to find that problems occur with automated sequencing. The peptide Leu-Arg-Arg-Ala-Ser-Leu-Gly is a substrate for guanidine-specific mono ADP-ribosyltransferase from muscle.[174] Automated sequencing of a mono ADP-ribosylated derivative obtained from a reaction catalyzed by cholera toxin gave low yields for PTH-arginine at the second and third cycles. Such low yields are consistent with some modification at both sites but do not prove modification, particularly because sequencing is not quantitative and because arginyl residues tend to give low yields of PTH-derivatives. No modified PTH-derivative of arginine was identified. To overcome problems of assignment associated with poor solubility of the putative ADP-ribosyl PTH-derivative, the filter has been pre-cut into several portions before sequencing and a protocol described for identifying phosphorylated sites has been followed.[162] After a cycle was finished in which a potential ADP-ribosylated residue could have been released, the filter was extracted with 50 percent formic acid. Reverse-phase

HPLC can be used effectively to identify the ADP-ribosylated derivative.[175]

Volumes 91, 106, 107, and 182 of *Methods in Enzymology* provide excellent descriptions of many of the general procedures used for the isolation and purification of peptides and proteins. And in some cases methods are available to facilitate the purification of peptides with specific modifying groups. For example, phosphorylated peptides can be separated from a mixture of peptides by affinity chromatography with an iminodiacetate agarose gel containing bound ferric ions.[176] The ferric ion binds phosphopeptides tightly under slightly acidic conditions—pH 5.0. After other peptides are washed away with suitable buffers, the phosphopeptide(s) can be eluted by ammonium acetate at pH 8.0. By this procedure,[177] a pure phosphopeptide has been obtained from an HPLC fraction of peptides derived from a protein of photosystem II of spinach chloroplasts. The separation of phosphopeptides from other peptides on HPLC can be achieved by utilizing an ion-pair reagent to interact with the negatively charged phosphate group.[178] An alternative is to convert the phosphopeptide to a new derivative possessing properties different from those of the other peptides in the mixture. β-elimination of a phosphopeptide and addition of ethanethiol yield a peptide with increased hydrophobicity. The change in property could allow its separation from other peptides on a reverse-phase HPLC system.[179] As described earlier, in specific instances, reverse-phase HPLC can distinguish modified peptides from unmodified peptides. For example, HPLC can be used to isolate the peptide containing γ-carboxyglutamate, ala-Asn-Ser-PheLeu-γGlu-γGlu-Met-Lys, derived from human-factor light chain by tryptic attack because it shows much weaker binding than its unmodified form does.[129] Since γ-carboxyglutamate residues in proteins decarboxylate readily by heating in a dry state, dry-heat could be used to change the retention properties of the modified peptide for further purification and identification, without destroying other amino acids. Another important way to purify peptides with specifically modified side chains is to use antibodies directed to the residue, for example, a phosphotyrosyl group, or to the modifying function itself, for example, a poly (ADP-ribose) chain. When the procedures described are used, however, it should not be assumed that they will provide a pure peptide. No single test can show that a peptide has been purified to homogeneity. A good index of purity of the peptide is a single peak in HPLC obtained under different chromatographic conditions, as well as an integral value of amino acids provided by amino acid analysis. Further evidence is obtained by structural analysis by Edman sequencing or by mass spectrometry.

Peptide Mapping

Two-dimensional electrophoretic methods separate modified peptides. This technique has been used extensively in the characterization of phosphorylation sites. Not only can phosphopeptides be separated, but if the amino acid composition and size of the peptide is known, evaluation of the charge of the parent peptide and its degradation products produced by specific proteases can define the site of phosphorylation, as described in a study of autophosphorylation sites in the insulin receptor.[180] Or alternatively, Edman sequencing can be performed to characterize the sites of phosphorylation.

Use of Mass Spectrometry in Peptide Sequencing

Fragmentation of protonated molecular ions of peptides in liquid secondary-ion mass spectrometry can occur in defined ways and can reveal the sequence.[181] The protonated-solvated peptide, when released from the liquid matrix, undergoes desolvation in the gas phase of the mass spectrometer. Charge delocalization occurs by internal solvation. The positive charge initially present on the amino terminus can be brought into juxtaposition with the amide groups in the chain and can lead to a protonation of peptide linkages throughout the structure. The protonated linkages are sensitive and can be cleaved if bombarded with low-energy argon atoms. Cleavage can occur by other methods—for example, by exposure to an ArF excimer laser in tandem mass spectometry. Two types of cleavages are shown in Figure 4.11.

Path 1 yields fragment ions (type B) with a charge on the carbonyl group, the ion being derived from the amino terminal portion of the peptide. Path 2 yields fragment ions (type Y) with a positive charge on the α-amino group, the ion being derived from the carboxyl part of the molecule.

A laser photodissociated mass spectrum of a peptide obtained in a tandem mass spectrometer illustrated below in Figure 4.12 shows the actual ions produced.

The peptide derived from the amino-terminus of a photosystem II protein of spinach chloroplasts contains an acetylated and phosphorylated threonyl residue.[182] Cleavage of the protonated molecular ion at m/z of 1386.4 at the thr-leu linkage shown in Figure 4.12 gives an ion at m/z of 1163.3, which comes from the loss of 223.1 mass units, the acetylated and phosphorylated threonyl function. The sequence of the entire structure can be deduced by examining the other Y-ions produced in this case. That type B-ions derived from the dephosphorylated peptide (m/z at

Mode of Cleavage of [M+H+] Ions

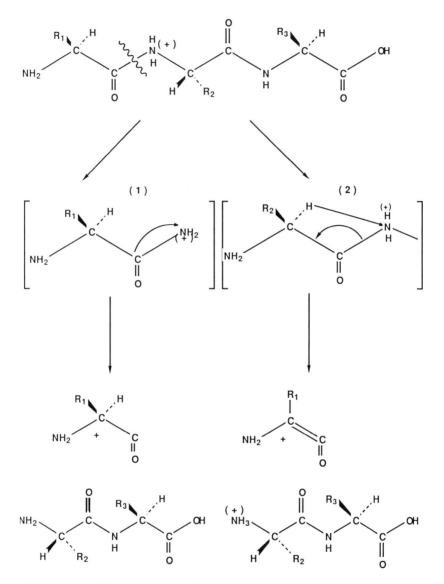

Figure 4.11. Chemistry of the cleavage of peptides in mass spectrometry. Reprinted with permission from *Proceedings of the National Academy of Sciences, USA* 83:6233 (1986).

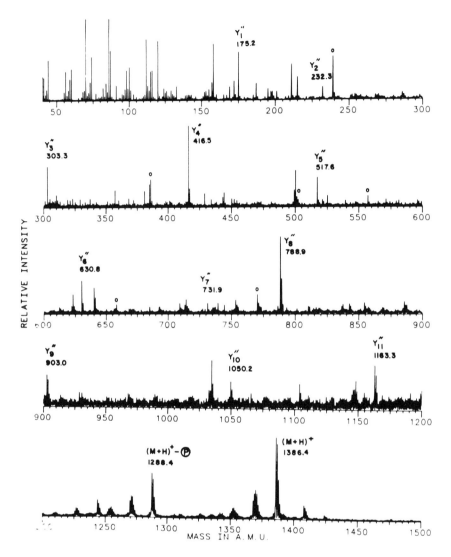

Figure 4.12. Laser photodissociated mass spectrum of a peptide of PS II protein. Reprinted with permission from *Journal Biological Chemistry* 263:1123 (1988). Copyright 1988, The American Society for Biochemistry & Molecular Biology.

1288.4) are also formed is indicated by the [O] signal in the mass spectrum and by the underlined ions in Figure 4.13, which are consistent with the primary structure of the peptide.

Mass spectral analysis of C-terminal fragments derived from neuron-specific class III b-tubulin provides information about post-translational modification that is not achievable by conventional amino acid sequencing alone.[183] Up to 21 charge variants of brain tubulin have been identified by isoelectric focusing, and it has been suggested these variants are due partly to developmentally regulated post-translational modification. Cleavage of the protein by CNBr, isolation of peptides, and mass spectral characterization shows that the peptides $(M+H)^+$ ions differ from each other by masses characteristic of forms with different amounts of glutamate. The collision activated spectra show the type Y and B ions expected for the sequence YEDDEEESEAQGPK, but the Y and B ions cannot account for the extra glutamates expected. Additional ions, however, are detected, and show that the extra residues, not in the linear sequence, are added to the side chain of the first glutamate. From one to six residues seem to be added on. Additionally, it was found that the seryl residue was phosphorylated and together phosphorylation and glutamylation can explain some of the charge variants of tubulin. Amino terminal sequencing by the Edman procedure showed tyrosine in the first cycle, but no PTH amino acid is found in the second cycle. This suggests a modification, but it does not define what modification has taken place. Analysis of the remaining peptide after the second cycle, the subtractive Edman degradation procedure, does show that several glutamates where removed by the Edman chemistry in support of multiple glutamylation of the first glutamate in the sequence.

Combined analysis of peptide derived from the C-terminus of α-tubulin by mass spectrometry and Edman sequencing shows how polyglutamylation occurs. Six residues may be linked to the γ-carboxyl group of a glutamate of the main chain, but how are these additional

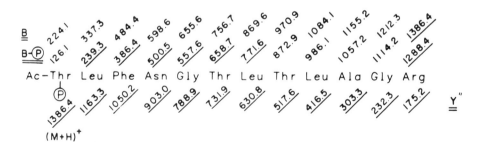

Figure 4.13. Amino acid sequence and fragment ion assignments for an acetylated and phosphorylated peptide. Reprinted with permission from *Journal Biological Chemistry* 263:1123 (1988). Copyright 1988, The American Society for Biochemistry & Molecular Biology.

residues linked? Is it through g or a linkages or some combination? The full answer is not known, but the first two residues are linked through the a-peptide bonds.[184]

Successful sequencing of modified peptides requires the use of many procedures. Clearly, mass spectrometry will find increasing use, but other methods, both enzymatic and chemical not addressed in this book are also of great importance in determining the primary structure of peptides. Information about these procedures can be found in the volumes of *Methods in Enzymology*.

REFERENCES

1. Schmidt, R.A., Schneider, C.J., and Glomset, J., *J. Biol. Chem.* 259, 10175–10180 (1984).
2. Kristiansen, T., *Methods in Enzymol.* 34, 331–341 (1974).
3. Gerard, C., *Methods in Enzymol.* 182, 529–539 (1990).
4. Adamietz, P., and Hilz, H., *Methods in Enzymol.* 106, 461–471 (1984).
5. Zhang, J., and Snyder, S.H., *Biochemistry* 32,2228–2233 (1993).
6. Klarlund, J.K., Latini, S., and Forchammer, J., *Biochem. Biophys. Acta* 971, 112–20 (1988)
7. Frackelton, A.R.,Jr., Posner, M., Kannan, B., Meremelstein, F., *Methods in Enzymol.* 201, 79–92 (1991).
8. Glenney, J.R., *Methods in Enzymol.* 201, 92–100 (1991).
9. Beletskii, I.P., Nelipovich, P.A., Umanskii, S.R., *Biokhimiya* 53, 1144–52 (1988).
10. Bofinger, D.P., Fucile, N.W., and Spaulding, S.W.,*Anal. Biochem.* 170,9–18 (1988).
11. Harris, W.H., and Graves, D.J., *Arch. Biochem. Biophysics* 276, 102–108 (1990).
12. Thomas, J.A., and Beidler, D., *Anal. Biochem.* 157, 32–38 (1986).
13. Rokutan, K., Thomas, J.A., and Sies, H., *Eur. J. Biochem.* 179, 233–239 (1989).
14. Ohsawa, K., Kimura, M., Kurosawa-Ohsawa, K., Takahashi, M., Koyama, M., Abiko, Y., Hirahara, K., Matsusihi, T., and Tanaka, S., *J. Chromatogr.* 597, 285–291 (1992).
15. Crabb,J.W., and Heilmeyer,L.M.J., *J. Biol. Chem.* 259, 6346–6350 (1984).
16. Schlesinger, D., Ed., *Macromolecular Sequencing and Synthesis: Selected Methods and Applications*, New York: Alan R. Liss, Inc., p. 35 (1988).
17. Martensen, T.M., *J. Biol. Chem.* 257, 9648–9652 (1982).
18. Martensen,T.M., and Levine,R.L.,*Methods in Enzymol.* 99, 402–405 (1983).
19. Meyer, H.E., Swiderek, K., Hoffmann-Posorske, E., Korte, H., and Heilmeyer, L.M.G.,Jr., *J. Chrom.* 397, 113–121 (1987).

20. Cooper, J.A., Sefton, B.M., and Hunter, T., *Methods in Enzymol.* 99, 387–402 (1985).
21. Duclos, B., Marcandiere, S., and Cozzone, A.J., *Methods in Enzymol.* 201, 10–21 (1991).
22. Boyle, W.J., Van Der Geer, P.V., and Hunter, T., *Methods in Enzymol.* 201, 110–149 (1991).
23. Towler, D.W., and Glaser, L., *Proc. Natl. Acad. Sci.* 83, 2812–2816 (1987).
24. Goddard, C., and Felsted, R.L., *Biochem. J.* 253, 839–843 (1988).
25. Chin, Q., and Wold, F., *Methods in Enzymol.* 106, 265–274 (1984).
26. Thotakura, N.R., and Bahl, O.P., *Methods in Enzymol.* 138, 350–359 (1987).
27. Tarentino, A.L., and Plummer, T.H., Jr., *Methods in Enzymol.* 138, 770–778 (1987).
28. Heinrikson, R.L., and Meredith, S.C., *Anal. Biochem.* 136, 65–74 (1984).
29. Paik, W.K., *Methods in Enzymol.* 106, 265–274 (1984).
30. Goff, C.G., *J. Biol. Chem.* 249, 6181–6190 (1974)
31. Payne, D.M., Jacobson, E.L., Moss, J., and Jacobson, M.K., *Biochemistry* 24, 7540–7549 (1985).
32. Smith, L.S., Kern, C.W., Halpern, R.M., and Smith, R.A., *Biochem. Biophys. Res. Commun.* 71, 459–465 (1976).
33. Clarke, S., McFadden, P.N., O'Connor, C.M., and Lou, L.L., *Methods in Enzymol.* 106, 330–344 (1984).
34. Paik, W.K. and Kim, S., in *The Enzymology of Post-translational Modification of Proteins,* (Freeman, R.B. and Hawkins, H.C., Ed.), New York:. Academic Press, p.187–228 (1985).
35. Degani, C., and Boyer, P.D., *J. Biol. Chem.* 248, 8222–8226 (1973).
36. Drakenburg, T., Fernlund, P., Roespstorff, P., and Stenflo, J., *Proc. Natl. Acad. Sci.* 80, 1802–1806 (1983).
37. Suzuki, H., Quesada, P., Farina, B., and Leone, E., *J. Biol. Chem.* 261, 6048–6055 (1986).
38. Koch, T.H., Christy, M.R., Barkley, R.M., Sluski, R., Bohemier, D., Van Buskirk, J.J., and Kusch, W.M., *Methods in Enzymol.* 107, 563–575 (1984).
39. Edge, A.S.B., Faltynek, C.R., Hof., L., Reichert, L.E. Jr., and Weber, P., *Anal Biochem.* 118, 131–137 (1981).
40. Sojar, H.T., and Bahl., O.P., *Archiv. Biochem. Biophys.* 259, 52–57 (1987).
41. Klotz, A.V., Leary, J.A., and Glazer, A., *J. Biol. Chem.* 261, 15891–15894 (1986).
42. Stenflo, J., Lundwall, A., Dahlback, B., *Proc. Natl. Acad. Sci.* 84, 368–372 (1987).
43. Thannhauser, T.W., Konishi, Y., and Scheraga, H.A., *Anal. Biochem.* 138, 181–188 (1984).
44. Damodaran, S., *Anal. Biochem.* 145, 200–204 (1985).
45. Pigiet, V., and Conley, R.R., *J. Biol. Chem.* 253, 1910–1920 (1978).
46. Guan, K.L., and Dixon. J.E., *J. Biol. Chem.* 266, 17026–17030 (1991).
47. Weis, J.B., and Lote, C.J., *Nature New Biol.* 234, 25–26 (1971).
48. Jacobson, M.K., Loflin, P.T., Abdoul-Ela, N., Mingmuang, M., Moss, J., and Jacobson, E.L., *J. Biol. Chem.* 265, 10825–10828 (1990).

49. Magee, A.I., Koyama, A.H., Malfer, C., Wen, D., and Schlesinger, M.J., *Biochem. and Biophys. Acta* 798, 156–166 (1984).
50. Kaufman, J.F., Krangel, M.S., and Stominger, J.L., *J. Biol. Chem.* 259, 7230–7238 (1984).
51. Lerch, K., *Methods in Enzymol.* 106, 355–359 (1984).
52. Hantke, K., and Braun, V., *Eur. J. Biochem.* 34, 284–296 (1973).
53. Nielsen, J.B.K., and Lampen, J.O., *J. Biol. Chem.* 257, 4490–4495 (1982).
54. Belz, R., Crabb, J.W., Meyer, H.E., Wittig, R., and Duntze, W., *J. Biol. Chem.* 262, 546–548 (1987).
55. Ishibashi, Y., Sakagami, Y., Isogai, A., and Suzuki, A., *Biochemistry.* 23, 1399–1404 (1984).
56. Rilling, H.C., Bruenger, E., Epstein, W.W., and Kandutsch, A.A., *Biochem. Biophys. Res. Commun.* 163, 143–148 (1989).
57. Farnsworth, C.C., Wolda, S.L., Gelb, M.H., and Glomset, J.A., *J. Biol. Chem.* 264, 20422–20429 (1989).
58. Renstrom, B., and DeLuca, H.F., *Biochim. Biophys. Acta.* 998, 69–74 (1989).
59. Takahashi, N., and Breitman, T.R., *Methods in Enzymology* 189, 233–238 (1990).
60. Stadtman, T., *Methods in Enzymol.* 107, 576–581 (1984).
61. Ogata, N., Ueda, K., Kagamiyama, H., and Hayaishi, O., *J. Biol. Chem.* 255, 7616–7620 (1980).
62. Price, P.A., *Methods in Enzymol.* 91, 13–17 (1983).
63. Nelsestuen, G.L., *Methods in Enzymol.* 107, 503–507 (1984)
64. Clarke, S., Sparrow, K., Panasenko, S., and Koshland, D.E., Jr., *J. Supramol. Struct.* 13, 315–328 (1980).
65. Piacentini, M., Martinet, N., Beninati, S., and Folk, J.E., *J. Biol. Chem.* 263, 3790–3794 (1988).
66. Lhoest, J., and Colson, C., *Mol. Gen. Genetics.* 154, 175–180 (1977).
67. Smith, R.A., Halpern, R.M., Bruegger, B.R., Dunlap, A.K., and Fricke, O., *Methods in Cell Biol.* 19, 153–159 (1978).
68. Wei, Y.F., and Matthews, H.K., *Methods in Enzymology* 200, 388–414 (1991).
69. Bodley, J.W., Dunlop, P.C., and Van Ness, B.G., *Methods in Enzymol.* 106, 378–387 (1984).
70. Van Ness, B.G., Howard, J.B., and Bodley, J.W., *J. Biol. Chem.* 255, 10717–10720 (1980).
71. Raghavan, M., Smith, C.K., and Schutt, C.E., *Anal. Biochem.* 178, 194–197 (1989).
72. Huszar, G., *Methods in Enzymol.* 106, 287–295 (1984).
73. Wolff, J., and Covelli, I., *European J. Biochem.* 9, 371–377 (1969).
74. Allfrey, V.G., Di Paola, E.A., and Sterner, R., *Methods in Enzymol.* 107, 224–240 (1984).
75. Cooper, H.L., Park, M.H., and Folk, J.E., *Methods in Enzymol.* 106, 344–351 (1984).
76. Rylatt, D.B., Keech, D.B., and Wallace, J.C., *Arch. Biochem. Biophys.* 183, 113–122 (1977).

77. Chin, C.Q.C., *Methods in Enzymol.* 106, 17–21 (1984).
78. Nakajima, T., and Volcani, B.E., *Biochem. Biophys. Res. Commun.* 39, 28–33 (1970).
79. Butler, W.T., *Methods in Enzymol.* 82, 339–346 (1982).
80. Conrad, S.M., in *The Enzymology of Post-Translational Modification of Protein,* (Freeman, R.B. and Hawkins, H.G., Ed.), New York: Academic Press, 339–363 (1985).
81. Busch, H., *Methods in Enzymol.* 106, 238–262 (1984).
82. Bornstein, P., and Piez, K.A., *Biochemistry* 5, 3460–3473 (1966).
83. Diedrich, D.L., and Schnaitman, C.A., *Proc. Natl. Acad. Sci.* 75, 3708–3712 (1978).
84. Brot, N., Fless, H., Coleman, T., and Weisbach, T., *Methods in Enzymol.* 107, 352–360 (1984).
85. Sliwkowski, M.X., *Methods in Enzymol.* 107, 620–623 (1984).
86. Lin, T.S., and Kolattukudy, P.E., *Archiv. Biochem. Biophys.* 196, 255–264 (1979).
87. Waite, J.H., and Benedict, C.V., *Methods in Enzymol.* 107, 397–413 (1984).
88. Yoshino, K., Takao, T., Suhara, M., Kitai, T., Hori, H., Nomura, K., Yamaguchi, M., Shimonishi, Y., and Suzuki, N., *Biochemistry* 30, 6203–6209 (1991).
89. Chrispeels, M.J., *Methods in Enzymol.* 107, 361–369 (1984).
90. Kornfeld, R., and Kornfeld, S., *The Biochemistry of Glycoproteins and Proteoglycans,* (Lennarz, W.J., Ed.), New York: Plenum Press, 1–34 (1980).
91. Lamport, D.T.A., *Biochemistry* 8, 1155–1163 (1969).
92. Martensen, T.M., *Methods in Enzymol.* 107, 3–23 (1984).
93. Ord, M.G., and Stocken, L.A., *Biochem. J.* 161, 583–592 (1977).
94. Adams, J.B., *Biochem. J.* 97, 345–352 (1965).
95. Gustafson, G.L., and Gander, J.E., *Methods in Enzymol.* 107, 172–183 (1984).
96. Drygin, Y.F., Zuklys, K.L., Terskich, A.V., and Bogdanov, A.A., *Eur. J. Biochem.* 175, 57–63 (1988).
97. Buko, A.M., Kentzer, E.J., Petros, A., Menon, G., Zuiderweg, E.R.P., and Sarin, V.K., *Proc. Natl. Acad. Sci.* 88, 3992–3996 (1991).
98. Huttner, W.B., *Methods in Enzymol.* 107, 200–223 (1984).
99. Rhee, S.G., Chock, P.B., and Stadtman, E.R., in *The Enzymology of Post-translational Modification of Proteins,* (Freeman, R.B., and Hawkins, H.C., Ed.), New York: Academic Press, 273–297 (1985).
100. Tse, Y-C., Kirkegaard, K., and Wang, J.C., *J. Biol. Chem.* 255, 5560–5565 (1980).
101. Hunt, S., *Methods in Enzymol.* 106, 413–438 (1984).
102. Malencik, D.A., Zhao, Z., and Anderson, S.R., *Anal. Biochem.* 184. 353–359 (1990).
103. Fry, S.C., *Methods in Enzymol.* 107, 388–397 (1984).
104. Lederer, F., Alix, J-H., and Hayes, D., *Biochem. Biophys. Res. Commun.* 77, 470–480 (1977).
105. Chen, R., Brosius, J., and Wittman-Liebold, B., *J. Mol. Biol.* 111, 173–181 (1977).

106. Pettigrew, G.W., and Smith, G. M., *Nature* 265, 661–662 (1977).
107. Tsunasawa, S., and Sakiyama, F., *Methods in Enzymol.* 106, 165–170 (1984).
108. Tsunasawa, S., and Sakiyama, F., *Methods in Enzymol.* 91, 84–92 (1983).
109. Carr, S.A., Biemann, K., Shoji, S., Parmalee, D.C., and Titani, K., Proc. Natl. Acad. Sci. 79, 6128–6131 (1982).
110. Aitken, A., and Cohen, P., *Methods in Enzymol.* 106, 205–210 (1984).
111. Henderson, L.E., Krutzsch, H.C., and Orslan, S., *Proc. Natl. Acad. Sci.* 80, 339–343 (1983).
112. Kolattukudy, P.E., *Methods in Enzymol.* 106, 210–217 (1984).
113. Doolittle, R.F., *Methods in Enzymol.* 25, 231–244 (1972).
114. Kreil, G., and Kreil-Kiss, G., *Biochem. Biophys. Res. Commun.* 27, 275–280 (1967)
115. Snell, E., *Trends in Biochem. Science* 2, 131–135 (1977).
116. Kapke, G., and Davis, L., *Biochemistry* 14, 4273–4276 (1975).
117. Kreil, G., *Methods in Enzymol.* 106, 218–223 (1984).
118. Tatemoto, K., and Mutl, V., *Proc. Natl. Acad. Sci.* 75, 4115–4119 (1978).
119. Kreil, G., in *The Enzymology of Post-Translational Modification of Proteins*, (Freeman, R.B., and Hawkins, H.C, Ed.) New York: Academic Press, p. 41–51 (1985).
120. Low,M.G., *Biochem. J.* 244, 1–13 (1987).
121. Clarke, S., Vogel, J., Deschenes, R.J., and Stock, J., *Proc. Natl. Acad. Sci.* 85, 4643–4647 (1988).
122. Coyle, D., and Jacobson, M., personal communication.
123. Cervantes-Laurean, D., Minter, D.E., Jacobson, E.L., and Jacobson, M.K., *Biochemistry* 32, 1528–1534 (1993).
124. Epstein, W.W., Lever, D., Leining, L.M., Bruenger, E., and Rilling, H.C., *Proc. Natl. Acad. Sci.* 88, 9668–9670 (1991).
125. Redman, K.L., and Rubenstein, P.A., *Methods in Enzymol.* 106, 179–192 (1984).
126. Gierasch, L.M., *Biochemistry* 28, 923–930 (1989).
127. Deutch, C.E., *Methods in Enzymol.* 106, 198–205 (1984).
128. Flavin, M., and Murofushi, H., *Methods in Enzymol.* 106, 223–237 (1984).
129. Walsh, K. A., and Sasagawa, T., *Methods in Enzymol.* 106, 22–29 (1984).
130. Concetti, A., and Gariboldi, P., *Biol.Metals* 3, 125–126 (1990).
131. Carr, S.A., and Biemann, K., *Methods in Enzymol.* 106, 29–58 (1984).
132. Barber, M., Bordoli, R.S., Sedgwick, R.D., and Tyler, A.N., *J. Chem. Soc. Chem. Commun.* 325– (1981).
133. Gibson, B.W., Falick, A.M., Burlingame, A.L., Nadasdi, L., Nguyen, A.C., and Kenyon, G.L., *J. Am. Chem. Soc.* 109, 5343–5348 (1987).
134. Hunt, D.F., Yates, J.R. III, Schabonowitz, J., Winston, S., and Hauer, C.R., *Proc. Natl. Acad. Sci.* 83, 6233–6237 (1986).
135. Morris, H.R., Panico, M., and Taylor, G.W., *Biochem. Biophys. Res. Commun.* 117, 299–305 (1983).
136. Petrilli, P., Pucci, P., Morris, H.R., and Addeo, F., *Biochem. Biophys. Res. Commun.* 140, 28–37 (1986).

137. Poulter, L., Ang, S-G., Gibson, B.W., Williams, D.H., Holmes, C.F.B., Caudwell, F.B., Pitcher, J., and Cohen, P., *Eur. J. Biochem.* 175, 497–510 (1988).
138. Pucci, P., and Sepe, C., *Biomedical and Environmental Mass Spectrom.* 17, 287–291 (1988).
139. Carr, S.A., Roberts, G.D., and Hemling, M.E., in *Mass Spectrometry of Biological Materials* 8, (McEwen, C.N., and Larsen, B.S.,Ed.), New York: Marcel Dekker, p. 87–133 (1990).
140. Biemann, K., *Ann. Rev. of Biochem.* 61, 977–1010 (1992).
141. Sharon, N., *Complex Carbohydrates: Their Chemistry, Biosynthesis and Functions*, Addison-Wesley Pub. Co., Reading, Mass., (1975).
142. Merkle, R., and Cummings, R., *Methods in Enzymol.* 138, 232–259 (1987).
143. Johnson, D.C., and LaCourse, W.R., *Analytical Chem.* 62, 589–597A(1990).
144. Hardy, M.R., and Townsend, R.R., *Proc. Nat. Acad. Sci.* 85, 3289–3293 (1988).
145. Townsend, R.R., Hardy, M.R., Hindsgaul, O., and Lee, Y.C., *Anal. Biochem.* 174, 459–470 (1988).
146. Naik, S., Oates, J.E., Dell, A., Taylor, G.W., Dey, P.M., and Pridham, J.B., *Biochem. Biophys. Res. Commun.* 132, 1–7 (1985).
147. Dell, A., Thomas-Oates, J.E., Rogers, M.E., and Tiller, P.R., *Biochimie* 70, 1435–1444 (1988).
148. Dell, A., *Adv. in Carbohydrate Chem. and Biochemistry* 45, 19–72 (1987).
149. Dell, A., Carmen, N.H., Tiller, P.R., and Thomas-Oates, J.E., *Biomedical and Environmental Mass Spectrom.* 16, 19–24 (1988).
150. Bierman, C.J., *Advances in Carbohydrate Chem. and Biochemistry* 46, 251–271 (1988).
151. Koerner, T.A.W., Prestegard, J.H., and Yu, R.K., *Methods in Enzymol.* 138, 38–59 (1987).
152. Sefton, B.M., *Mol. Biol. of Intracell. Protein Sorting and Organelle Assembly*, 215–219 (1988).
153. O'Brien, P.J., St. Jules, R.S., Reddy, T.S., Bazan, N.G., Zatz, M., *J. Biol. Chem.* 262, 5210–5215 (1987).
154. Fergurson, M.A.J., and Williams, A.F., *Annu. Rev. of Biochem.* 57, 285–320 (1988).
155. Fergurson, M.A.J., and Cross, G.A.M., *J. Biol. Chem.* 259, 3011–3015 (1984).
156. Fergurson, M.A.J., Haldar, K., and Cross, G.A.M., *J. Biol. Chem.* 260, 4963–4968 (1985).
157. Fergurson, M.A.J., Homans, S.W., Dwek, R.A., and Rademacher, T.W., *Science* 239, 753–759 (1988).
158. Schmitz, B., Klein, R.A., Duncan, I.A., Egge, H, Gunawan, J., Peter-Katalinie, J., Dabrowski, U., and Dabrowski, J., *Biochem. Biophys. Res. Commun.* 146, 1055–1063 (1987).
159. Holder, A.A, *Biochem. J.* 209, 261–262 (1983).

160. Takio, K., Blumenthal, D.K., Walsh, K.A., Titani, K., and Krebs, E.G., *Biochemistry* 25, 8049–8057 (1986).
161. Meyer, H.E., Hoffmann-Posorske, E., Korte, H., Heilmeyer, L.M.G.,Jr., *FEBS Lett.*, 204, 61–66 (1986).
162. Wang, Y., Fiol, C.J., DePaoli-Roach, A.A., Bell, A.W., Hermodson, M.A., and Roach, P.J., *Analytical Biochem.* 174, 537–547 (1988).
163. Tornqvist, H.E., Pierce, M.W., Frackelton, A.R., Nemenoff, R.A., and Avruch, J., *J. Biol. Chem.* 262, 10212–10219 (1987).
164. Tonks, N.K., Charbonneau, H., Diltz, C.D., Kumar, S., Cicirelli, M.F., Krebs, E.G., Walsh, K.A., and Fischer, E.H., *Advances in Protein Phosphatases* 5, 149–180 (1989).
165. Turck, C.W., Hermann, J., Escobedo, J.A., and Williams, L.T., *Peptide Research* 4, 36–39 (1991).
166. Mahrenholz, A.M., Votaw, P., Roach, P.J., Depaoli-Roach, A.A, Zioncheck, T.F., Harrison, M.L., and Geahlen, R.L., *Biochem. Biophys. Res. Commun.* 155, 52–58 (1988).
167. Wei, Y.F., and Matthews, H.K., *Methods in Enzymol.* 200, 388–414 (1991).
168. Fernlund, P., and Stenflo, J., *J. Biol. Chem.* 257, 12170–12171 (1982).
169. Cairns, J.R., Williamson, M.K., and Price, P.A., *Anal. Biochem.* 199, 93–97 (1991).
170. Hauschka, P.V., Henson, E.B., and Gallop, P.M., *Analytical Biochem.* 108, 57–63 (1980).
171. Paxton, R.J., Mooser, G., Pande, H., Lee, T.D., and Shiveley, J.E., *Proc. Natl. Acad. Sci.* 84, 920–924 (1987).
172. Robb, R.J., Kutny, R.M., Panico, M., Morris, H.R., and Chowdhry, V., *Proc. Natl. Acad. Sci.* 81, 6486–6490 (1984).
173. Hsi, K-L., Chen, L., Hawke, D.H., Zieske, L.R., and Yuan, P-M., *Anal. Biochem.* 198, 238–245 (1991).
174. Kharadia, S., and Graves, D.J., *J. Biol. Chem.* 262, 17379–17383 (1987).
175. Kharadia, S., Ph.D Dissertation, Iowa State University, 1990.
176. Anderson, L., and Porath, J., *Anal. Biochem.* 154, 250–254 (1986).
177. Michel, H.P., and Bennett, J., *FEBS Lett.* 212, 103–108 (1987).
178. Haystead, T.A.J., Campbell, D.G., and Hardie, D.G., *Eur. J. Biochem.* 175, 347–354 (1988).
179. Holmes, C.F.B., *FEBS Lett.* 215, 21–24 (1987).
180. Tavare', J.M., and Denton, R.M., *Biochem. J.* 252, 607–615 (1988).
181. Hunt, D.F., Yates, J.R.,III, Shabinowitz, J., Winston, S., and Hauer, C.R., *Proc. Natl. Acad. Sci.* 83, 6233–6237 (1986).
182. Michel, H., Hunt, D.F., Shabanowitz, J., and Bennett, J., *J. Biol. Chem.* 263, 1123–1130 (1988).
183. Alexander, J.E., Hunt, D.F., Lee, M.K., Shabonowitz, J., Hanspeter, M., Berlin, S.C., MacDonald, T.L., Sundberg, R.J., Rebhun, L.I., and Frankfurter, A., *Proc. Natl. Acad. Sci.* 48, 4685–4689 (1991).
184. Redeker, V., Le Caer, J-P., Rossier, J., and Prome, J., *J. Biol. Chem.* 266, 23461–23466 (1991).

5

Physical and Chemical Properties of Modified Proteins

The incorporation of new chemical groups into proteins can have important effects on the biological properties of proteins. For example, modification may influence the turnover of the protein, its localization in the cell, its regulatory function, and the catalytic character of an enzyme. To understand the molecular basis for these changes, it is necessary to evaluate the chemical and physical properties of the modified proteins. What new properties are introduced because of the modification? How do we learn about what these modifications do?

A good place to start is to find out what the physical properties of the modified amino acids are and how they differ from those of the free amino acids. The solubilities of amino acids partitioning between polar and apolar phases can be used to describe the apolarity of the side chain. The logarithm of the partition coefficient (P) is is a measure of hydrophobicity and is proportional to a free energy of phase transfer.[1] Values may be obtained experimentally or effectively logarithm P, may be derived from hydrophobic fragmental constants by the following equation;[2]

$$\log (P) = \sum_{n=1}^{N} a_n . f_n$$

where f_n is the hydrophobic fragmental constant of the nth atom or group in the structure[3] and a_n represents the number of times the moiety is found in the structure. For example, to calculate log P for the side chain of phenylalanine, the fragmental constants for the side chain groups

(CH_2) and (C_6H_6) are simply added. Values have been obtained for all of the 20 naturally occurring amino acids in proteins and for some of the post- and co-translation modification forms (Table 5.1).[2]

Table 5.1 Calculation of side chain hydrophobicity values for amino acids and modified forms

Amino acid	Log P	Scaled
arginine	-2.061	0.000
aspartic acid	-1.935	0.028
glutamic acid	-1.868	0.043
histidine	-1.321	0.165
asparagine	-1.003	0.236
glutamine	-0.936	0.251
lysine	-0.790	0.283
serine	-0.453	0.359
threonine	-0.042	0.450
glycine	0.184	0.501
alanine	0.702	0.616
cysteine	0.987	0.680
proline	1.128	0.711
methionine	1.246	0.738
valine	1.640	0.825
tryptophan	1.878	0.878
tyrosine	1.887	0.880
phenylalanine	2.423	1.000
Modified amino acids		
asn-(GlcNac)$_2$.(Man)$_9$	-34.586	-7.254
γ-carboxyglutamate	-4.281	-0.495
O3-glycosylserine	-3.498	-0.320
5–hydroxylysine	-1.614	0.100
N-ε-trimethyllysine	-1.526	0.119
N-ε-dimethyllysine	-1.384	0.151
N-ε-monomethyllysine	-1.022	0.283
N-ε-acetyllysine	1.436	0.780
4-hydroxyproline	0.304	0.527
S-palmitoylcysteine	7.993	2.242
Modified amino terminus		
formylation	-0.028	0.453
acetylation	0.490	0.569
myristylation	6.814	1.979

Reprinted with permission from *Analytical Biochemistry* 193, 72. Copyright 1991, Academic Press, Inc.

A positive value of log P indicates the side chain partitions from the organic to the aqueous phase and is a measure of hydrophobicity. Scaled values are shown in Table 5.1 for the amino acids; these values range from 0.0 for arginine (most hydrophilic) to 1.0 for phenylalanine (most hydrophobic).

Note that the log P values for the modified amino acid can differ significantly from those of its parent amino acids and that the scaled values extend outside the range for the free amino acids. That is, some modifications increase the hydrophilicity and others the hydrophobicity. It was estimated that an asparaginyl residue carrying an N-linked oligosaccharide could be about 10^{32} times more water soluble than an arginyl side chain.[2] Knowledge of the physical properties of the modified amino acids provides some basis on which to understand, at least, in part how modification of the side chains and of the end groups influences protein structure.

MODIFICATION CAN CHANGE THE CHARGE OF THE PROTEIN

The incorporation of phosphate, sulfate, or carboxyl groups are direct ways of introducing new negative charges into the protein. Addition of glutamic acid to the side chain γ-carboxyl of a glutamyl residue in tubulin[4] also brings new negative charge to the protein, and addition of arginine at the n-terminus in reactions catalyzed by amino acyl tRNA synthase introduces positive character to the protein. The negative charge of a carboxylate group may be lost by esterification or amidation. Likewise, the positive charge of an amino group may be removed by an acylation reaction. Consider some effects of specific types of modification.

Methylation

The modification of the ε-amino group of lysine alters the positive character of the side chain and can influence its interactions with anionic ligands. Mono-, di-, and trimethylated forms of lysyl residues exist and possess different properties in proteins.

A trimethylated lysyl residue in a protein would maintain its positive charge at all pH values, whereas an unmodified amino at pH's above its pKa value would not. Monomethylated and dimethylated amino groups have different pKa values from the unmodified residue and could have a different charge from that of lysine depending upon the pH. But at physiological pH values, most lysines, mono-, and di- substituted forms would possess a positive charge. This does not mean, however, that the positive character of these sites are identical. ^{13}C-NMR spectroscopy of

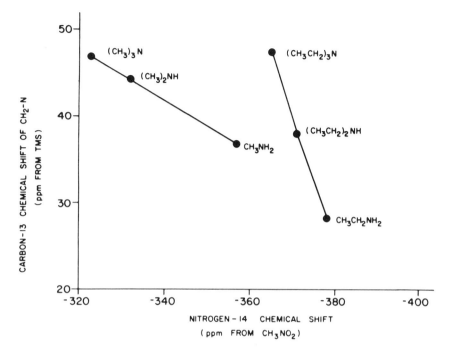

Figure 5.1. Relationship between ^{14}N and ^{13}C chemical shifts in substituted amines. Reprinted with permission from *Biochemical Biophysical Research Communications* 64:514 (1975). Copyright 1975, Academic Press.

substituted amines shows that a progressive decrease in electron density of adjoining carbon atoms to nitrogen occurs upon methylation.[5] Chemical shifts of ^{13}C and ^{14}N resonances of simple amines (Figure 5.1) change proportionally for methylated and ethylated amines. The correlation has been interpreted to mean that the electron density at nitrogen and neighboring carbon atoms decreases proportionally with methylation, that is, progressive methylation increases the positive character of the N atom.

Note that an increase in methylation of lysine causes an increase in hydrophilicity of the side chain (Table 5.1). That increase corresponds with the increase in positive character of the nitrogen atom observed here.

It could be expected that because of the increase in positive character of the N-atom, anionic ligands, for example, phosphoryl groups in DNA and RNA, may interact more strongly with residues containing more methyl groups. Elongation Factor 1α contains di- and trimethyl lysines,[6] and although the functions of modification have not been defined, it has been suggested that modification influences the interactions of the protein with nucleic acid.[7] Histones having a greater proportion of N-methylated lysine interact more strongly with DNA in chromatin than do unmodified histones.[8]

Although methylation can change pK_a values and charge density, effects of modification may be manifested through an alteration of the protein's tertiary structure. The isoelectric point of ε-N-trimethyl-lysine is higher than that of unmodified lysine, but trimethylation of lysine in apo- and holocytochrome c decreases the pI values. This apparent anomaly is explained by a "global" effect of modification. Evaluation of the Stokes radii and sensitivities to proteolysis suggest that methylation of cytochrome C causes the structure to become more compact.[9]

Dimethyl arginine in protein is suggested to have a role in protein-RNA interaction.[10] Modified-nucleolar proteins contain a repetitive sequence Gly-DMArg-Gly-Gly-Phe-Gly-Gly-DMArg-Gly. The modified arginyl residues may interact with the phosphoryl groups of RNA; the glycine stretch could provide flexibility needed for the interaction.[11] Arginine in a basic sequence of HIV-1 Tat protein is believed to make a fork by H-bonding with the phosphoryl groups in a stem loop of tRNA. It has been suggested that methylation, by influencing H-bonding characteristics of arginyl residues, may regulate specific and nonspecific modes of binding.[12] It has not been established, however, whether arginyl residues in a basic segment can be methylated.

When yeast iso-1-cytochrome c is modified by protein methylase III, mono-, di-, and trimethyl derivatives can be identified,[13] but the trimethyl derivative is the principal species formed. An approach taken to evaluate the role of this charged group in its interactions in proteins is ^{14}N-NMR spectroscopy. The isotope is 100 percent naturally abundant and can provide reasonable signals, but NMR of N atoms of amino groups in proteins gives extremely broad lines because of quadrapole relaxation mechanisms and thus limits the use of NMR spectroscopy. An exception is with the trimethylated derivative, which gives sharp lines because of symmetry of the modified group.[14] Studies with calmodulin show that the trimethyl group does not interact strongly with other groups in the protein, that is, it is exposed and has considerable rotational freedom. Although its chemical role in calmodulin is not understood, it does prevent ubiquitination and may have an important function in regulating its turnover.

Phosphorylation

The phosphate group may alter the charge by one or two units, depending on the pKa of the phosphoryl group and on the pH of the medium. But how might we determine exactly the properties and charge state of the bound

Table 5.2. Chemical shifts and pKa values for amino acids and phospho-proteins

Derivatives	Chemical Shift (ppm) dianion	Chemical Shift (ppm) monoanion	pKa
phosphoserine	4.6	0.6	5.8
phosphothreonine	4.0	0.0	5.9
phosphotyrosine	1.0	-3.3	5.8
Phosphoarginine	-3.0	-5.4	4.3
N^3-phosphohistidine	-4.5	No change	
N^1-phosphohistidine	-5.5	No change	
Phosphoproteins			
pepsin (p-ser.)			
ovalbumin			
• p-ser.$_{68}$	5.0	—	6.0
• p-ser.$_{344}$	4.75	—	6.0[17]
phosphorylase a	4.2	—	
CNBr fragment(1-91)	4.6	—	5.45[18]
phos. a (p-ser.)			
histone			
• N1-p-his.$_{18}$	-5.3	—	
• N1-p-his.$_{75}$	-4.9	—	

Reprinted with permission from *Methods in Enzymology* 177:263. Copyright 1989, Academic Press, Inc.

enphosphoryl group? The application of ^{31}P-NMR spectroscopy can be particularly informative. Helpful reviews have been prepared by Vogel.[15,16]

The chemical shift of the phosphoryl group depends upon many factors but principally upon atoms bound to phosphorous, bond geometry, and the extent of π-bonding. A change in charge may influence the chemical shift for some phosphate esters (Table 5.2). Note that chemical shifts occur in the conversion of monoanionic forms to dianionic forms for seryl, threonyl, tyrosyl, and arginyl, but not for histidyl, esters.

Studies with phosphorylated proteins (excluding those containing p-his.) suggest that the phosphoryl groups are titratable and that a chemical shift accompanies the titration. The microenvironment may influence the pKa values and may alter the steepness of the titration curve. Positive charges close to the phosphorylated residue could lower the pKa and increase the steepness (Hill coefficient > 1.0). On the other hand, negative charges could raise the pKa value and decrease the Hill coefficient (n < 1.0). The possibility exists that neighboring groups with similar pKa values may titrate in the same pH range and that the extent of the ^{31}P-chemical shift may reflect partial protonation of several residues.[19] Thus, ^{31}P-NMR may give insight about the nature of the modified group, its charge state, and microenvironment. The application of Fourier transform infrared-spectroscopy (FTIR) has provided direct proof that phosphoryl groups in phosvitin and ovalbumin undergo ionization with pKa values of about 6.0.[20]

The charged phosphoryl groups can promote structural rearrangements in the protein and influence the properties of regions of the protein that are distant from the phosphorylation site. That is the case for glycogen phosphorylase. Activation of muscle glycogen phosphorylase occurs by phosphorylation of serine$_{14}$, and in the conversion of the enzyme from its inactive T state to the active R state, contacts are made between the phosphoryl group and Arg$_{69}$ of its own subunit and Arg'$_{43}$ of its neighbor (Figure 5.2).[21,22] Note that multiple interactions occur between

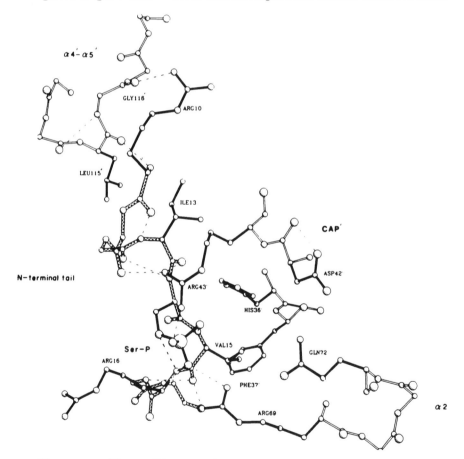

Figure 5.2.   The cap'N-terminal in R state glycogen phosphorylase a. Reprinted with permission from *Journal of Molecular Biology* 218:233 (1991). Copyright 1991, Journal of Molecular Biology.

the phosphoryl moiety and the guanidinyl groups. The phosphorylation event promotes a structural change in the amino terminal region, which in turn affects the active site region, which is some 40 Å removed. The phosphoryl group does more than simply just introduce negative charges. That it can also form multiple interactions may explain its important and diverse role in biological structures.

Yet phosphorylation of isocitrate dehydrogenase of *E. coli* provides a quite different scenario. This phosphorylation occurs directly in the active site region. Serine$_{113}$, the phosphorylation site, makes an H-bond with the γ-carboxylate of isocitrate. Phosphorylation inactivates the enzyme by blocking the binding of isocitrate.[23] A comparison of the 3-D structure of the phosphorylated enzyme with the dephosphorylated enzyme and the inactive mutant forms (in which serine has been converted to aspartate or glutamate) suggests that minimal changes of conformation occur. Calculations of electrostatic potential change, and changes in free energy of binding due to modification[24] support the view that the inactivation is due to the negative charge and to the size of the phosphate moiety, which restricts isocitrate binding by electrostatic and steric hindrance effects. Further evidence of a steric effect is that a smaller substrate analog lacking a third carboxylate, 2R malate, can fit and be acted upon by the phosphorylated enzyme.[25]

Sulfation

A sulfate group incorporated into proteins might be expected to have effects similar to those of a phosphate group. Both groups are anionic and possess tetrahedral geometry. The sulfate ester is somewhat larger than the phosphate ester, but a major difference between them is that the former has a single negative charge at all physiological pH values. Hence, a sulfate ester may interact differently from a phosphate ester. It can participate in electrostatic interactions and act as hydrogen-bond acceptor. The phosphate ester with a single negative charge also can participate in electrostatic interactions, but it can act as both an acceptor and donor for the formation of hydrogen bonds.[26]

Antibodies prepared to a phosphorylated derivative of tyrosine illustrate the importance of the ionization of the side chain in ionic interactions.[27] Binding occurs at pH's where the dianion of phosphorylated tyrosine is the predominant ionic form binding occurs, but not at pH's where the monanion is the prevalent ionic species. Tyrosyl sulfate, like $H_2PO_4^-$, does not bind presumably because of its single negative charge.

The sulfate dianion is known to affect the properties of glycogen phosphorylase and glycogen synthase in a way related to the effects caused by covalent phosphorylation, that is, formation of R- and T-like states, respectively. X-ray analysis of glycogen phosphorylase b shows that free sulfate interacts with the same side-chain groups shown for the phosphoryl group in phosphorylase a.

The protein hirudin contains a sulfated tyrosyl residue in the flexible carboxy-terminal region. This protein makes a strong complex (Kd = 10^{-14}M) with thrombin, and serves as a physiological mechanism with which to prevent blood coagulation when the leech (Hiruda medicinulis) ingests blood. The protein can be prepared in desulfated and sulfated forms, and kinetic studies of its interaction with thrombin,[28] 2D-NMR spectroscopy,[29] and x-ray crystallography[30] reveal structural aspects of the protein and its interaction with thrombin.

Ionic interactions account for approximately 32 percent of the binding energy of hirudin with thrombin. A strong correlation exists between the ionic interactions of genetically engineered forms of hirudin, including a desulfated form, with thrombin. The linear relationship between binding energy and charge state of hirudin suggests that in the flexible C-terminal region, the sulfate moiety and the carboxylate groups, also singly negatively charged, each contribute about -4kJ mol^{-1} to the binding energy.

The desulfated form of hirudin binds more slowly and dissociates more readily than does the sulfated form. Overall, the binding is reduced ten-fold by the absence of the sulfate group.[31] X-ray crystallography suggests that tyrosine$_{63}$ sulfate of hirudin may interact with tyrosine$_{76}$ in thrombin. A phosphorylated variant of hirudin has been prepared and found to interact with essentially the same kinetics as does the sulfated form with thrombin.[32] Thus, it seems that an additional negative charge contributed by a dianionic form of phosphate is not needed for optimal binding.

Carboxylation

Carboxylation, like sulfation, introduces a new negative charge into a site. But the consequences may be quite different inasmuch as the sites of modifications are the β and γ methylene groups of aspartyl and glutamyl side chains, respectively. The products bearing two neighboring carboxyl groups can have two negative charges close by, and a carboxyl or a carboxylate group may have a profound effect on the acid-base characteristics of its neighbor. For example, the two carboxyl groups of methylmalonic acid, an analogue of g-carboxyglutamate, titrate with widely separated pKa values of 2.7 and 5.1. ^{13}C NMR spectroscopy of g-carboxyglutamate as a function of pH yields pKa values of 2.0 and 4.4 for the two side-chain groups.[33] It seems that a nonionized carboxyl group in g-carboxyglutamate(Gla) increases the acidity of the first titrating group. A negatively charged carboxylate produced from the

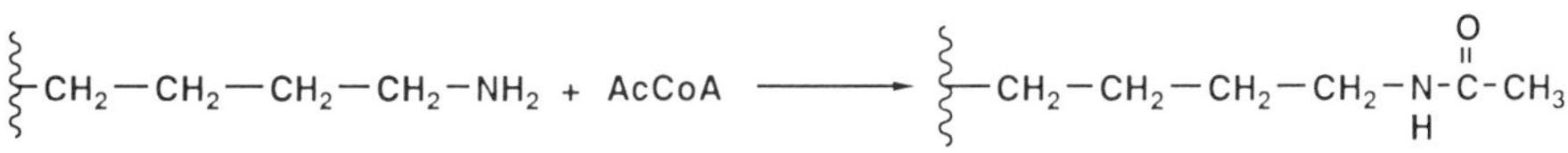

$$\xi\text{-CH}_2-\text{CH}_2-\text{CH}_2-\text{CH}_2-\text{NH}_2 \ + \ \text{AcCoA} \ \longrightarrow \ \xi\text{-CH}_2-\text{CH}_2-\text{CH}_2-\text{CH}_2-\overset{\text{O}}{\underset{\text{H}}{\text{N-}}}\overset{||}{\text{C}}\text{-CH}_3$$

Figure 5.3. Enzymatic reaction of acetyl CoA with ε-amino groups in proteins.

first ionization decreases slightly the acidity of the second group, pKa 4.4, as compared with the γ-carboxyl group of unmodified glutamyl residue, which has a pKa value of 4.1.

A Gla residue with its two carboxylate groups may enter into specific interactions with guanidinium groups not possible with Glu. A hydrogen bond pattern between both amino nitrogens of the guanidinium group with the two dicarboxylate groups has been defined in model compounds.[34] Such interactions may be important in protein structures containing Gla residues. It has also been suggested that such an interaction of Gla, a disubstituted malonate, with arginine may keep Gla from decarboxylating.[35]

Acetylation

The charge state of an ε-amino group of a lysyl residue can be modified by the enzymatic reaction with acetyl CoA (Figure 5.3).

If the protein contains a stretch of lysyl residues, as found in specific histones, charge neutralization could have an important effect on electrostatic interactions of this segment of the protein with anionic groups, for example, of DNA in nucleosomal particles, or of nucleoplasmin in Xenopus laevis oocyte nuclei.

Although it may be anticipated that charge neutralization follows acetylation, the precise effect depends upon the chemical nature of the lysyl residues. A cluster of lysyl or other positively charged groups could lower the pKa of certain residues so that at physiological pH, certain residues could exist in the deprotonated form. In this instance acetylation would change the chemical character of the side chain but not its charge state.

The molecular mechanisms by which acetylation of histones influence chromatin structure are not resolved. Acetylation and its reverse process, deacetylation, can influence deposition of histone, structure of the nucleosomal particle, and higher order structures of chromatin.[36]

Arginylation

Proteins possessing negatively charged amino acids at the amino terminus are not targeted for modification by the ubiquitin-protein ligase. But an addition of the basic amino acid, arginine, to the end catalyzed by the action of arginyl t-RNA synthetase can sensitize specific proteins.[37] Is the sensitivity because ligase has an absolute requirement for arginine in the protein substrate for binding or because the charge of the amino terminus is neutralized? That chemical amidation of carboxyl groups of glutamate and aspartate at the amino terminal ends of lactalbumin and soybean trypsin inhibitor induces sensitivity to the ligase suggests the merit of the latter explanation.[38] A protein possessing an acidic end amino acid with pKa values of the α-amino group and of a side chain carboxyl group of 8.0 and 3-4, respectively, would exhibit a partial negative charge at its amino end at pH 7.0 and above. Arginylation would induce a positive character to this region.

MODIFICATION CAN CHANGE THE HYDROPHOBICITY OF THE PROTEIN

The addition of a fatty acid such as myristic or palmitic acid or of an isoprenoid like farnesine into a protein obviously can change the hydrophobic character of a protein. First a lipid-like material is incorporated, and second the modification may block the ionization of the functional group to which the lipid is attached. Acylation of an amino group or isoprenylation of a sulfhydryl group blocks ionization. Methylation introduces a substituent with some hydrophobicity. If the reaction is on a carboxl group, the loss of charge can be significant. Methylation of the amino terminus would not alleviate charge. Hence, hydrophobic character depends not only upon the group that is added but upon the amino acid residues and the type of reaction involved. Enzymatic reduction of a methionyl sulfoxide group in a protein would increase hydrophobicity and may constitute an important physiological mechanism with which to repair oxidized protein. For example, elastase inhibitor damaged by smoke involves formation of a sulfoxide. The repair may bring back normal function. Lack of reduction may contribute to the uncontrolled proteolysis associated with emphysema (Chapter 1). An increase in hydrophobic character was determined for post-translational modification of P_{21} Ras protein by measuring its partition into 1 percent Triton X-114 as a consequence of modification.[39,40] This method may apply to other proteins with increased hydrophobicity.

Fatty Acid Acylation of the Amino Terminus

Myristoylation

In 1982, tetradecanoic acid (myristic acid) was identified as the amino

Table 5.3. Retention times of fatty acid-acylated peptides

Fatty acid formula	Retention time (min)	Fatty acid
$CH_3(CH_2)_8COOH$	14	Decanoic acid
$CH_3(CH_2)_{10}COOH$	21	Dodecanoic acid
$CH_3(CH_2)_{12}COOH$	26	Myristic acid
$CH_3(CH_2)_7O(CH_2)_4COOH$	16	6-Oxymyristic acid
$CH_3(CH_2)_2O(CH_2)_{10}COOH$	11-12	11-Oxymyristic acid
$CH_3O(CH_2)_{11}COOH$	12-13	13-Oxymyristic acid

Acylated peptides of GNAAS(^{125}I)YRR. Reprinted with permission from *Proceedings of the National Academy of Sciences* 85:8795, 1988.

terminal blocking group in the catalytic subunit of cyclic-AMP dependent protein kinase in cardiac muscle.[41] Since then, proteins such as pp60[v-src], the gag protein of murine leukemia virus, the β-subunit of calcineurin, G proteins, cytochrome b_5 reductase, and many others have been reported to contain this functional group.

Myristoylation of pp60[v-src] is necessary for the association of the protein with the plasma membrane but alone it is insufficient to account for a stable association.[42,43] Seemingly, the myristoyl group and specific parts of the amino-terminal region are necessary for binding. Saturation kinetics and competition of binding by specific myristoylated peptide fragments are in keeping with a specific myristoyl-src receptor in the plasma membrane.[44]

The hydrophobic features of the myristoyl moiety may promote associative behavior with other proteins and/or other lipid-containing substances. But it is also possible that specific recognition of the chain length of the myristoyl moiety in the modified protein is important for binding. Oxyanalogs of myristic acid, 6-oxymyristic, 11-oxymyristic, and 13-oxymyristic acids have stereochemical structures and chain lengths very similar to those of myristic acid but differ from myristic acid in hydrophobicity. The use of these analogs has helped sort out whether binding is more related to length or to hydrophobicity.

In Table 5.3, the retention times of acylated peptides containing fatty acids and the oxyanalogs of myristic acid obtained in reverse phase HPLC are given.[45] The retention times, which are correlated with hydrophobicity, increase with chain length. Note that the oxyderivatives have similar binding characteristics and that their retention times are similar to those of decanoic acid, which is four carbons shorter.

If the binding of specific proteins, for example, pp60[v-src], to membranes is more related to the hydrophobicity of the myristoyl moiety than to its chain length, it might be expected that the oxyanalog-protein derivatives may show weaker binding of the modified proteins to membranes than the myristoylated form. Redistribution of specific proteins from the

Table 5.4 Glycoprotein (percent of bound fatty acids)

Fatty acids	Native	Cystic fibrosis
C16:0	47.0	36.5
C18:0	22.0	48.7
C20:0	0.5	0.1
C22:0	0.6	1.5
C24:0	5.9	1.8
C16:1	2.7	1.1
C18:1	14.5	8.0

Reprinted with permission from *Journal of Biological Chemistry* 258:8535, 1983.

membrane to the cytosol has been found following incorporation of oxyanalogs, for example, with pp60[v-src] and the model protein p22-WT-Ser186. This finding suggests that hydrophobicity of the myristoyl moiety is more important than chain length for targeting the binding of these proteins to membranes.[46]

Fatty Acid Acylation of Side Chains

Modification may occur on threonyl, seryl, and cysteinyl residues. The O and S-esters can be distinguished by their sensitivity to hydroxylamine and thiols. A thiol ester may be cleaved by 0.05M NH_2OH or by 2 percent mercaptoethanol but an oxyester is insensitive to thiols and requires greater amounts of hydroxylamine (1M) for effective cleavage.[47,48] Whether the modification involves one or two residues in a protein or is extensive, a marked effect on the hydrophobicity of the protein is expected.

Palmityl, stearyl, oleoyl, and miscellaneous other fatty acid residues have been found covalently bound to proteins. Myelin proteolipid protein from brain consists of 2 percent by weight of palmitic, oleic, and stearic acids.[49] Gastric mucin, a protein thought to provide a protective role to gastric mucosa against damage by chemicals, enzymes, or physical agents, contains approximately twenty moles of these three fatty acid/mole of protein.[50] In individuals with cystic fibrosis, bound fatty acids can increase three-fold. The distribution of fatty acids in mucin differs from normal individuals from that from individuals with cystic fibrosis (Table 5.4). Note that many different fatty acids are present, but that palmitate, stearate, and oleate account for the major fraction of the total fatty acids. The increase in fatty acids contributes to the increased viscosity of the mucin and may be partly responsible for the dysfunction of the protein in this disease.[51]

The human β_2 adrenergic receptor,[52] bovine rhodopsin,[53] and specific Ras proteins[54] are examples of proteins containing only one-two moles of palmitic acid/mole. The effect of modification is pronounced as it alters specific interactions of these proteins with membranes. The adrenergic receptor illustrated in Figure 5.4 shows that the protein interacts with the plasma membrane through seven membrane spanning segments, and palmitoylation at Cys_{341} is believed to promote a new interaction with the membrane leading to the formation of the fourth intracellular loop necessary for the coupling of the receptor to Gs for adenyl cyclase action.[53]

Mutation of Cys at position 341 to Ser causes a loss in sensitivity of the receptor to isoproterenol. A cysteine residue is found in equivalent positions in other adrenergic receptors and in rhodopsin. In the latter instance, however, there are two adjacent cysteinyl residues, both of which can be palmitoylated,[50] and thus can lead to important contacts with photoreceptor membranes. The exact biological and chemical roles that these acyl groups play in signal transduction await further investigation.

Prenylation

Mevalonic acid, the isoprenoid precursor, is converted to farnesyl pyrophosphate in various cellular systems, thereby leading to the the incorporation of a prenyl nucleus into specific proteins such as lamin B,[55,40,56] Ras,[54] and various mating peptide hormones from fungi and yeast.[57,58] These polypeptides all contain a C_{15} farnesyl moiety attached via a thiol ether to a cysteinyl residue. This fact was nicely proved in Lamin B. Here, a release of radiolabeled material derived from mevalonic acid was caused by Raney-nickel, a reagent that is known to catalyze the hydrogenolysis of carbon-sulfur bonds. Alcoholic KOH, a reagent that cleaves esters, was ineffectual. The carbon fragment identified as 3,7,11-trimethyl-2,6,10-dodecatriene by gas chromatography and mass spectrometry shows that a farnesyl group is present in Lamin B.

Raney-nickel may be quite useful for examining other proteins for similar modifications because cleavage can occur in intact proteins with this reagent. In fact, the use of Raney-nickel of prenylated proteins of Chinese hamster ovary cells and Hela cells led to the discovery that these proteins contain the geranylgeranyl group.[59,60] This function has also been identified in the γ-subunit of G-proteins.[61,62]

It has not yet been established what these isoprenoids do in protein structures or what precise interactions they have with other biomolecules. The all trans-isoprenoid side chain might intercalate effectively with hydrophobic patches in proteins or in lipid bilayers. A geranylgeranyl group (20 carbons) versus a farnesyl group (15 carbons) would be expected to have additional hydrophobic character and to interact more strongly with lipid bilayers.

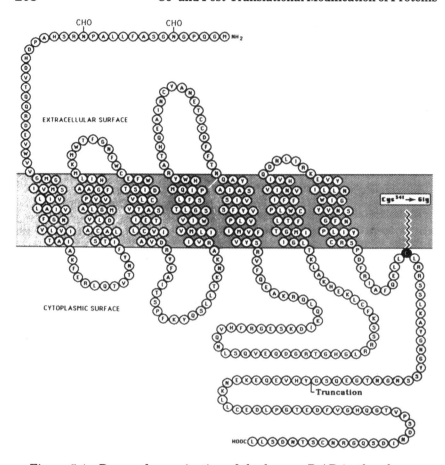

Figure 5.4.   Proposed organization of the human B_2AR in the plasma membrane. The palmitoylation at Cys^{341} and the truncation in T^{365} are indicated. Palmitoylated Cys^{341} may anchor the amino-terminal portion of the carboxy-tail in the plasma membrane, thereby creating a fourth intracellular loop critical for formation of a functional G_s binding site. Reprinted with permission from *Journal of Biological Chemistry* 258:8535 (1983). Copyright 1983, American Society for Biochemistry.

Studies with lamin B and Ras proteins suggest specific roles for the modified protein. Nuclear lamina serve to provide structure to the nuclear envelope and are involved in associative processes with chromatin. In mitosis, depolymerization of lamins occur, and lamin A is released from the inner nuclear membrane, while lamin B remains attached. Lamin B binding to membranes is saturable, and experiments suggest a specific protein receptor may be involved.[63],[64] Although not proven, the farnesyl moiety along with the polypeptide constituents may be bound to a specific receptor as suggested for interactions of myristylated proteins. The interaction of Ras proteins with membranes is a means by which Ras can influence biological processes.[65] Specific point mutations of these proteins can lead to cellular transformation.[66] Farnesylation of a cysteinyl residue in the carboxy terminal sequence of the Ras proteins is necessary for their association with membranes. Palmitoylation of a nearby cysteinyl residue increases the avidity of the modified protein for the membrane.[54]

Although prenylation introduces hydrophobic character to the protein, it alone is insufficient for effective association of the modified protein with the membrane. Proteins like p21[K-ras(B)] bearing the same C-terminal motif are prenylated, proteolyzed to yield a carboxyl terminal cysteine, and then methylated. Both proteolysis and methylation appear essential for effective membrane binding of prenylated Ras. Methylation of the carboxy end removes a negative charge and introduces some hydrophobicity to the region, which could influence how the prenyl nucleus interacts with the lipid bilayer or a receptor protein.[67]

MODIFICATION CAN INFLUENCE PROTEIN CONFORMATION, FOLDING, AND STABILITY

Structural organization of proteins induced by enzymatic processes can occur as a result of both co- and post-translationally directed reactions. For example, disulfide bond formation, glycosylation, hydroxylation, proteolytic processing, and prolylpeptidyl isomerization can influence the attainment of a native-folded state. Phosphorylation-dephosphorylation and a variety of other reactions may alter the conformation of a completed polypeptide chain.

Disulfide bonds

The bonding of two sulfhydryl groups via a disulfide bridge influences protein structure and stability.[68] An analysis of 22 protein structures containing 72 disulfide bonds shows that 23 of the disulfides link extended β-strands, 4 bonded helices, 12 linked helical regions to β-strands, and 6 were found in β-turns. Although disulfides link extended strands, a plurality (but < 50 percent) is found in irregular regions of the protein structures.[69]

Torsion angles, bond distances, and conformational parameters of 72 disulfides have been evaluated in proteins whose structures have been resolved to < 2.0Å resolution. The most common torsion angle for the disulfide bond is around -90° or +90°. Eight disulfide types occur; a

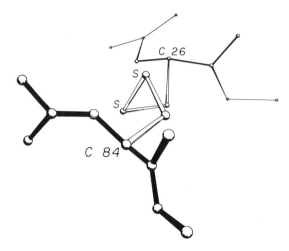

Figure 5.5. A left-handed spiral disulfide from ribonuclease S, viewed end-on. Reprinted with permission from *Advances in Protein Chemistry* 34:167 (1987). Copyright 1987, Academic Press.

prominent structure is the "left-handed spiral" conformation[70] shown in Figure 5.5 for the protein, RNase S.

Disulfide bonds in proteins have different stabilities, and a relation exists between their stabilities and their effects on the conformational stability.[71] Table 5.5 shows such a relation for bovine trypsin inhibitor.[72]

The effective concentration of protein thiol is taken as a measure of the differences in free energies of the protein disulfide and an intermolecular disulfide. The higher this concentration is, the more stable is the bond. Hence, the disulfide bond 5–55 is more stable than 30-51, which is more stable than 14-38. The latter disulfide shows greater than 50 percent exposure to solvent and influences its stability. Note that there exists the same order of protein stability, which is measured by the temperature required for unfolding of protein forms missing a particular disulfide.

Ribonuclease A contains a β-turn between residues 66-68 that is stabilized by H-bonding, and a disulfide between cys. residues 65 and 72 gives inflexibility to the region. Asparagine at position 67 can be deamidated by heating the protein. A cyclic imide is believed to be an intermediate in the process, requiring the ψ torsional angle for the single bond of C-C = O of the asparagine residue to rotate so that N atom of the asparaginyl peptide linkage is oriented correctly for optimal reaction with the carboxyamide function. At 37° and pH 8, little or no deamidation occurs of the native protein, but deamidation of the reduced protein occurs effectively under the same conditions. It thus seems that the local conformation influenced by disulfide bonding is important and limits the

Table 5.5. Correlation between stability of a disulfide and its contribution to the conformational stability of BPTI[a]

Disulfide	Stability in folded conformation (effective concentration of thiols, M)	Temp °C at which folded conformation without disulfide disappears
5–55	3.7×10^5	~38
30–51	1.4×10^3	~55
14–38	1.9×10^2	74

Reprinted with permission from *Journal of Physical Chemistry* 89:2452. Copyright 1985, American Chemical Society.

deamidation reaction.[73]

Glycosylation

Glycans bound to protein are often described by their mode of linkage to the protein, for example, the O-linked sugars and the N-linked sugars. Bound carbohydrate can influence protein structure and vice versa, and effects depend upon types of sugars, the linkages, stereochemistry, size of the bound saccharides, and characteristics of the protein.

Chains of O-linked sugars bound to protein are formed by a stepwise addition of carbohydrate to a completed polypeptide chain. These glycans are found densely clustered in stretches of the primary structures of cell surface glycoproteins and mucins, and the oligosaccharides in these packed regions influence the properties of the polypeptide chain.[74] Physical studies of mucins show that the structure can be described generally as a random coil conformation, but that the bound carbohydrate influences the dynamics of the structure. Determination of the radius of gyration, which is a measure of the statistical average distance of the end of the chain to its center, shows that a mucin chain has a 2.5–3-fold larger radius of gyration (Figure 5.6) than does a denatured polypeptide chain of equal number of amino acid residues.[75]

Deglycosylated mucin has a radius of gyration similar to that of other denatured proteins showing that the large radius of gyration of mucin is not due to amino acid and sequence differences between the polymers but is related to the presence of carbohydrate, which tends to stiffen the polypeptide backbone. Light scattering and NMR studies[76] suggest that a whole chain of bound carbohydrate is not needed for effects on structure and that chain stiffness is influenced mainly by steric interactions with the first sugar, an O-linked GalNAc residue. Calculation of the persistence length, the distance that a chain moves before a significant change in direction occurs, yields a value of 145Å in comparison to an unsubstituted peptide value of about 10Å. Seemingly, the structure of

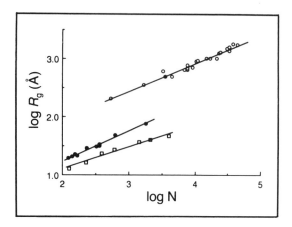

Figure 5.6. The comparison of N (the number of amino acids in the peptide chain) to the radii of gyration of mucins (open circle), proteins denatured in guanidine hydrochloride (closed circle) and globular proteins (open square). Reprinted with permission from *Trends in Biochemical Science* 15:291 (1990). Copyright 1990, Elsevier Trends Journals.

mucin approaches that of a rigid rod. This effect of bound carbohydrate on conformational stability is important for the orientation and function of membrane-bound glycoproteins.

Leuokosialin is a glycoprotein also containing O-linked carbohydrates. In the extracellular domain one of every three residues is glycosylated. It was found that the glycosylated segment can be described as an unfolded structure with a length of 45 nm, from transmission electron microscopy studies.[77]

Studies of model compounds can provide insight into carbohydrate-protein interactions. For example, circular dichroism measurements of model glycopeptides suggest that bound carbohydrate may interact directly by hydrogen bonding with the peptide backbone, thereby stabilizing a particular structure.[78] A β-turn is proposed, and Figure 5.7 shows how an acetyl function at C'2 may interact with an amide H and influence structural stability.

And last, X-ray analysis of glucoamylase, a glycoprotein containing O-linked and N-linked carbohydrate, provides further insight about the structure of bound carbohydrate and protein structure. Crystals of a proteolyzed form of glucoamylase derived from the fungus, Aspergillus awamori, have been analyzed and its structure has been resolved to 2.2 Å (Figure 5.8).[79]

The region containing 10 O-linked sugars is a well defined structure with the polypeptide chain in an extended conformation. The carbohy-

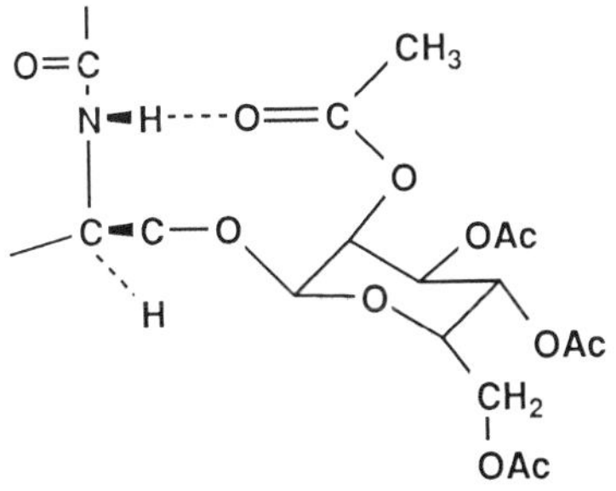

Figure 5.7. Schematic representation of a possible type of a glyco-turn. Type 1(III) β-turn with a ten-membered (C10) glyco-turn ββ. Reprinted with permission from *Biopolymers* 29:1549 (1990). Copyright 1990, John Wiley & Sons, Ltd.

drate, a-mannose (single units) in the C1 conformation are linked to seryl and threonyl side chains. It is believed these carbohydrates impart rigidity to the chain and this region of the protein containing clustered O-linked sugars is like a belt surrounding the "a/a-barrel" connecting the catalytic domain with a starch binding domain, not present in this proteolyzed form of the enzyme.

N-linked carbohydrate chains, in contrast to O-linked carbohydrate chains, are begun in a cotranslational process by the transfer of the oligosaccharide unit of $Glc_3 Man_9 GlcNAc_2$ from dolichol phosphate. The added structure can influence folding and stability of the protein.[80]

One approach to evaluating the roles of sugars is to remove them from the protein and to examine the structures of the partly deglycosylated proteins. Enzymes were used with $β_2$-glycoprotein, and the sequential loss of NeuAc, Gal, GlcNAc, and Man had virtually no effect on the secondary structure of the protein, as assessed by circular dichroism measurements. But significant changes in structure—an increase of β-turns and a decrease of random coil—were found after the removal of Man and then of GlcNAc from the bound trisaccharide Man GlcNAc GlcNAc.[81] The first N-linked sugar, GlcNAc, may not influence structure. An implication of this finding is that the addition of multiple sugars as one unit, which occurs during biosynthesis of the N-linked glycoprotein, may ensure proper folding of the peptide chain.

A second approach is to use site-directed mutagenesis to alter the glycosylation sites in known proteins so that N-glycosylation cannot occur, and evaluate what effects these changes have on the properties of biosynthesized proteins. Such studies done with simian viral hemagglutinin neuraminidase[82] and yeast acid phosphatase[83] suggest N-glycosylation is needed for proper folding.

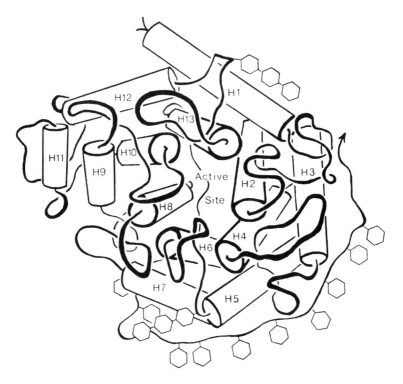

Figure 5.8. Schematic of the glucoamylase-470. α-helices are cylinders, sites of O-glycosylation are represented by single hexagons and sites of N-glycosylation are symbolized by chains of three hexagons. Reprinted with permission from *Journal of Biological Chemistry* 267, 1921–19298 (1992). Copyright 1992, The American Society for Biochemistry & Molecular Biology.

The gene for yeast acid phosphatase has 12 sequons for Asn-X-Ser/Thr and all of the putative glycosylation sites in the wild-type protein are N-glycosylated. The completed enzyme is secreted into the periplasmic space and culture medium, but not all of the sites need be modified for secretion. This was demonstrated by the use of hypoglycosylation mutants of the 12 sites. At 30° six glycosylation events are required, at 20° one is necessary, and at 9° none is needed to form active secreted enzyme. Hence, glycosylation is not needed to form active enzyme. Immunoprecipitation shows that protein expressed in the hypoglycosylation mutants can be found in an inactive state in the endoplasmic reticulum, and suggests that these forms that accumulate have not folded properly due to underglycosylation. Also, the position of glycosylation influences the

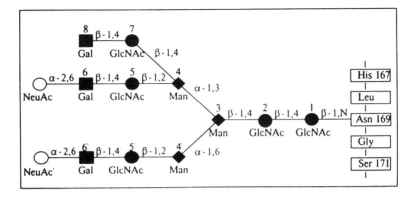

Figure 5.9. Structure of carbohydrate portion of phosvitin from NMR studies. Reprinted with permission from *Biochemistry* 29: 5574 (1990). Copyright 1990 American Chemical Society.

effectiveness of folding and glycosylation of sites at the C-terminus appear more influential than those in the middle or at the N-terminus.

The dynamics of bound carbohydrate to protein can be obtained by NMR spectroscopy. Early studies of ^{13}C-NMR spectroscopy suggested that the carbohydrate chain was not held tight by the protein but was quite flexible.[84,85,86] A powerful tool is 2-D proton NMR spectroscopy. A study of carbohydrate bound to phosvitin reveals its composition, structure (as illustrated in Figure 5.9), and flexibility.[87]

A possible interaction of the core region may occur, but in comparison with those of model compounds, the results suggest that a bound chain has the flexibility of a free carbohydrate chain.

Phosphorylation

Different mechanisms exist by which phosphorylation can alter protein structure. The introduction of a phosphate group with one or two negative charges can change ionic interactions in the protein, promote hydrogen bonding with other functional groups, or alter hydrophobic interactions; these processes can change the secondary, tertiary, and quaternary structures of proteins. These effects, in contrast to some other types of covalent modification, are completely reversible because of the action of protein phosphatases.

X-ray structural analysis of glycogen phosphorylase *b* and *a* shows how a single phosphate group in a monomeric chain of 841 residues influences protein conformation. The amino terminal region bearing the phosphorylatable serine in phosphorylase b bound by intramolecular interactions protein moves after phosphorylation, forming new contacts across the subunit interface of the dimer (Chapter 3). Phosphorylation of glycogen phosphorylase from rabbit muscle activates the enzyme, decreases its solubility, increases its stability, and causes the association of its dimeric form of the enzyme to the tetramer. But activation, which is caused by phosphorylation, is due to a conformational change of the dimer and not to a changed quaternary structure, which actually lowers enzyme activity. X-ray analysis of the R-state of tetrameric phosphorylase a shows that a portion of the glycogen storage domain participates in the tetramer interface and could account for the lower activity of the tetramer in comparison to the dimer. Other important contacts involved in tetramer formation involve the C-terminal subdomain and the tower helix. Polar and charged residues and a high number of H-bonds are involved in the interactions. Thus, from x-ray crystallography we have learned how phosphorylation changes both tertiary and quaternary structures of a protein.[88,89]

Phosphorylation can also promote the dissociation of multimeric structures. Neurofilament L protein,[90] nuclear lamins,[91,92] intermediate filaments containing vimentin,[93] and Tau, a microtubular associated protein,[94] are known instances in which phosphorylation promotes disassembly. Electron microscopy shows that the Tau protein becomes longer and stiffer after phosphorylation and that dephosphorylation shortens the protein and increases elasticity.^{13}C-NMR spectroscopy and other physical studies suggest that portions of keratin intermediate filaments proteins also become more rigid after phosphorylation.[95]

Assembly-disassembly of intermediate filaments consisting of vimentin or desmin can be influenced by dephosphorylation and phosphorylation reactions, respectively. These proteins consist of three domains—the head, rod, and tail. Double stranded coils are caused by specific interactions of the the α-helical rod domains; the head region consisting of basic residues influences filament stability and polymerization. Multiple phosphorylation of the head regions by protein kinase C or cyclic AMP-dependent protein kinase can cause disassembly of the intermediate filaments.[96,97] The fact that enzymatic dephosphorylation reverses the process is consistent with a physiological control mechanism. Though the precise molecular mechanism is not understood, the introduction of negative charges in the head region could overcome the positive charge of the region and influence ionic interactions of the head region which in turn could affect structures involved in coil-coil interactions. Filament disassembly also may be promoted by another post-translational modification reaction, arginine deimination.[98] Here citrulline is formed from arginyl groups in proteins and deimination of vimentin to about eight moles of citrulline/mole of protein produces the structural changes. The head domain was the major region of modification and results support the concept that the basic character of this region is important for formation and stablility of the filamentous structure.

Phosphorylation-dephosphorylation reactions may change hydrophobic interactions and influence protein structure. This attractive idea proposed on the basis of studies with polypeptide analogs of elastin may partly explain why the entanglements of specific neural proteins occur in Alzheimer's disease, may apply to the disassembly of filamentous proteins mentioned above, and may promote local conformational changes in regions where hydrophobic interactions exist.

An important aspect of hydrophobic bonding is the change of water structure occurring upon the interaction of apolar groups. Disorganization of bound clathrate H_2O occurs, and the release of water molecules into the bulk solvent contributes significantly to the positive change in entropy associated with formation of hydrophobic interactions.

The heating of polypeptide analogs of elastin, rich in hydrophobic amino acids, in aqueous solutions induces a conformational transition leading to the formation of a shortened chain of β-spirals. Because increased order of these peptides occurs upon heating, this phenomenon is referred to as an inverse temperature transition.[99] The favorable energetics for the process are explained by the overall positive entropy change of the process driven by the dissolution of the clathrate H_2O molecules around the individual apolar groups in the extended chain as new hydrophobic interactions occur. Introduction of charged groups, a carboxylate or a phosphate, which would increase hydrophilicity, favors formation of the extended structure and influences the temperature required to induce the conformational transition to the β-spirals.

The peptide, poly[15(IPGVG), (RGYSLG)] contains phosphorylatable sites for cyclic AMP-dependent protein kinase. Upon increasing temperature, the peptide becomes more ordered and forms an viscoelastic phase, and changes in its associative state are measured by following the increase in turbidity at 300 nm (Figure 5.10).

Note the unmodified peptide has an inverse temperature transition at 18° but after phosphorylation of only one serine/360 residues the transition temperature is raised to 32°. And alkaline phosphatase dephosphorylates the peptide and reverses the effect.[100] If the unmodified peptide is at 25°, that is, below the transition temperature, the chains are contracted. If the phosphorylated peptide is at 25°, that is, below the transition temperature, the chains are extended. Thus a change in chemistry can promote a mechanical movement of the chain and is referred to a chemomechanical transduction.[101] This interesting sugges

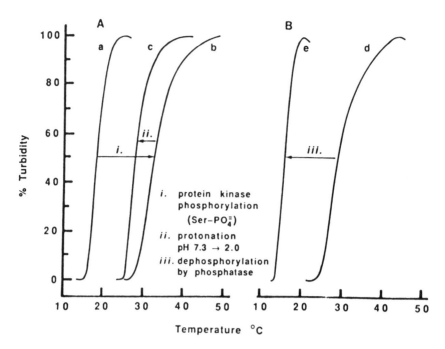

Figure 5.10. Folding and aggregational transitions of poly15-(lPGVG),(RGYSLG): **A.** Curve a. Polypeptide with protein kinase at time 0 minutes; Curve b. Polypeptide with protein kinase at time 48 hours; Curve c. Protonation of 48 hours phosphorylated sample. pH 2.0. **B.** Curve d. Control phosphorylated polypeptide (24 hours) reaction mixture: at time 4 hours; Curve e. Phosphorylated polypeptide (24 hours) treated with alkaline phosphatase: at time 4 hours. Complete dephosphorylation is achieved. Reprinted with permission from *Biochemical Biophysical Research Communications* 178:539 (1991). Copyright 1991, Academic Press.

tion can provide an explanation for some conformational changes and a mechanism by which signals may be transduced.

A region of hydrophobic amino acids containing a dephosphorylated serine, threonine, or tyrosine could have a different local conformation than if the hydroxyamino acid is phosphorylated because water structure about hydrophobic residues and anionic groups differ, and the two water structures are mutually incompatible.[102] Changes in solvent H_2O interactions brought about by phosphorylation-dephosphorylation may result in different ϕ and ψ angles of neighboring residues and change the orientation of the side chains, for example, of valine to isoleucine in the phosphorylatable sequence, val-ser-ile, found in glycogen phosphorylase. A local structural change based on changes in hydrophobicity could serve to trigger other structural changes in the protein.

Hydroxylation

It is well documented that the hydroxylation of the four position of prolyl residues in collagen influences the stability of the triple helix. If hydroxylation of the chains are prevented during biosynthesis, then formation of the triple helix is severely impaired. A model suggesting how hydroxylation could induce a conformational change favorable for the winding of the chains was presented in Chapter 3.

Hydroxylation of prolyl residues in tropoelastin has a different effect on folding different from than that of collagen and can inhibit fiber formation.[103] The use of polypentapeptide analogs of elastin (Val-Pro[or HydroxyPro]-Gly-Val-Gly)$_n$ reveals how hydroxylation can change structure.[104,95] Both peptides undergo an inverse temperature transition as measured by the effects of heat on aggregation, but the hydroxylated form requires much higher temperatures to undergo the structural alteration. Thus, an increase in hydrophilicity associated with hydroxylation is believed to cause changes leading to an unwinding of chains and to decreases in elasticity. Hydrophilicity increases when elastin becomes partly oxidized, and these changes may contribute to the loss of elastic recoil in elastic fibers of the lung in emphysema.[105]

A mechanism by which hydroxylation of lysines in collagen can influence structure is by leading to a stable cross-linked derivative.[106] The aldehyde derived from hydroxylysine can first make a Schiff base with an amino group of a lysyl residue followed by a rearrangement of the double bond to generate a stable keto form (Figure 1.7 of Chapter 1). The aldehyde of unhydroxylated lysine cannot make this stable structure.[107]

Myristoylation

Bound fatty acids can influence associative interactions but the exact function of the modification likely depends upon the type of group being introduced and its placement in the molecule. Myristoylation occurs at the amino terminus on a glycine residue (Chapter 4). To determine whether N-myristoylation could influence the protein stability of cAMP-dependent protein kinase, recombinant forms were expressed in *E. coli* with and without this modification.[108] Heat denaturation and sensitivity to a nonionic detergent (Triton x-100) showed that the myristoyl group in the catalytic subunit or in the reconstituted holoenzyme enhances stability but not enzymatic activity. The surface region near the amino terminus is quite hydrophobic and x-ray crystallography of the catalytic subunit shows that α-helical structure is initiated at residue 15.[109] It was suggested that the myristoyl group might serve to tether the first portion 14 amino acid to the surface of the large lobe of the catalytic subunit.[108] Exactly how the modification affects interactions and stability is not yet known.

Aspartyl Isomerization

A change in the linkage of amino acids in the polypeptide backbone would be expected to alter the folding and properties of a protein molecule. Such a change in linkage has been found in peptide bonds bearing an aspartic acid or asparagine linked to glycine or other amino acids with small side chains. In these cases, it is believed a peptide bond is formed between the β-carboxyl function and the α-amino group of the next amino acid (Chapter 3). Whether such modifications may be involved in the formation of abnormal proteins *in vivo* is of great interest. An evaluation has been made of aspartyl isomerization in β-amyloid proteins isolated from individuals who died of Alzheimer's disease.[110] Analyses of βA protein from brain parenchyma and leptomeningeal microvasculature show that modifications occur at position 1 and 7 in the chain with the main change being the formation of an isoaspartate linkage. As this modification is thought to influence proteoyltic processing, proteins with such modifications may accumulate in the β-amyloid deposits associated with the disease. But the relationships of these changes to Alzheimer's disease is not understood presently.

FLEXIBILITY AND FUNCTION

A modification of a protein can introduce to the structure a flexible side chain that is essential for the function of the protein. Or a protein may contain a flexible peptide segment, later removed, that can form a defined structure necessary for a biological process. An example of the latter is found in the involvement of signal peptides in the translocation of proteins across cell membranes.

A signal peptide consists of a hydrophilic segment (1-5 residues) with a net positive charge followed by a hydrophobic core (7-15 residues). At the end of the signal peptide sequence a hydrophilic region, the cleavage site, is found. The peptide is released by cleavage at this site when the process of translocation is finished.

Signal peptides do not appear to have well defined structures in aqueous solution, but they can form organized secondary structures in nonpolar environments. One model based on stereochemical analysis suggests that the signal peptide region adopts the structure of an α-helix as it encounters the phospholipid of the lipid bilayer. An α-helix allows for the formation of a compact structure with the complementary phosphorylated region of the phospholipid. Neutralization of the positive charge of signal peptide and negative charge of the phosphate could be an early step in the translocation of the protein across the membrane (Figure 5.11).[111]

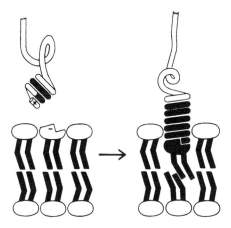

Figure 5.11. Scheme of the molecular mechanism of initiation of protein translocation across the membrane. Black color denotes hydrophobic regions of the molecule, white color-hydrophillic regions. Reprinted with permission from *Journal of Biomolecular Structure and Dynamics* 9:143 (1991). Copyright 1991, Adenine Press, Inc.

Flexibility of structures and effects on function due to added groups is seen as a result of the attachment of specific coenzymes to proteins. Some common features are shared by covalently bound lipoic acid, biotin, and pantetheine. Figure 5.12 shows their structures and modes of attachment to side chain groups of proteins. All three structures serve as "swinging arms" and facilitate catalytic processes in multienzyme complexes.

Acetyl CoA carboxylase from prokaryotes contains three proteins: (1) biotin carboxylase, (2) a transcarboxylase, and (3) a protein containing the bound biotin. The flexible side chain containing the bound biotin can shuttle between the active sites of the two enzymes and effect CO_2 fixation and its transfer to acetyl CoA. Similarly, phosphopanthetheine bound to a fatty acid synthase, a complex with a mass of 2.3 million daltons, serves to carry acyl CoA derivatives from the active centers of the various enzymes involved in the synthesis of fatty acid.

α-keto acid dehydrogenases catalyzing the general reaction exist
$$RCOCO_2^- + CoASH + NAD^+ - RCOSCoA + CO_2 + NADH$$
for reactions with pyruvate, a-ketoglutarate, and branched chain a-keto acids of amino acids, valine, leucine, and isoleucine. Each reaction is carried out by a multienzyme complex in which the dihydrolipoamide acyl transferase is present in the core of the enzyme. The bound lipoic acid of the enzyme can swing between the active sites as pictured in Figure 5.13 for the pyruvate dehydrogenase complex.[112]

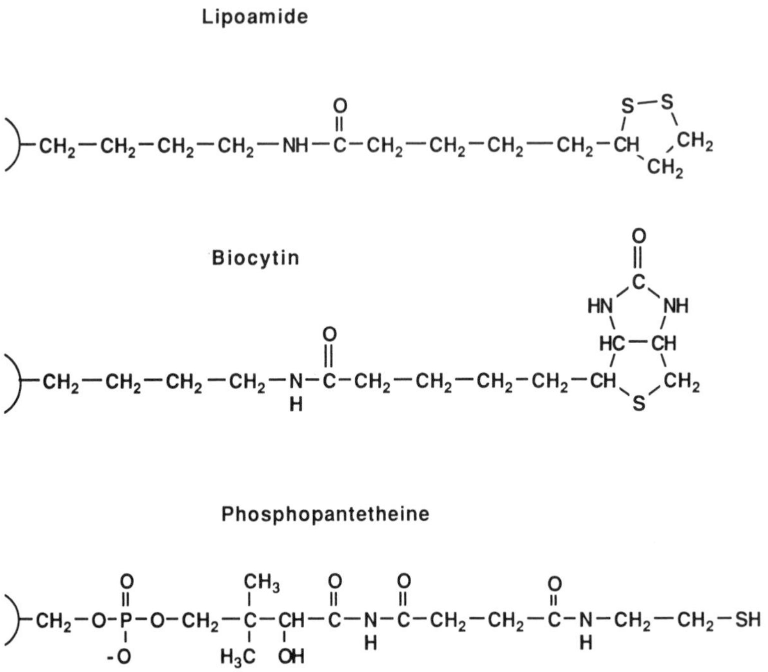

Lipoamide

$$\rangle\!\!-CH_2-CH_2-CH_2-CH_2-NH-\overset{\overset{\textstyle O}{\|}}{C}-CH_2-CH_2-CH_2-CH_2-\underset{\underset{\textstyle CH_2}{|}}{CH}\overset{\overset{\textstyle S-S}{\diagup\quad\diagdown}}{}CH_2$$

Biocytin

$$\rangle\!\!-CH_2-CH_2-CH_2-CH_2-\underset{\underset{\textstyle H}{|}}{N}-\overset{\overset{\textstyle O}{\|}}{C}-CH_2-CH_2-CH_2-CH_2-CH\quad CH_2$$

Phosphopantetheine

$$\rangle\!\!-CH_2-O-\overset{\overset{\textstyle O}{\|}}{\underset{\underset{\textstyle -O}{|}}{P}}-O-CH_2-\overset{\overset{\textstyle CH_3}{|}}{\underset{\underset{\textstyle H_3C}{|}}{C}}-\overset{\overset{\textstyle }{|}}{\underset{\underset{\textstyle OH}{|}}{CH}}-\overset{\overset{\textstyle O}{\|}}{C}-\underset{\underset{\textstyle H}{|}}{N}-\overset{\overset{\textstyle O}{\|}}{C}-CH_2-CH_2-\overset{\overset{\textstyle O}{\|}}{C}-\underset{\underset{\textstyle H}{|}}{N}-CH_2-CH_2-SH$$

Figure 5.12. Structures of protein-bound coenzymes .

The biotin-flexible arm has a reach of approximately 14Å, and physical studies show that bound lipoyl-lysine can rotate with a correlation time of 10^{-9} sec., which is consistent with its migration between sites during the catalytic events.[113] But to explain how the coenzyme can effectively reach these sites in the multienzyme complexes, it is necessary to invoke a movement of a part of the dihydrolipoamide acyltransferase, the lipoyl domain, and the swinging arm of the lipoyl-lysine. Physical studies, including the pioneering studies of the Reed laboratory using electron microscopy,[114,115] 1-NMR spectroscopy,[116] fluorescence energy transfer measurements,[117] and transient dichroism measurements[118] show how flexibility of the protein and of the bound coenzyme influence function.[119]

Myosin II filaments from *Acanthemobes castellani* can be phosphorylated on the nonhelical tail piece, and this phosphorylation causes inactivation of the actin activated Mg^{+2} ATPase associated with the N-terminal globular domain of myosin.[120,121] How modification at one end of the molecule can have such a profound effect at the opposite end is of much interest. In the rod-like section of myosin, a hinge occurs about 40 percent down the chain from the globular head.[122] One model suggests

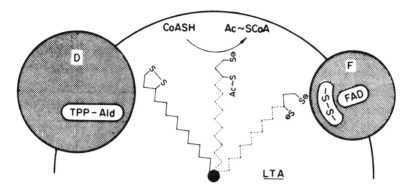

Figure 5.13. A schematic representation of the possible rotation of a lipoyllysyl moiety between a-hydroxyethylthiamine pyrophosphate (TPP-Ald) bound to pyruvate dehydrogenase (D), the site for acetyl transfer to CoA, and the reactive disulfide of the flavoprotein (F). The lipoyllysyl moiety is an integral part of dihydrolipoyl transacetylase (LTA). Reprinted with permission from *Accounts of Chemical Research* 7:40 (1974). Copyright 1974, American Chemistry Society.

that the tail region can be spacially adjacent to the hinge region. The tail piece contains 5 arginine, 2 glutamic acid, and 3 serine residues, and phosphorylation of the tail piece causes a change in the net charge of this region. Phosphorylation of a single residue, serine1489, in a mysoin derivative is sufficient to cause inactivation of filamentous myosin.[123] Modification could influence structure of the hinge region, which in turn could change the orientation of the N-terminal region and its interaction with actin which promotes ATPase activity.[124] To study the molecular dimensions of myosin and structural alterations induced by phosphorylation, electrical birefingence measurements were made.[125,126] When the electric field was turned off, relaxation of the oriented molecules occurred, and information was obtained about the overall structure and some internal motions. The basic structures of phosphorylated and dephosphorylated myosin filaments are thought to be very similar, but the dephosphorylated filament is believed to be fifty-fold stiffer. Hence, it was suggested that function is changed by altering the flexibility of the hinge region by phosphorylation.[126]

MODIFICATION CAN CHANGE LIGAND BINDING

Two general mechanisms exist by which modification can influence ligand binding. First, the modified residue may itself participate directly in an interaction with the ligand; for example, a phosphoryl group may provide a site for a metal ion. Second, the modified residue may not be directly involved in the interaction but cause a conformational change in the protein, which either enhances or diminishes ligand binding.

The Modified Group is Directly Involved in Ligand Binding

Phosphorylation

The phosphoryl group can provide a site for metal attachment. Phosvitin, a protein with an Mw of approximately 34,000 found in egg yolk, contains up to 10 percent phosphorous on a weight basis. The protein contains a great percentage of seryl residues, and up to 80 percent of these may be phosphorylated.[127] Segments of the protein may have as many as eight residues phosphorylated in a row. Ca^{+2} and Mg^{+2} bind with near stoichiometric equivalence to the phosphorylated residues, and Fe^{+3} may bind in a ratio of 0.5:1.0. These facts implicate a direct role for the phosphate group in binding, and spectroscopic studies based on the magnetic properties of Fe^{+3} provide substantive evidence for an interaction of the metal ion with the oxygen atoms of the phosphoryl groups.[128] The binding is approximately 14 orders of magnitude tighter than with Ca^{+2} or Mg^{+2} ($Ka \sim 10^4$ M). The interaction with Fe^{+3} depends upon the degree of phosphorylation of the protein, and different types of complexes may be formed but their exact structures have not been elucidated.[129] Although the exact biological role for phosvitin is not proved, it carries a high content of metal ion in egg yolk. Further studies on binding and of the relation of phosphorylation to the release of bound metal may provide insight into its biological role.

Carboxylation

It was recognized early that an abnormal form of prothrombin is present in serum of animals that received vitamin K antagonists, and that this form of prothrombin did not show the same interactions with calcium as did natural prothrombin.[130,131] The absence and presence of γ-carboxyglutamate[132,133] in abnormal and natural prothrombin, respectively, can account for these differences. But is γ-carboxyglutamate involved directly in metal ion binding? If so, how does this involvement relate to interactions of prothrombin with phospholipid surfaces and the activation of prothrombin by activated Factor X in the intrinsic pathway of blood coagulation?

γ-carboxyglutamate(Gla) is found in a highly conserved amino terminal region of various blood-clotting proteins, for example, prothrombin, factors VII, IX, and X, and proteins C, S, and Z. From 9-12 Gla residues are present, and these proteins are known to contain several binding

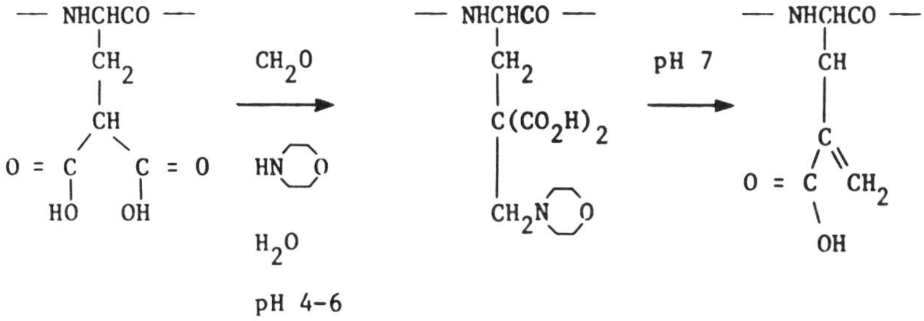

— NHCHCO —   — NHCHCO —   — NHCHCO —
    | | |
   CH$_2$ CH$_2$O   CH$_2$ pH 7   CH
    | → | |
   CH   C(CO$_2$H)$_2$ C
  / \   HN( )O | / \\
O = C C = O   H$_2$O   CH$_2$N( )O   O = C CH$_2$
  | | \
 HO OH   pH 4-6   OH

    1   2   3

Figure 5.14. Reaction of γ-carboxyglutamate in proteins with morpholine. Reprinted with permission from *Analytical Biochemistry* 139:82 (1984). Copyright 1984, Academic Press.

sites for multivalent cations. Human and bovine prothrombin have 7 and 10 binding sites for calcium, respectively.[134] The near equivalence of metal binding sites and γ-carboxyglutamate suggests a direct interaction between them, and the little binding in chemically modified prothrombin, in which γ-methylene glutamyl residues replace γ-carboxyglutamate, provides further evidence for its involvement in metal ion binding. The chemical change is accomplished by treating the protein with morpholine and formaldehyde as depicted in Figure 5.14.[135]

Valuable information about properties of these metal binding sites have been derived by physical and chemical studies of the proteins and model compounds. Binding of calcium measured by various methods, for example, fluorescence,[136] circular dichroism,[137] NMR spectroscopy,[138] and equilibrium dialysis show that there are several high-affinity binding sites.[139] These binding sites are believed to be involved in important conformational changes in the amino terminal region that are promoted by the binding of calcium. A second class of sites is thought to serve as a bridge between the protein and the phospholipid membrane surface. Particularly informative in regard to how this interaction might occur was obtained from the spectroscopic anlaysis of bound lanthanides to γ-carboxyglutamate and malonate derivatives.[33] A major conclusion to be drawn from these findings is that half of the coordination sphere of the metal ion is used up in binding two carboxylate groups, which would make the second half available for other ligands, for example, the anionic components of phospholipid. Studies of the dissociation of Protein S and other γ-carboxyglutamate-containing proteins from membranes by high calcium supports this view of protein-membrane bridging.[140]

The results of x-ray crystallography of Ca^{+2} prothrombin fragment 1 confirm some of the early concepts about metal ion binding and show the nature of certain complexes.[141] The Gla domain resembles a discoid surface with the γ-carboxyglutamate residues exposed to solvent. The paired residues, 7-8, 20-21, and 26-27, are oriented for participation in phospholipid binding. Solution conformations evaluated by molecular dynamics simulations show specific calcium-Gla interactions.[142]

Glycosylation

Fibrinogen, a glycoprotein, contains high- and low-affinity binding sites for calcium. In blood coagulation, it is known that calcium ions influence fibrin polymerization, but the mechanism is not fully understood. The low-affinity binding sites with Kd of mM value are thought to be involved, because effective polymerization occurs with calcium in the mM range, the physiological concentrations of free calcium. Sialic acid bound to the biantennary carbohydrate chains seems to constitute a low affinity binding site. Early work established the nature of the interaction of sialic acid with calcium.[143] More recent studies show that the binding interactions of sialofibrinogen are consistent with interactions of calcium with free sialic acid and that asialofibrinogen does not contain the low affinity sites.[144] Thus, the negative charge of the bound sialic acid is believed to hinder associative processes involving fibrin monomers. Occupancy by calcium neutralizes the charge facilitating new protein-protein interactions.

Sialic acid in glycophorin in erythrocytes has a role in the binding of viruses to receptors. Chemical amidation of its carboxyl group, which removes the negative charge, blocks association of encephalomyocarditis virus. A direct role for sialic acid in attachment is suggested.[145]

The Modified Group is not Directly Involved in Ligand Binding

Glycosylation

The Mr 46,000 mannose 6-phosphate receptor functions to target newly synthesized lysosomal enzymes into lysosomes.[146] Four N-glycosylated segments, two that contain complex-type oligosaccharides and two that contain high mannose oligosaccaharides, are present in the receptor. Site-directed mutagenesis studies showed that mutants deleted at three of the sites bound to a phosphomannan-Sepharose affinity matrix with binding characteristics of the wild type receptor. Mutation at asparagine 113, a mannose-binding site, to threonine results in loss of binding activity. But binding changes are attributed to substitution of asparagine for threonine, not loss of glycosylation. The nonglycosylated forms of the receptor, however, are less stable to freeze-thawing and lose their ability to form dimers. Hence, the oligosaccharides are not necessary for binding but are important for stabilization of the high affinity conformation.[147]

It seems oligosaccharides N-linked to a specific asparagine, asn_{238}, in human corticosteroid binding protein are essential for high affinity binding of cortisol.[148] This conclusion was reached on the basis of Scatchard binding curves of proteins obtained by site-directed mutagenesis. These mutant forms were made to block specific N-glycosylation reactions. Alteration of asn_{238} to gln blocked glycosylation and affected binding. Based on a related structure in α1-antitrypsin,[149] it is presumed that the oligosaccharide interacts with a tryptophan residue in the protein, creating the steroid binding site. It was not established whether oligosaccharide, in this case, is part of the binding site for steroid.

Phosphorylation

The phosphorylation of a single seryl residue in glycogen phosphorylase in each polypeptide chain of the dimeric form changes the property of the protein so that its affinity for the allosteric activator, AMP, is increased by nearly two orders of magnitude.[150] Clearly, it would not be expected that a direct interaction of the phosphoryl seryl residue with the nucleotide could explain these large differences in affinity for the dephospho- and phospho-forms of the enzyme. But what are the interactions? X-ray crystallographic studies provide the answers. The main points are the following: (1) AMP and the phosphorylated site, although bound in between the α2-helix of one subunit and a cap of the second subunit, are at distinct regions about 12Å apart; (2) the conformation of the bound nucleotide differs in the two forms of phosphorylase. The anticonformation of the nucleotide is bound to both forms, but the ribose ring changes from C2' endo- to the C3' endo-conformation in the phosphorylated form. This change results from a change in torsional angles of the C4' and C5' bonds. In phosphorylase b, the conformation is gauche and in phosphorylase a, trans; (3) new and altered contacts of the protein occur with the bound nucleotide. Phosphorylation causes the α2-helix to contract and promotes new side chain interactions. The altered conformation of the nucleotide binding site and a changed nucleotide conformation allows for closer stacking interaction between the adenine ring and $tyrosine_{75}$. New hydrogen bonds occur with the ribosyl group and with the side chains of Asn_{44} and Gln_{71}. More solvent-accessible area is buried in the nucleotide complex of phosphorylase a than of phosphorylase b. All these factors contribute then to the the tighter interaction of the allosteric activator with the phosphorylated form. Interestingly, the changes in protein structure induced by phosphorylation weaken the binding of the allosteric inhibitor, glucose-6-P, which shares a common phosphoryl binding site with AMP. Figure 5.15 illustrates the interactions with the two forms of the protein.[151]

The activated R state of phosphorylase b resembles closely the structure of the R state of phosphorylase a. Thus, phosphorylation provides a structure that ensures tight and specific binding of AMP. IMP does not bind as tightly as AMP due to the lack of H-bonding between the N-1 of the purine with $Asn_{44'}$.[152]

Interactions in Receptors

There is considerable interest in understanding (1) how the binding of hormones or growth factors to a specific receptor on the surface of a cell can influence an enzymatic reaction of a distant portion of its own receptor in the cytosol, and (2) what effect the enzymatic reaction has on processes at the receptor site. Consider the action of the insulin receptor. The protein is encoded by a single gene and is post-translationally modified and proteolytically processed to form a transmembrane unit of two polypeptide chains linked by disulfide bonds. The α-subunits comprising the insulin binding site have a molecular weight of 135kDa. The β-subunits of 95kDa possess tyrosyl kinase activity, and it is accepted that insulin binding promotes autophosphorylation of the β-chains activating subsequent phosphorylation events as a part of insulin action. But how are the processes of insulin binding and autophosphorylation linked, and does phosphorylation influence insulin binding?

One approach taken to provide an answer to the first question was to use gel filtration, electrophoresis under nondenaturing conditions, and ultracentrifugation to analyze the structural alterations of different forms of the purified insulin receptor in response to insulin.[153] The determination of the Stokes radius and sedimentation coefficients showed that the $(\alpha\beta)_2$ receptor undergoes a significant conformational change upon insulin binding. It is assumed that autophosphorylation occurs in this transformed state. Upon dissociation of insulin from the phosphorylated receptor, the conformational change is reversed. Insulin has little effect on the Stokes radius and sedimentation coefficient of the $(\alpha\beta)$ form of the receptor, and it is suggested that the association of the two $(\alpha\beta)$ receptors are required for enhanced autophosphorylation. Binding of the chains brought about by insulin could promote a transphosphorylation event.

Another approach taken to evaluate insulin activation, associative interactions of subunits, and phosphorylation was to determine whether phosphorylation of the subunits undergo cis- or transphosphorylation

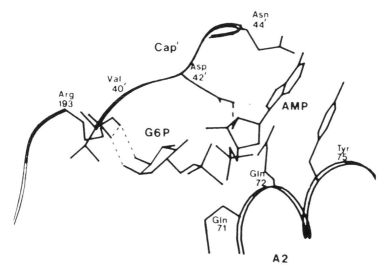

Figure 5.15. Structure of the AMP binding site in phosphorylase *b*. Reprinted by permission for *Nature* 336::221 (1988); Copyright (c) (1988) Macmillan Magazines Limited.

reactions.[154] Truncated forms of the receptor may be derived by trypsin treatment of Fao hepatoma cells. The modified receptor contains an intact β-subunit and is connected to a small remaining fragment of the α-subunit by an α-disulfide bond.[155,156] The truncated receptor is not responsive to insulin, but it can undergo phosphorylation. Kinetic studies show that phosphorylation is independent of receptor concentration. The studies show that reactions can occur by an intramolecular process; therefore, association of subunits is not required absolutely for autophosphorylation. Insulin stimulation of the αβ heterodimers is, however, dependent upon receptor concentration. The linear relationship of phosphorylation to the square of the receptor concentration suggests a bimolecular trans event. The initial rate of phosphorylation of the holoreceptor is 2-3-fold higher than the truncated form. Effective transmission of the signal requires intact αβ-subunits and their associative interaction. The process of insulin binding and activation of the receptor may involve initial intramolecular autophosphorylation reactions that serve to trigger other phosphorylation events, including the intermolecular reactions.

Photoactivation of an insulin analog containing benzoylphenylanine can provide high efficiency cross-linking of the insulin derviative to its receptor. The cross-linked receptor containing a single insulin derivative linked to each $\alpha_2\beta_2$ receptor is able to autophosphorylate.[157] This cross-linked receptor was used to determine whether phosphorylation could only occur on the subunit pair containing the bound insulin derivative (cis phosphorylation), on the subunit pair without insulin, or on both pairs. The results showed that the β-subunits of occupied and unoccupied pairs were phosphorylated, but the form without insulin was phosphorylated more effectively. The sites of phosphorylation are believed to be the same in both cases. One model to explain the asymmetric phosphorylation proposes that the insulin bound unit phosphorylates the unbound form (trans phosphorylation) and then the autophosphorylated form acts to phosphorylate the β-subunit of the αβ containing the insulin derivative.[158]

It seems the process of insulin binding and activation of the receptor involves initial intramolecular autophosphorylation reactions which serve to trigger other phosphorylation events, including the intermolecular reactions.

The β-adrenergic receptor is a transmembrane protein. Seven segments of the chain intertwine through the membrane, and it is believed that the binding site for hormone exists in the transmembrane region. Disulfide bonding is important for binding of the receptor. The "competitive" inhibitory effect of dithiothreitol, DTT, a reducing agent, on ligand binding to the receptor site could be interpreted to mean the DTT sensitive bonds are directly involved in the binding site.[159] And there are cysteinyl residues in transmembrane regions that could be involved in disulfide bonding, which is critical for ligand binding. Nevertheless, a "competitive "pattern does not prove that one distinct site is involved. It only means, as in enzyme kinetics, that processes are mutually exclusive. Chemical modification at one site could change conformation and influence the characteristics of a distant ligand binding site. Site-directed mutagenesis suggests that the latter explanation is correct and that extracellular disulfide is important for ligand binding. A similar involvement for disulfide in rhodopsin is suggested for binding of 11-cis-retinal.[160]

REFERENCES

1. Nozaki, Y., and Tanford, C., *J. Biol. Chem.* 246, 2211–2217 (1974).
2. Black, S.D., and Mould, D.R., *Anal. Biochem.* 193, 72–82 (1991).
3. Rekker, R.F., *The Hydrophobic Fragmental Constant*, Amsterdam: Elsevier, (1977).
4. Edde, B., Rossier, J., Le Caer, J-P., Desbruyeres, E., Gros, F., and Denoulet, P., *Science* 247, 83–84 (1990).
5. Baxter, C.S., and Byvoet, P., *Biochem. and Biophys Res Commun.* 64, 514–518 (1975).

6. Dever, T.E., Costello, C.E., Owens, C.L., Rosenberrt, T.L., and Merrick, W.C., *J. Biol. Chem.* 264, 20518–20525 (1989).

7. Hiatt, W.R., Garcia, R., Merrick, W.C., and Sypherd, P. S., *Proc. Natl. Acad. Sci.* 79, 3433–3477 (1982).

8. Byvoet, P., unpublished results.

9. Paik, W.K., and Kim, S., *J. Theor. Biol.* 155, 335–342 (1992).

10. Lestourgeon, W.M., Beyer, A.L., Christensen, M.E., Walker, B.W., Poupore, S.M., and Daniels, L.P., *Cold Spring Harbor Quant. Biol.* 42, 885–898 (1977).

11. Christensen, M.E., and Fuxa, K. P., *Biochem Biophys. Res. Commun.* 155, 1278–1283 (1988).

12. Calnan, B.J., Tidor, B., Biancalana, S., Hudson, D., and Frankel, A.D., *Science* 252, 1167–1171 (1991).

13. Paik, W.K., Park, K.W., Frost, B.F., and Kim, S., in *Advances in Post-Translational Modification of Proteins and Aging* (Edited by Zappia, V., Galletti, P., Porta, R., and Wold, F.) New York: Plenum Press 231, 317–325 (1988).

14. Hugne, E., Ph.D. Dissertation, U. of Calgary (1989).

15. Vogel, H.J., *Methods in Enzymol.* 177, 263–282 (1989).

16. Vogel, H.J., in "Phosphorous-31NMR, Principles and Applications" (D.G. Gorenstein, ed.), p.104–154, Academic Press, New York, (1984).

17. Vogel, H.J., and Bridger, W.A., *Biochemistry* 21, 5825–5831 (1982).

18. Takrama, J.F., and Graves, D.J., *Biochim Biophys. Acta* 1077, 371–378 (1991).

19. Shrager, R.I., Cohen, J.S., Heller, S.R., Sachs, D.H., and Schechter, A.N., *Biochem.* 11, 541–547 (1972).

20. Sanchez-Ruiz, J.M., and Martinez-Carrion, M., *Biochem.* 27, 3338–3342 (1988).

21. Barford, D., and Johnson, L.N., *Nature* 340, 609–616 (1989).

22. Barford, D., Hu, S.-H, and Johnson, L.N., *J. Mol. Biol.* 218, 233–260 (1991).

23. Dean, A.M., Lee, M.H.I., and Koshland, D.E., Jr., *J. Biol. Chem.* 264, 20482– (1989).

24. Hurley, J.H., Dean, A.M., Sohl, J.L., Koshland, D.E., Jr., and Stroud, R.M., *Science* 249, 1012–1016 (1990).

25. Dean, A.M., and Koshland, D.E., Jr., *Science* 249, 1044–1046 (1990).

26. Fries, D.C., Sunderalingam, M., *Acta Cryst.* B27, 401–410 (1971).

27. Jones, J.A., Wood, A., and Cushley, W., *Bioscience Reports* 6, 265–273 (1986).

28. Stone, S.R., Dennis, S., and Hofsteenge, J., *Biochemistry* 28, 6857–6863 (1989).

29. Haruyama, H., and Wuthrich, K., *Biochemistry* 28, 4301–4312 (1986).

30. Grutter, M.G., Priestle, J.P., Rahuel, J., Grossenbacher, H., Bode, W., Hofsteenge, J., and Stone, S.R., *EMBO J.* 9, 2361–2365 (1990).

31. Braun, P.J., Dennis, S., Hofsteenge, J., and Stone, S.R., *Biochemistry* 27, 6517–6522 (1988).

32. Hofsteenge, J., Stone, S.R., Donnella-Deana, A., and Pinna, L.A., *Eur. J. Biochem.* 188, 55–59 (1990).

33. Sperling, B., Furie, B.C., Blumenstein, M., Keyt, B., and Furie, B., *J. Biol. Chem.* 253, 3898–3906 (1978).

34. Yokomori, Y., and Hodgson, D.J., *Int. J. Peptide Protein Res.* 31, 289–298 (1988).
35. Gray, A.L., Hoke, R.A., Deerfield, D.W.II, and Hiskey, R.G., *J. Org. Chem.* 50, 2189–2191 (1985).
36. Csordas, A., *Biochem. J.* 265, 23–38 (1990).
37. Ferber, S., and Ciechanover, A., *Nature* 326, 808–811 (1987).
38. Elias, S., and Ciechanover, A., *J. Biol. Chem.* 265, 15511–15517 (1990).
39. Bordier, C., *J. Biol. Chem.* 256, 1604–1607 (1981).
40. Beck, L.A., Hosick, T.J., and Sinensky, M., *J. Cell Biology* 107, 1307–1316 (1988).
41. Carr, S.A., Biemann, K., Shoji, S., Parmelee, and Titani, K., *Proc. Natl. Acad. Sci. USA* 79, 6128–6131 (1982).
42. Garber, E.A., Cross, F.R., and Hanafusa, H., *Mol. and Cell Biol.* 5, 2781–2788 (1985).
43. Buss, J.E., and Sefton, B.M., *J. Virol.* 53, 7–12 (1985).
44 Resh, M.D., *Cell* 58, 281–286 (1989).
45. Heuckeroth, R.O., Glaser, L., and Gordon, J.I., *Proc. Natl. Acad. Sci.* 85, 8795–8799 (1988).
46. Johnson, D.R., Cox, A.D., Solski, P.A., Devadas, B., Adams, S.P., Leimburger, R.M., Heuckeroth, R.O., Buss, J.E., and Gordon, J.I., *Proc. Natl. Acad. Sci.* 87, 8511–8515 (1990).
47. Schmidt, M.F.G., and Lambrecht, B., *J. Gen. Virol.* 66, 2635–2647 (1985).
48. O'Brien, P.J., St.Jules, R.S., Reddy, T.S., Bazan, N.G., and Zatz, M., *J. Biol. Chem.* 262, 5210–5215 (1987).
49. Bizzozero, O.A., McGarry, J.F., and Lees, M.B., *J. Biol. Chem.* 262, 2138–2145 (1987).
50. Towler, D.A., Gordon, J.I., Adams, S.P., and Glaser, L., *Ann. Rev. Biochem.* 57, 69–99 (1988).
51. Slomiany, A., Liau, Y.H., Carter, S.R., Newman, L.J., and Slomiany, B.L., *Biochem. Biophys. Res. Commun.* 132, 299–306 (1985).
52. O'Dowd, B.F., Hnatowich, M., Caron, M.G., Lefkowitz, R.J., and Bouvier, M., *J. Biol. Chem.* 264, 7564–7569 (1989).
53. Ovchinnikov, Y.A., Abdulaev, N.G., and Bogachuk, A.S., *FEBS Lett.* 230, 1–5 (1988).
54. Hancock, J.F., Magee, A.I., Childs, J.E., and Marshall, C.J., *Cell* 57, 1167–1177 (1989).
55. Wolda, S.L., and Glomset, J.A., *J. Biol. Chem.* 263, 5997–6000 (1988).
56. Farnsworth, C.C., Wolda, S.L., Gelb, M.H., and Glomset, J.A., *J. Biol. Chem.* 264, 20422–20429 (1989).
57. Sakagami, Y., Yoshida, M., Isogai, A., and Suzuki, A., *Science* 212, 1525–1526 (1981).
58. Anderegg, R.J., Betz, R., Carr, S.A., Crabb, J.W., and Duntze, W., *J. Biol. Chem.* 263, 18236–18240 (1988).
59. Rilling, H.C., Breunger, E., Epstein, W.W., and Crain, P.F., *Science* 247, 318–320 (1990).
60. Farnsworth, C.C., Gelb, M.H., and Glomset, J.A., *Science* 247, 320–322 (1990).

61. Yamane, H.K., Farnsworth, C.C., Xie, H., Howald, W., Fung, B.K-K., Clarke, S., Gelb, M.H., and Glomset, J.A., *Proc. Natl. Acad. Sci.* 87, 5868–5872 (1990).
62. Mumby, S.M., Casey, P.J., Gilman, A.G., Gutowski, S., and Sternweiss, P.C., *Proc. Natl. Acad. Sci.* 87, 5873–5877 (1990).
63. Worman, H.J., Yuan, J., Blobel, G., and Georgatos, S.D., *Proc. Natl. Acad. Sci.* 85, 8531–8534 (1988).
64. Senior, and Gerace, L., *J. Cell Biol.* 107, 2029–2036 (1988).
65. Willumsen, B.M., Norris, K., Papergo, A.G., Hubbert, N.L., and Lowry, D.R., *EMBO J.* 3, 2581–2585 (1984).
66. Barbacid, M., *Ann. Rev. Biochem.* 56, 779–827 (1987).
67. Hancock, J.F., Cadwallader, K., and Marshall, C.J., *The EMBO Journal* 10, 641–646 (1991).
68. Wetzel, R., *Trends in Biochem. Sci.* 12, 478–482 (1987).
69. Srinivasan, N., Sowdhami, R., Ramakrishnan, C., and Balaram, P., *Int. J. Peptide Protein Res.* 36, 147–155 (1990).
70. Richardson, J.S., *Adv. in Protein Chem.* 34, 167–339 (1981).
71. Creighton, T.E., *Methods in Enzymol.* 107, 305–329 (1984).
72. Creighton, T.E., *J. Phys. Chem.* 89, 2452–2459 (1985).
73. Wearne, S.J., and Creighton, T.E., *Proteins* 5, 8–12 (1989).
74. Jentoft, N., *Trends in Biochem. Sci.* 15, 291–294 (1990).
75. Shogren, R., Gerken, T.A., and Jentoft, N., *Biochemistry* 28, 5525–5536 (1989).
76. Gerken, T.A., Butenhof, K.J., and Shogren, R., *Biochemistry* 28, 5536–5543 (1989).
77. Cyster, J.G., Shotton, D.M., and Williams, A.F., *EMBO J.* 10, 893–902 (1991).
78. Hollosi, M., Perczel, A., and Fasman, G.D., *Biopolymers* 29, 1549–1564 (1990).
79. Aleshin, A., Golubev, A., Firsov, L.M., and Honzatko, R.B., J. Biol. Chem. 267, 19291–19298 (1992).
80. Paulson, J.C., *Trends in Biochem Sci.* 14, 272–275 (1989).
81. Walsh, M.T., Watzlawick, H., Putnam, F.W., Schmid, K., and Brossmer, R., *Biochemistry* 29, 6250–6257 (1990).
82. Ng, D.T.W., Hiebert, S.W., and Lamb, R.A., *Mol. Cell Biology* 10, 1989–2001 (1990).
83. Riederer, M.A., and Hinnen, A., *J. of Bacteriology* 173, 3539–3546 (1991).
84. Dill, K., and Allerhand, A., *J. Biol. Chem.* 254, 4524–4531 (1979).
85. Berman, E., Walters, D.E., and Allerhand, A., *J. Biol. Chem.* 256, 3853–3857 (1981).
86. Goux, W.J., Perry, C., and James, T.L., *J. Biol. Chem.* 257, 1829–1835 (1982).
87. Brockbank, R.L., and Vogel, H.J., *Biochemistry* 29, 5574–5583 (1990).
88. Barford, D., and Johnson, L.N., *Protein Science* 1, 472–493 (1992).
89. Graves, D.J., and Wang, J.H., in *The Enzymes*, (Boyer, P.D., Ed.), New York: Academic Press, pp. 435–482, (1972).

90. Hisanaga, S., Gonda, Y., Inagaki, M., Ikai, A., and Hirokawa, N., *Cell Regul.* 1, 237–248 (1990).
91. Peter, M., Nakagawa, J., Doree, M., Labbe, J.-C., and Nigg, E.A., *Cell* 61, 591–602 (1990).
92. Ward, G.E., and Kirschner, M.W., *Cell* 61, 561–577 (1990).
93. Chou, Y-H., Bischoff, J.R., Beach, D., and Goldman, R.D., *Cell* 62, 1063–1071 (1990).
94. Hagestedt, T., Lichtenberg, B., Wille, H., Mandelkow, E.-M., and Mandelkow, E., *J. of Cell Biol.* 109, 1643–1651 (1989).
95. Yeagle, P.L., Frye, J., and Eckert, B.S., *Biochemistry* 29, 1508–1514 (1990).
96. Inagaki, M., Nishi, Y., Nishizawa, K., Mastuyama, M., and Sato, C., *Nature* 328, 648–652 (1987).
97. Inagaki, M., Gonda, Y., Matsuyama, M., Nishizawa, K., Nishi, Y., and Sato, C., *J. Biol. Chem.* 263, 5970–5978 (1988).
98. Inagaki, M., Takahara, H., Nishi, Y., Sugawara, K., and Sato, C., *J. Biol. Chem.* 264, 18119–18127 (1989).
99. Urry, D.W., *J. Protein Chem.* 7, 1–34 (1988).
100. Pattanaik, A., Gowda, D.C., and Urry, D.W., *Biochem. Biophys. Res. Commun.* 178, 539–545 (1991).
101. Urry, D.W., in *Protein Folding: Deciphering the Second Half of the Genetic Code*, (Lila Gierasch and Jonathan King, Eds.) American Association for the Advancement of Science, pp 63–71 (1990).
102. Urry, D.W., Chang, D-K., Zhang, H., and Prasad, K.U., *Biochem. Biophys. Res. Commun.* 153, 832–839 (1988).
103. Barone, L.M., Faris, B., Chipman, S.D., Toselli, P., Oakes, B.W., and Franzblau, C., *Biochem. Biophys. Acta* 840, 245–254 (1985).
104. Urry, D.W., Sugano, H., Prasad, K.U., Long, M.M., and Bhatnagar, S., *Biochem. Biophys. Res. Commun.* 90, 194–198 (1979).
105. Clark, J.G., Kuhn, C., and Mecham, R.P., *Int. Rev. Connect. Tissue Res.* 10, 249–331 (1983).
106. Bailey, A.J., Robins, S.P., and Balian, G., *Nature* 251, 105–109 (1974).
107. Kivirikko, K.I., and Myllyla, R. in *The Enzymology of Post-translational Modification of Proteins* 1, (Freedman, R.B., and Hawkins, H.C., Eds.), New York: Academic Press, 54–91 (1980).
108. Yonemoto, W., McGlone, M.L., and Taylor, S.S., *J. Biol.Chem.* 268, 2348–2352 (1993).
109. Knighton, D.R., Zheng, J., Ten Eyck, L.F., Ashford, V.A., Xuoang, N-h., Taylor, S.S., and Sowadowski, J.M., *Science* 253, 407–414 (1991).
110. Roher, A.E., Lowenson, J.D., Clarke, S., Wolkow, C., Wang, R., Cotter, R.J., Reardon, I.M., Zurcher-Neely, H.A., Heinrikson, R.L., Ball, M.J., and Greenberg, B.D., *J. Biol.Chem.* 268, 3072–3083 (1993).
111. Kajava, A.V., Bogdanov, M.V., and Nesmeyanova, M.A., *J. of Biomolecular Structure and Dynamics* 9, 143–157 (1991).
112. Reed, L.J., *Accounts of Chem. Res.* 7, 40–46 (1974).
113. Ambrose, M.C., and Perham, R.N., *Biochem. J.* 155, 429–432 (1976).
114. Oliver, R.M., and Reed, L.J., in *Electron Microscopy of Proteins* (Harris, J.R., ed.) 2, 1–48 (1982).

115. Reed, L.J., and Hackert, M.L., *J. Biol. Chem.* 265, 8971–8974 (1990).
116. Perham, R.N., Duckworth, H.W., and Roberts, G.C.K., *Nature* 292, 474–477 (1981).
117. Sheperd, G., and Hammes, G.G., *Biochemistry* 16, 5234–5241 (1977).
118. Harrison, J.P., Morrison, I.E.G., and Cherry, R.J., *Biochemistry* 29, 5596–5604 (1990).
119. Perham, R.N., *Biochemistry* 30, 8501–8512 (1991).
120. Cote, G.P., Collins, J.H., and Korn, E., *J. Biol. Chem.* 256, 12811–12816 (1981).
121. Collins, J.H., Cote, G.P., and Korn, E., *J. Biol. Chem.* 257, 4529–4534 (1982).
122. Kuznicki, J., Cote, G.P., Bowers, B., and Korn, E., *J. Biol. Chem.* 260, 1967–1972 (1985).
123. Sathyamoorthy, V., Atkinson, M.A.L., Bowers, B., and Korn, E., *Biochemistry* 29, 3793–3797 (1990).
124. Atkinson, M.A.L., Lambooy, P.K., and Korn, E.J., *J. Biol. Chem.* 262, 15809–15811 (1987).
125. Wijmenja, S.S., Atkinson, M.A.L., Rau, D., and Korn, E.D., *J. Biol. Chem.* 262, 15803–15808 (1987).
126. Rau, D.C., Ganguly, C., and Korn, E., *J. Biol. Chem.* 268, 4612–4621 (1993).
127. Taborsky, G., *Adv. Inorg. Biochem.* 5, 235–279 (1983).
128. Webb, J., Mulatani, J.S., Saltman, P., Beach, N.A., and Gray, H.B., *Biochemistry* 12, 1797–1802 (1973).
129. Grogan, J., and Taborsky, G., *J. Inorg. Biochem.* 29, 33–47 (1987).
130. Niehlin, J.E., and Ganrot, P.O., *Scand. J. Clin. Lab. Invest.* 22, 17–22 (1968).
131. Ganrot, P.O., and Niehlin, J.E., *Scand. J. Clin. Lab. Invest* 22, 23–28 (1968).
132. Stenflo, J., Ferlund, P., Egan, W., and Roepstorff, P., *Proc. Natl. Acad. Sci.* 71, 2730–2733 (1974).
133. Nelsesteun, G.L., Zytkovicz, T.H., and Howard, J.B., *J. Biol. Chem.* 249, 6347–6350 (1974).
134. Deerfield, D.W.II, OLson, D.L., Berkowitz, P., Koehler, K.A., Pedreson, L.G., and Hiskey, R.G., *Biochem. Biophys. Res. Commun.* 144, 520–527 (1987).
135. Wright, S.F., Bourne, C.D., Hoke, R.A., Koehler, K.A., and Hiskey, R.G., *Anal. Biochem.* 139, 82–90 (1984).
136. Nelsestuen, G.L., *J. Biol. Chem.* 251, 5648–5656 (1976).
137. Bloom, J.W., and Mann, K.G., *Biochemistry* 17, 4430–4438 (1978).
138. Furie, B.C., Blumenstein, M., and Furie, B., *J. Biol. Chem.* 254, 1251–12530 (1979).
139. Deerfield, D.W.II, Berkowitz, P., Olson, D.L., Wells, S., Hoke, R.A., Koehler, K.A., Pedersen, L.G., and Hiskey, R.H., *J. Biol. Chem.* 261, 4833–4839 (1986).
140. Schwalbe, R.A., Ryan, J., Stern, D.M., Kisiel, W., Dahlback, B., and Nelsestuen, G.L., *J. Biol. Chem.* 264, 20288–20296 (1989).

141. Soriano-Garcia, M., Park, C.H., Tulinsky, A., Ravichandran, K.G., and Jankun-Skrzypczak, E., *Biochemistry* 28, 6805–6810 (1989).

142. Charifson, P.S., Darden, T., Tulinsky, A., Hughey, J.L., Hiskey, R.G., and Pedersen, L.G., *Proc. Natl. Acad. Sci.* 88, 424–428 (1991).

143. Jaques, L.W., Brown, E.B., Barret, E.B., Barret, J.M., Brey, W.S.,Jr., and Weltner, W.,Jr., *J. Biol. Chem.* 252, 4533–4538 (1997).

144. Dang, C.H., Shin, C.K., Bell, W.R., Nagaswami, C., and Weisel, J.W., *J. Biol. Chem.* 264, 15104–15108 (1989).

145. Tavakkol, A., and Burness, A.T.H., *Biochemistry* 29, 10684–10690 (1990).

146. Stein, M., Zijderhand-Bleekemolen, J.E., Geuze, H., Hasilik, A., and von Figura, K., *EMBO J.* 6, 2677–2681 (1987).

147. Wendland, M., Waheed, B., Schmidt, B., Hille, A., Nagel, G., von Figura, K., and Pohlmann, R., *J. Biol. Chem.* 266, 4598–4604 (1991).

148. Avvakumov, G.V., Warmels-Rodenhiser, S., and Hammond, G.L., *J. Biol. Chem.* 268, 862–866 (1993).

149. Powell, L.M., and Pain, R.H.J., *J. Mol. Biol.* 224, 224–252 (1992).

150. Helmreich, E., Michaelides, M.C., and Cori, C.F., *Biochemistry* 6, 3695–3710 (1967).

151. Sprang, S.R., Acharya, K.R., Goldsmith, E.J., Stuart, D.I., Varvill, K., Fletterick, R.J., Madsen, N.B., and Johnson, L.N., *Nature* 336, 215–221 (1988).

152. Barford, D., Hu, S.-H., and Johnson, L.N., *J. Mol. Biol.* 218, 233–260 (1991).

153. Florke, R-R., Klein, H.W., and Reinauer, H., *Eur. Biochem J.* 191, 473–482 (1990).

154. Shoelson, S.E., Schnetzler-B.M., Pilch, P.F., and Kahn, C.R., *Biochemistry* 30, 7740–7746 (1991).

155. Shoelson, S.E., White, M.F., and Kahn, C.R., *J.Biol. Chem.* 263, 4852–4860 (1988).

156. Xu, Q.-X., Paxton, R.J., and Fujita-Yamaguchi, Y., *J. Biol. Chem.* 265, 18673–18681 (1990).

157. Shoelson, S.E., Lee, J., Lynch, C.S., Backer, J.M., and Pilch, P.F., *J. Biol. Chem.* 268, 4085–4098 (1993).

158. Lee, J., O'Hare, T., Pilch, P.F., and Shoelson, S.E., *J. Biol. Chem.* 268, 4092–4098 (1993).

159. Dohlman, H.G., Caron, M.C., DeBlasi, A., Frielle, T., and Lefkowitz, R.J., *Biochemistry* 29, 2335–2342 (1990).

160. Karnik, S.S., Sakmar, T.P., Chen, H.-B., and Khorana, H.G., *Proc. Natl. Acad. Sci.* 85, 8459–8463 (1988).

6

Inhibitors and Activators of Cellular Modification Reactions

A study of the enzymology of covalent modification reactions in purified systems provides information on how these reactions occur and are regulated. The studies of protein chemistry help us define factors influencing the reaction and how we can indentify the reaction products. But how can we use this information to learn more about these reactions in the intact cell? Consider the use of enzyme inhibitors and activators.

ENZYME INHIBITORS

General Principles for the Design of Inhibitors

Two general classes of inhibitors exist: reversible and irreversible. Both types can be used in cellular systems, but several factors need to be considered in using them. First, and most important, how specific is the inhibitor? For example, if the compound is an inhibitor of one enzyme, will it not inhibit other enzymes? If the selectivity isn't complete, is there a range of concentration where the inhibitor would work on one system but not the other. In this case, knowledge of the level of the compound in cells could help the investigator interpret the results. Second, the inhibitor should be able to reach its cellular target. For example, an inhibitor that could act on a process in the cytosol may prove ineffective in cell cultures because it cannot penetrate the cellular membrane. In some cases, the solubility of the compound may limit the applications.

Third, the compound should have a sufficient chemical and biological half-life to influence cellular reactions. An inhibitor may contain features making it unstable at neutral pH and 37° or susceptible to enzymatic modification and inactivation. But these chemical or enzymatic reactions may be slowed down if the inhibitor is bound tightly to its specific target.

One important use of inhibitors is to help identify what proteins are modified in a cell in response to a biological signal. In reversible modification, the extent of modification is due to a balance of activities of enzymes which catalyze the modification with those which catalyze the removal of the modifier. The effect of the inhibitor will depend on what step is being inhibited and what process is rate-limiting. For example, assume the forward reaction is slower than the reverse process. An inhibitor of the forward reaction may have little effect on the amount of the modified protein because once it is formed it is rapidly turned over. On the other hand, an inhibitor of the reverse process could change the steady state concentration of the modified protein if this step now became rate-limiting. In fact, the use of an inhibitor might even lead to the identification of new target proteins if these modified proteins were turned over rapidly in the absence of the inhibitor. A specific inhibitor for one direction of a reversible process can provide insight into how the modification is linked to a biological response. A second and specific inhibitor of the reverse process could help substantiate the view that this cycle of modification is important.

Slow-Acting Inhibitors

An inhibitor binding avidly to its target protein may act rapidly or display a kinetic pattern described by the progress curves shown in Figure 6.1. In this case, the inhibitor may have little effect on the initial rate of the reaction, but at later stages the reaction rate will fall off as the inhibitor forms a tight complex with the enzyme. These effects are characteristics of slow-acting inhibitors.[1] An inhibitor resembling the transition state may also have these characteristics because the conformational change of the enzyme needed to accommodate tight binding of the transition state inhibitor can be rate limiting. Or an inhibitor may contain some structural features, for example, hydrophobic regions that do not bind at full strength until the enzyme has undergone a time-dependent conformational change. An important review of this subject has already appeared.[2]

A kinetic scheme that can account for many slow-acting inhibitors is shown in Figure 6.2. This scheme suggests that interaction of the inhibitor occurs in stages. After a portion of the inhibitor binds (forma-

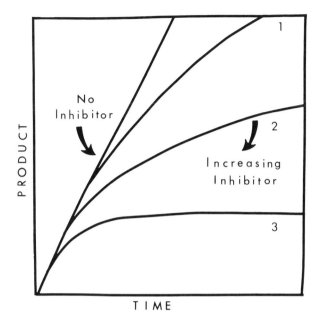

Figure 6.1. Progress curves for the action of a slow-acting inhibitor.

tion of an EI complex), a conformational change in the enzyme may occur to accommodate a more complete binding of the inhibitor (formation of the EI* complex). This latter stage may account for the slowness of the response to the inhibitor. If the ratio of k_5/k_6 is small, the $t_{1/2}$ for dissociation of the inhibitor may be seconds, minutes, hours, or even weeks. Thus, the biological half-life for this type of inhibition may allow for certain experiments in cells not possible with other inhibitors.

An inhibitor resembling the ground state may bind avidly to the enzyme, as do slow-acting inhibitors, but it may dissociate more quickly than some other inhibitors. Hence, knowledge obtained from the time course (Figure 6.1) and from other kinetic studies could lead to a

$$E + S \underset{k_2}{\overset{k_1}{\rightleftharpoons}} ES \overset{k}{\longrightarrow} E + P$$

$$E + I \underset{k_4}{\overset{k_3}{\rightleftharpoons}} EI \underset{k_6}{\overset{k_5}{\rightleftharpoons}} EI^*$$

Figure 6.2. Reaction scheme for slow-acting inhibitors.

rationale for the design of new inhibitory molecules. The results could influence how experiments are done in cellular systems, for example, how long cells should be incubated with the inhibitor before a growth factor or hormone is added, and what time course should be used to evaluate the effect of the inhibitor.

A specific inhibitor may contain two features: a part resembling the substrate (in the ground state or transition state), and in some cases a part that would allow for extensive contacts with the protein. An apolar part may lead to the slow burying and entrapment of the inhibitor. The concept of extra binding (exo binding) was championed by Baker in the design of specific inhibitors of enzymes.[3] The action of hirudin on thrombin represents such a case. The tight binding of hirudin arises from multiple contacts, some of which occur in the active site region, but tight binding is also derived from interactions at the exo-site anion binding region of thrombin.[4,5]

A proteinase inhibitor of 83 residues from barley shows kinetics typical of slow-acting inhibitors with subtilisin BPN', subtilisin Carlsberg, chymotrypsin, and pancreatic elastase. A dissociation constant, K_i, of 29

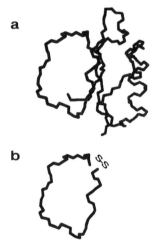

Figure 6.3. Design of the peptide inhibitor. The backbone structures of (a) CI-2 and (b) the peptide inhibitor are shown. The residues that are incorporated into the peptide are shown in black; the remainder of the CI-2 residues are shaded. The region that is chosen to create the peptide consists of the binding loop, together with most of the residues that are responsible for maintaining the structure of this loop. A disulfide bridge is introduced into the peptide to create a closed cyclic molecule. Reprinted with permission from *Biochemistry* 30:10717 (1991). Copyright 1991, American Chemical Society.

x 10^{-12} M was evaluated for its interaction with subtilisin BPN'. Because crystal structures are known for the free inhibitor and the inhibitor bound to the subtilisin BPN',[6,7] a rationale is possible for the design of small inhibitors.

Based on the fact that only the loop region of the proteinase inhibitor was involved in the interactions, molecular modeling studies suggested a peptide of 18 residues enclosed into a loop form by a disulfide bond could serve to mimic the structure of the active portion of the natural material (Figure 6.3).[8] Residues 54–63 of the proteinase inhibitor, found in the cyclic inhibitor, are believed to interact with subtilisin BPN'. Indeed it was the case as the peptide showed similar inactivation kinetics and a K_i value of 28 x 10^{-12} M was found. From the k_{off} constant, it can be calculated that the $t_{1/2}$ is for the release of the peptide is 10.7 hrs. Because of the tight binding and slow release, peptides or other mimetics may find use as inhibitors of biological processes.

Reversible Inhibitors

Several classes of reversible inhibitors are known to influence co- and post-translational modification reactions in cellular systems: (1) natural proteins or peptides, (2) natural products, and (3) chemically derived materials. Table 6.1 illustrates how some of these materials influence specific covalent modification reactions.

Prenylation, proteolysis, and carboxymethylation

These post-translational modifications (the isoprenylation pathway) that occur in eukaryotes can influence protein interactions and membrane binding. How these modifications influence cellular functions can be studied by using specific inhibitors of the different enzymatic reactions. For example, mevinolin is a potent inhibitor of HMG-CoA reductase and by blocking the formation of isoprenoid pyrophosphates in cells, this would curtail subsequent prenylation of proteins. In addition, mevinolin by inhibiting prenylation also would inhibit proteolysis and C-terminal carboxymethylation because these reactions will not occur unless the cysteinyl side chain is prenylated first. However, mevinolin would also impair sterol biosynthesis. Hence, an inhibitory effect of mevinolin in cells may not be simply related to an effect on the isoprenylation pathway.

An effect on carboxymethylation, the last reaction in the reaction sequence, might be easier to interpret. The enzymological studies in vitro have provided a rationale for in vivo studies. First, studies of prenylated analogs show that a whole peptide is not needed for enzyme recognition. N-acetyl-S-farnesyl-L-cysteine(L-AFC) and S-(farnesyl-3-thio) propionic

Table 6.1. Action of reversible inhibitors in cellular systems

Modification	Inhibitor	Effect
Deacetylation of ε-amino groups of lysine	butyrate	inhibits deacylation[9,10,11] and promotes hyper-acetylation of chromatin
	trichostatin A	inhibits histone deacetylase[12] induction of Friend cell differentiation[13]
Carboxylation of γ-carboxyl glutamate	dicoumarol (chloro K)	blocks vitamin K- dependent group of carboxylation[14,15] impairs blood clotting
	2, 3 ,5, 6–tetra-chloro-pyridinol	inhibition of vitamin K-epoxide reductase[16]
N-glycosylation	tunicamycin (antibiotics)	inhibits coupling[17,18] of dol-P and UDP- GlcNAc
	kifunensine	inhibits mannosidase I glycoprotein processing[19]
Adenylation of	glutamine	inhibits deadenylation[20] tyrosine and stimulates adenylation of GS
	α-ketoglut	inhibits adenylation and stimulates deadenylation of E. coli glutamine synth.
Prenylation of methyl cysteine	mevinolin	inhibits hydroxy- methylglutaryl-CoA reductase[21]
	limonene	inhibits isoprenylation of 21-26kDa proteins in NIH3T3 cells[22]
Peptidyl prolylcis-trans isomerization	cyclosporin A	inhibits peptidyl prolyl isomerase[23] slows down triple helix formation of collagen.
Hydroxylation of proline and lysine	α,α-dipyridyl	chelates Fe^{+2} inhibiting hydroxlases in collagen biosynthesis[24,25]
ADP-ribosylation	benzamide	general inhibitor at low concentration. (μM) inhibits poly-ADP ribosylation and at mM conc. inhibits mono-ADP ribosylation[26]
arg-ADP ribosylation	meta-iodo benzylguanidine	inhibits mono-ADP ribosylation[27]
Transglutamination	cystamine	inhibits transglutaminase (glu-lys isopeptide) in parallel with macrophage fusion[28]
Myristoylation	2-hydroxy myristic acid	Acyl-CoA form inhibits N-myristoyl transferase in LSTRA cells[29]
	10-(propoxy) decanoic acid	highly toxic to trypanosomes in culture-blocks incorporation of myristic acid into glycan anchor of variant surface glycoprotein[30]

acid can be effectively methylated.[31] S-farnesylthioacetic acid cannot be methylated, but it is a potent competitive inhibitor (K_i of $1.2\mu M$) with respect to L-AFC.[32] Second, the kinetic studies suggest that the reaction proceeds in an ordered sequence, with S-adenosyl-L-methionine (AdoMet) binding first and S-adenosyl-L-homocysteine(AdoHcy) being released last. Hence, AdoMet is needed for binding of a farnesyl derivative.[32] If inhibition by the S-farnesyl derivatives is to be effective in a cell, sufficient amounts of AdoMet or perhaps AdoHcy need be present to saturate the binding site of the isoprenylated protein methyltransferase. Interestingly, farnesyl-cysteine analogs have been found to inhibit carboxymethylation of p21[ras] in vivo and suggest that carboxymethylation has a role in signal transduction in eukaryotes.[33] Recently it was shown that concentrations of AFC (10–$50\mu M$) that inhibit carboxymethylation of Ras inhibit agonist response in human platlets.[34] An important review on prenylation and carboxymethylation is available.[35]

Protein Phosphorylation-Dephosphorylation

Natural proteins. Generally the proteins inhibiting kinase and phosphatase reactions are heat stable and of low molecular weight. Peptides can mimic some of their actions effectively; thus, a complex organized structure may not be needed for their actions. An important characteristic of these inhibitors is that binding occurs with high affinity (in the nanomolar range), suggesting that multiple interactions exist between the inhibitor proteins and their target enzymes.

The heat-stable inhibitor of cyclic AMP-dependent protein kinase has a molecular weight of 11,300. It binds competitively with respect to protein substrates with a dissociation constant of about 2 nM.[36] Mg-ATP influences the interaction of the inhibitor.[37] If the enzyme is not preincubated with Mg-ATP, inhibition is less and the progress curve is suggestive of a slow-binding inhibitor. Effective synthetic peptides [PKI -(6–22)-amide] of the inhibitor protein contain two parts, a pseudosubstrate region and a region at the amino terminus containing phenylalanine 10, which is important for high affinity binding.[38] Circular dichroism measurements of the binding of a peptide substrate and effective inhibitors suggest that binding occurs in different stages.[39]

An ATP-dependent phosphoprotein phosphatase[40] is made up of a catalytic subunit and a regulatory protein, inhibitor 2,[41] which is a potent inhibitor of the phosphatase. Activation of the phosphatase *in vitro* may be regulated by phosphorylation of the inhibitor protein (M_r 22kDa). Phosphorylation of threonine 72 by glycogen synthase kinase 3 leads to the activated state.[42] Phosphorylation of other site(s) on the inhibitor by casein kinase II may influence the phosphorylation and activation by glycogen synthase kinase 3.[43] Dephosphorylation of the bound inhibitor occurs, but this occurrence does not immediately inactivate the enzyme. In fact, dephosphorylation activates the reaction of the catalytic subunit on exogenous substrates.[44] A conformational change in the catalytic subunit (activation) induced by phosphorylation of the inhibitor may be reversed slowly by the action of the dephosphorylated inhibitor causing inactivation. This system (Figure 6.4) shows some characteristics of slow-acting inhibitors (Figure 6.1).[45]

A different scenario exists for inhibitor 1(M_r 18.6kD). This protein inhibits phosphoprotein phosphatase PP-1, but only in its phosphorylated state, which is achieved by the phosphorylation of a threonyl residue by cyclic AMP-dependent protein kinase.

But what happens in the cell? Are the inhibitor proteins modified? One way of evaluating function and changes of inhibitors is to treat an animal, a perfused organ or tissue, or cells in culture with an effector of cellular function, for example, a hormone, and then determine whether the inhibitor has changed. For example, injection of epinephrine into rabbits leads to phosphorylation of inhibitor 1 and combination with PP-1, which is released from the glycogen particle by phosphorylation. A scheme for the changes occurring in response to epinephrine is shown in Figure 6.5.[46]

Injection of insulin or epinephrine in rabbits leads to changes of glycogen synthase associated with phosphatase activity, but no changes in the phosphorylated state of threonyl or seryl residues of inhibitor 2 were detected.[47] Based on in vitro studies showing a rapid turnover of phosphate of the bound inhibitor in the Mg-ATP-dependent phosphatase, a change in covalent state may not be expected. Thus, the in vitro studies of inhibitor 1 and 2 served nicely as a guide for interpreting in vivo effects.

Another way to investigate the biological action of inhibitors is to microinject these molecules directly into cells. The role of protein phosphorylation-dephosphorylation in maturating *Xenopus* oocytes was nicely addressed by using the inhibitor of cyclic AMP-dependent protein kinase and inhibitors 1 and 2 of protein phosphatase-1. The phosphorylated form of inhibitor 1 (the active form) and the dephosphorylated form of inhibitor 2 (the active form) increase the time required for maturation induced by progesterone,[48,49] whereas the protein kinase inhibitor[50,51] increases the rate of maturation. The results obtained with inhibitors provide good evidence of cyclic AMP mediated protein phosphorylation regulating meiotic cell division of oocytes.

Microinjection of enzymes or antibodies also provide important information about cellular regulation. Injection of antibodies directed to inhibitor 2 in fibroblasts led to results suggesting that the inhibitor could alter the activity of the protein kinase p34[cdc2] influencing mitotic induction.[52]

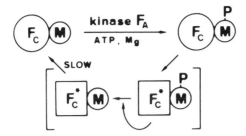

Figure 6.4. Scheme for activation of Mg(II) ATP-dependent phosphatase activity by F_A. Reprinted with permission from *Journal of Biological Chemistry* 259:5864 (1984). Copyright 1984, The American Society for Biochemistry & Molecular Biology.

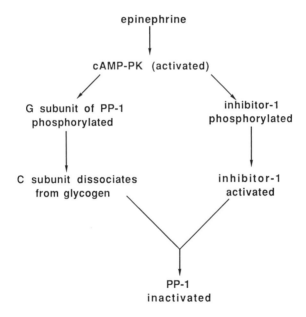

Figure 6.5. Scheme for inactivation of PP-1 by epinephrine. Reproduced, with permission, from the *Annual Review of Biochemistry* 58:453 (1989). Copyright 1989 by Annual Reviews, Inc.

Natural products. Complex nonprotenoid-organic molecules have been isolated from marine sponges, blue-green algae, bacteria, fungi, and plants that can affect specific phosphorylation and dephosphorylation reactions.

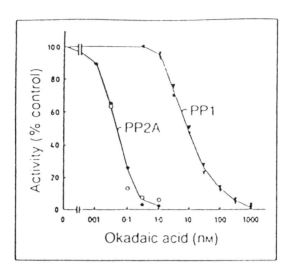

Figure 6.6. Inhibition of PP-1 and PP-2A by okadaic acid. Reprinted with permission from *Trends in Biochemical Science* 15:98 (1990). Copyright 1990, Elsevier Trends Journals.

Okadaic acid obtained from *Halichondria okadaii*, a marine sponge, is a potent inhibitor of protein phosphatases. It acts with an IC_{50} below 0.1nM on protein phosphatase 2A and with an IC_{50} of 10nM on phosphatase 1 (Figure 6.6) but has no effect on other phosphatases studied at these concentrations.[53]

Kinetic studies suggest that the compound acts as a mixed inhibitor[54] although the results do not eliminate competitive inhibition. The compound, which is a polyether derivative of a 38 carbon fatty acid, can penetrate into cells and, because of its specificity, can be used to evaluate the biological roles of PP-1 and PP2A.

Experiments with hepatocytes and adipocytes show that okadaic acid influences protein phosphorylation[55] and can mimic the action of insulin in isolated adipocytes.[56] In this instance, okadaic acid stimulated the activity of specific protein kinases presumably by blocking their dephosphorylation and deactivation by PP-1 or PP 2A.

Microcystin LR, a cyclic peptide isolated from blue green algae, is lethal when injected into mice in small amounts (1–2mg).[57] It is a potent and specific inhibitor of PP-1 and PP2A.[54,58] and, like okadaic acid, contains a hydrophobic structural element, 3-amino-9-methoxy-2,6,8-trimethyl-10-phenyldeca-4,6–dienoic acid. Interestingly, microcystin LR, okadaic acid, and inhibitors 1 and 2 seem to act at common sites on the phosphatases.[32] Knowledge of structural features influencing strong binding and specificity could lead to the design of other in vivo inhibitors of phosphatases.

Genistein, an isoflavanoid from pseudomonas; erbastin, a hydroquinone from streptomyces; and staurosporine, a microbial alkaloid from streptomyces are among a growing list of natural compounds known to inhibit the activity of various tyrosyl kinases, for example, epidermal growth factor receptor, insulin receptor kinase, v-src, c-src, and v-abl. These compounds have been used in cell cultures to probe the involvement of tyrosyl phosphorylation in differentiation, cell growth, and carcinogenesis. But how specific are these inhibitors? Staurosporine, which also inhibits tyrosine kinase, is a potent inhibitor of protein kinase C and of Ca^{+2}/ calmodulin dependent-protein kinase II,[59] so effects of inhibitors should be interpreted cautiously.

Chemically derived materials. Tyrphostins, benzylidene derivatives related to the natural product erbastin, are competitive inhibitors of the phosphorylatable tyrosyl containing substrate and have solubility characteristics allowing their penetration into cells. They act 1,000-fold more effectively with the EGF receptor than with the insulin receptor. Tyrphostin can block epidermal growth factor-induced cell proliferation by blocking the tyrosyl kinase of the EGF receptor.[60] Thiazolidonediones, potent inhibitors of the EGF receptor tyrosyl kinase activity, block the action of EGF on the growth of BALB/MK and A431 cell lines.[61]

Bisindolylmaleimides,[62,63] analogs related to the staurosporine, have been made and these show more selective inhibition of protein kinase C than staurosporine in vitro and in whole cells. A K_i of 14 nM was reported for one derivative, GF109203X, and this compound interfered with protein kinase C phosphorylation reactions in human platelets and Swiss 3T3 fibroblasts. For example, in platelets phosphorylation of P47, caused by protein kinase C, is inhibited by GF 109203X, but phosphorylation of P20, caused by myosin light chain kinase, is only slightly affected, whereas phosphorylation of both proteins is affected similarly by staurosporine.

An, isoquinolinesulfonamide derivative, KN-62, is a potent inhibitor of Ca^{+2}/calmodulin-dependent protein kinase with little or no effect on myosin light chain kinase, cAMP-dependent protein kinase, and protein kinase C. Because of it specificity it has been used to probe the physiological role(s) of Ca^{+2}/CaM kinase reactions in cultured PC D 12 pheochromocytoma cells.[64] Also, a more water soluble compound, KN-63, which has essentially the same specificity of inhibition of kinases as KN-62 has been used. The reduction of dopamine with KN-63 in cultured cells was interpreted to mean that phosphorylation of tyrosine hydroxylase by

Ca^{+2}/CaM kinase was inhibited specifically by the methoxybenzenesulf-onamide derivative.[65]

Metabolites

The use of these compounds has had an important role in the discovery of covalent modification reactions. The discovery of the adenylation of glutamine synthetase represents a good example.

Early studies suggested the synthesis of glutamine synthetase in *E. coli* is inhibited by NH$_4$$^+$, but another effect of the cation was suggested by the rapid inactivation of the enzyme (90% loss of activity in 2 minutes at 37°) caused by the incubation of E. coli cells with NH$_4$$^+$.[66] The latter effect can be explained in part by the formation of glutamine, which promotes an enzyme catalyzed modification of the enzyme by ATP. Glutamine, which causes inactivation,[67,68,69] not only influences the transfer of an adenyl group to a tyrosyl residue, but it also inhibits enzymatic deadenylation. But the regulation is more complex than these explanations suggest and is explained by a bicyclic cascade shown in Figure 6.7. Note that glutamine also influences another nucleotidylation reaction, uridylation. Here it inhibits the transfer of a uridyl group from UTP and inhibits deuridylation, just the opposite of adenylation. The uridylated protein can activate the adenylation process. An additional effect is seen on protein synthesis. The P$_{II}$ protein (the deuridylated form) which may accumulate in the presence of glutamine has an indirect effect in inhibiting the biosynthesis of glutamine synthetase.[70]

The level of glucose in animal cell cultures influences gene expression. One response of cells deprived of glucose is the synthesis of glucose-regulated proteins.[71] Another response is a decrease in the synthesis of the insulin receptor.[72] Are these processes regulated by covalent modification reactions? Yes, they are, and it seems that glucose acts by influencing the glycosylation of transcription factors influencing gene expression.[73,74] This view is supported by the fact that the glycosylation inhibitors, tunicamycin and deoxyglucose, have effects similar to that of the starved glucose state. The mechanisms by which glycosylation regulates expression are not yet known.

A rationale used for inhibition of methylation in cells is based on the fact that S-adenosylhomocysteine, the product of methylation reactions, is a potent inhibitor of methyl transferases. Compounds in cell cultures that elevate S-adenosylhomocysteine may block protein methylation and lead to the discovery of endogenous substrates and roles of methylation.

A function of methylation reactions in the phagocytic and chemotactic action of leucocytes was probed by using an adenosine deaminase inhibitor, erythro-9-(2-hydroxy-3-nonyl) adenine, plus adenosine and L-

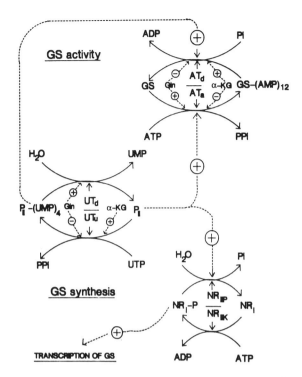

Figure 6.7. The cyclic cascade of glutamine synthetase regulation. Inter-relationship between the uridylylation cycle, the adenylylation cycle, and the phosphorylation cycle; the reciprocal controls of these interconversions by L-glutamine (Gln) and a-ketoglutarate (-KG) are shown; (+) indicates stimulation, (-) indicates inhibition. Abbreviations: GS, glutamine synthetase; P_{II}, regulatory protein; AT_a and AT_d, adenylyltransferase catalyzing adenylylation and deadenylylation, respectively; UT_d, uridylyl-removing enzyme or deuridylylation enzyme; UT_u, uridylyltransferase; NR_I, glnG product also known as NTRC; NR_{IIK} and NR_{IIP}, glnL product (also known as NTRB) catalyzing phosphroylation and dephosphorylation of NR_I, respectively. Reprinted with permission from *Advances in Enzymology* 62:37 (1989). Copyright 1989, John Wiley & Sons, Ltd.

homocysteine lactone with human monocytes.[75] S-adensoylhomocysteine in the cell was elevated 1,500-fold; carboxymethylation reactions of proteins was decreased, and chemotactic activity was reduced, but phagocytosis was not. Thus, the use of the inhibitor helped point out a relation between methylation and chemotactic activity and differences in regulation of phagocytosis and chemotaxis.

Irreversible Inhibitors

Studies with affinity labeling agents and suicide inhibitors, molecules that initially are chemically inert but become activated by the action of a specific enzyme, have been used to provide insight into post- and co-translational reactions occurring within cells or on their surfaces. Adenosine dialdehyde was useful in identifying endogenous proteins for N-methylation reactions in pheochromo-cytoma cells (PC12).[76] N-methylation is irreversible; as with many covalent modification reactions, this may result in full or nearly full methylated proteins in the intact cell. Hence, the study of methylation reactions in cellular extracts may give negative results because the protein substrates are already modified. Adenosine dialdehyde inhibits S-adensyl-L-homocysteine hydrolase from building up the concentration of S-adenosylmethionine, which competitively inhibits methylation. Cells treated with adenosine dialdehyde yielded proteins, which could be identified in the hypomethylated state. Incubation of extracts with radioactive S-adenosylmethionine led to the detection of over 50 different protein substrates. The major products of N-methylation were proteins containing NG, N'G-dimethylarginine. The use of inhibitors in other systems could find use for identifying endogenous substrates.

Fluorosulfonylbenzoyladenosine (FSBA) is an affinity-labeling reagent used to modify ATP-binding sites in kinases, including protein kinases. An early study suggested that an ectoprotein kinase activity on the surface of myoblasts was important for the process in muscle cell development in which mononucleated myoblasts fuse to form the multi-nucleated myotubes.[77] Incubation of myoblasts (L_6 line) with FSBA blocks cell fusion and is correlated with inactivation of the ectoprotein kinase activity; this correlation shows a linkage between the processes.[78] The use of affinity-labeling reagents to modify enzymes on cell surfaces may find applications in the study of other modification reactions in signal transduction.

3,4-dehydroproline was found to be a potent inhibitor of prolyl hydroxylase in L929 fibroblasts.[79] The free amino acid is not effective in vitro, but peptides containing it can cause a time-dependent inactivation of prolyl hydroxylase.[80] It serves to inhibit formation of collagen in animal cells and formation of hydroxyproline-rich proteins in plants. This amino acid analog of proline can be incorporated into protein in place of proline and bind to the hydroxylase.[81] The modified proteins when bound to the active site of prolyl hydroxylase are believed to be acted upon in the active site by an oxy radical at the C4 position of the dehydroproline ring, producing a radical that can irreversibly inactivate and bind the enzyme.[82] Hence, dehydroproline has the characteristics of a suicidal inhibitor.

Table 6.2. Action of irreversible inhibitors in cellular systems

Modification	Inhibitor	Effect
Peptide bond cleavage	Leupeptin	Inhibitor of cysteine proteases; influences protein degradation in muscle[83]
	Chymostatin	Inhibitor of serine proteases (related to chymotrypsin)[84]
Dephosphorylation of tyrosine	phenylarsene oxide	inhibits tyrosine phosphatase[85,86] increase of tyr. phosphorylation inhibits insulin action of glucose uptake in adipocytes, T cell function
cross-linking of collagen	β-aminopropio- (lathyrism)	inhibits lysyl nitrile oxidase[87]
hydroxylation of proline	3,4 dehydro proline	suicidal inhibitor of prolyl hydroxylase[81] inhibits hydroxylation and growth of soybean cell culture[88]
	Oxaproline- containing peptides	collagen synthesis in fibroblasts[89]

Some examples of the action of other irreversible inhibitors are shown in Table 6.2. A particularly novel way of evaluating phosphorylation reactions in intact cells is through the use of thiophosphate.[90] Hela cells grown in the presence of thiophosphate yield both ATP and GTP with the thio group in the γ-position. Thiophosphorylation of proteins then occurs, but because dethiophosphorylation by protein phosphatases is limited,[91,92] thiophosphoproteins accumulate. Dethiophosphorylation is slower than dephosphorylation probably because of differences in the chemical properties of the two esters. Sulfur, which is less electronegative than oxygen, could reduce the effectiveness of a nucleophilic reaction at the central phosphorous atom of the ester. The use of thiophosphate can facilitate characterization of newly derivitized protein; for example, in the study with hela cells, thiophosphorylated histones could be recovered after cell disruption by binding the proteins to an organic mercurial Sepharose derivative.

Phosphorylation reactions can be influenced in vivo by the use of metal ions or metal derivatives. Aluminum ions have potent effects on phosphorylation reactions in the brain and may provide insight about neurological disorders, including Alzheimer's disease.[93] Use of other effectors of protein phosphorylation in PC12 cells supports the view of a linkage between abnormal phosphorylation and cerebral amyloidosis associated with the disease.[94] Figure 6.8 illustrates the cellular effects of aluminum ions.

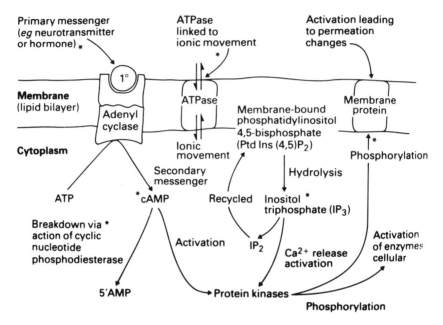

Figure 6.8. Cellular events affected by aluminum. Reprinted with permission from *Chemistry in Britain* 26:1169 (1990). Copyright 1990, The Royal Society of Chemistry.

The asterisks indicate sites of action. Binding of aluminm ions to inositol triphosphate is exceedingly tight and with a dissociation, which is 10^7 times slower than with Ca^{+2}, it can alter phosphorylation reactions mediated through this effector.[95]

ENZYME ACTIVATORS

Protein modification reactions can be activated by numerous agents including metal ions, growth factors, hormones, drugs, and so on. If the reactions are reversible, the effect could be related to (1) a stimulation of the modifying enzyme (2) an inhibition of the enzyme reversing the modification or (3) a combination of these two processes. Experiments with whole cells and cellular fractions can help define how these agents work.

Determination of Sites of Activation

Consider the action of zinc ions and its influence on protein phosphorylation reactions in particulate fractions of LSTRA lymphoma cells.[96] Zinc ions activate autotyrosylphosphorylation of kinase p56[lck], but as zinc ions can inhibit certain tyrosyl phosphatases the immediate cause of activation is not certain. No phosphorylation occurs in the presence of 2mM EDTA, but when μM zinc ions are added phosphorylation does occur. Mg^{+2} in the μM range has no effect. A Zn-ATP complex for phosphorylation of the autophosphorylation site is suggested. The effect of other metal ions and the lack of effect of orthovanadate, a phosphotyrosine phosphatase inhibitor, further suggest a direct effect of zinc on the kinase. A second binding of zinc ions is believed due to the presence of zinc binding sites (cysteine motifs) in the kinase and membrane constituents, for example, CD 4 or CD 8.[97] It is suggested that binding could directly activate autophosphorylation of the lck kinase.

Similarly, zinc ions activate protein kinase C and influence its binding to plasma membranes in T lymphocytes.[98] Zinc ions in the presence of pyrithione, a zinc ionophore, influences binding of lymphocytes to red blood cells (rosette formation).[99] Free zinc has no effect because it does not penetrate into cells. Hence to test a role for zinc in cellular systems, attention need be paid to mechanisms of transport. The effect in lymphocytes, due to the loss of the human B cell receptor, is believed to be related to translocation and activation of protein kinase C in the membranes. It is proposed that Zn is involved by binding to the zinc fingerlike structures in protein kinase C[100] which promote the associative interactions with membrane constituents.

Tumor promoting-phorbol esters and related compounds are utilized in cellular systems as probes to evaluate how growth factors, hormones, neurotransmitters, and other signaling agents might elicit their biological effects. The phorbol esters are used because they serve to activate protein kinase C. Hence a common effect of the phorbol esters and a biological agent is suggestive of an involvement of protein kinase C.

But how do the phorbol esters cause activation? Protein kinase C, a calcium-phospholipid dependent enzyme, is activated by diacylglycerol. This latter agent promotes a conformational change in the enzyme dislodging a pseudosubstrate region from the active site. In a sense this makes the enzyme catalytic site open for business. Kinetic studies show that phorbol esters can substitute for diacylglycerol and activate the enzyme by increasing its affinity for calcium.[101] The structures of phorbol esters and other agents of similar activity with respect to diacylglycerol are shown in Figure 6.9.

How these compounds are related is not immediately evident from examining their structures alone. Analysis of the effects of isomers, derivatives, and molecular modeling does provide answers about common key features in the structures. For example, 1,2-sn-diolein is active but its enantiomer 2,3-sn-diolein is not.[102] A hydroxyl group at C3 of DAT might mimic the hydroxyl group of diacylglycerol, but since 3-deoxy-DAT

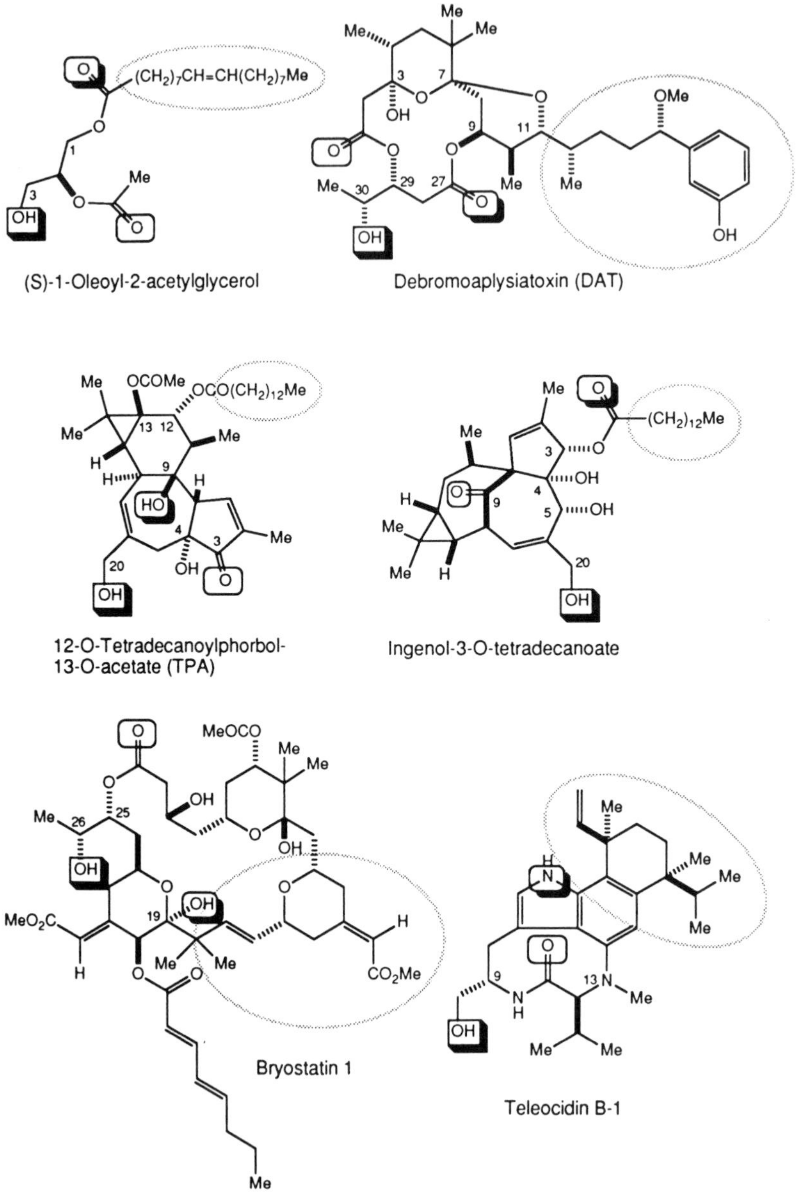

Figure 6.9. Proposed PKC activation model. Identical symbols indicate equivalent atoms: proposed spacial correspondence of hydrophobic area is indicated by a dotted circle. For bryostatins, the C20 fatty acid residue is unlikely to be the hydrophobic region as C20 acetyl and C20 deacetylated bryostatins retain the biological activity. Reprinted with permission from *Proceedings of the National Academy of Sciences, USA* 86:9672 (1989).

has the same activity as the parent form, this OH group is not critical. It is suggested that three hydrophilic atoms in all of the activating molecules are key features in molecules which can activate protein kinase C.[103,104] In the above figure this is indicated by atoms encircled by rounded squares, shadowed circles, and shadowed squares.

Phorbol esters bind in the regulatory amino domain of protein kinase C and the analysis of binding to mutant forms shows that two binding sites exist.[105] The sequence of the sites that contain the cysteine-rich regions is consistent with the involvement of putative zinc fingers in binding of phorbol esters.

An important aspect of activation involves a migration of protein kinase C from the cystosol to the particulate fraction.[106] Specific intracellular binding proteins (RACKS) have been identified in cytoskeletal elements.[107,108] These proteins, by anchoring activated protein kinase C to specific sites, could influence what protein substrates could be phosphorylated in response to a biological signal. A model showing the difference between inactive and active protein kinase C, interactions of lipids, Ca^{+2}, and the RACK protein is shown in Figure 6.10.

What Enzymes Might Cause the Modification in Intact Cells?

It may be learned that an activator can influence a specific type of covalent modification reaction, but how do we learn what enzymes are involved in the cellular response? Knowledge of reactions in vitro can help, because these can define what proteins can be modified, the residues that can be modified, and whether specific sequences are reacted on by one enzyme and not another. Also, information of the chemical properties of the modified sites can be useful in distinguishing specific reactions, such as in ADP-ribosylation of glutamyl, cysteinyl, or arginyl residues, or in phosphorylation of seryl or tyrosyl residues (see Chapter 4). But, we should also realize that an activator or an inhibitor may influence new modification reactions for which no background information exists.

Epinephrine activates a phosphorylation of acetyl-CoA carboxylase in adipocytes causing enzyme inactivation. Because of the vast information linking epinephrine action with the second messenger cyclic AMP, it might be expected that cyclic AMP-dependent protein kinase would be involved. But does this enzyme modify acetyl-CoA carboxylase directly or is some other mechanism involved? Sequence analysis of modified sites of acetyl-CoA carboxylase from ^{32}P-labeled adipocytes was quite revealing. The putative site of modification ser-77 in the phosphopeptide, ser-ser[77]-met-ser[79]-gly-leu-his-leu-val-lys, derived from the protein by tryptic digestion turned out not to be phosphorylated. Phosphorylation

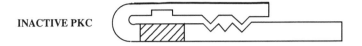

INACTIVE PKC

ACTIVATED PKC

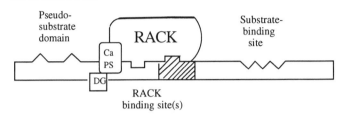

Figure 6.10. Model of PKC with A RACK binding site, DG, diacyl-glycerol; PS, Ptdser. Reprinted with permission from *Proceedings of the National Academy of Sciences, USA* 88:3997 (1991).

was found at ser-79, and this phosphorylation was increased by epinephrine treatment. It has been concluded that the effect of cyclic AMP-dependent protein kinase is not directly involved and that phosphorylation of ser-79 likely occurs by AMP-activated protein kinase.[109] These results emphasize the need to analyze the structures of the modified site to define how modification occurs.

A similar case exists with glycogen synthase. Epinephrine, which is an activator of glycogenolysis, also serves as an inhibitor of glycogen synthesis. Phosphorylation of glycogen synthase can cause inactivation, and because cyclic AMP-dependent protein kinase can phosphorylate the enzyme, it was thought earlier that inactivation of glycogen synthase in vivo was caused by a direct phosphorylation of this enzyme. More recent studies show that other kinases are involved and that inactivation by epinephrine in vivo is initiated through a phosphorylation of a regulatory subunit of protein-phosphatase 1 bound to glycogen. The phosphatase released from the glycogen particle and bound by phosphorylated inhibitor 2 in the cytosol may not act on the bound glycogen synthase. Phosphorylation of glycogen synthase that occurs now is not counterbalanced by the action of protein phosphatase 1.

But what kinases are involved, what sites are phosphorylated, and what levels of phosphorylation occur? Nine serine residues can be phosphorylated in glycogen synthase. Two sites are found in the amino terminus and the rest are in the c-terminal region. One account deals with changes in the amino terminal region.[110] To define what events occur in the amino terminal region, glycogen synthase was isolated from animals injected with either epinephrine or propanolol and digested with trypsin. Peptides of the phosphorylated region were isolated. Di-, mono-, and unphosphorylated forms were defined by FAB-mass spectrometry, and quantitative amino acid analysis.

An important point illustrated in this study is the need to define structures, discussed in Chapter 4. FAB-mass spectrometry can be done on mixtures, but it can give results which underestimate stoichiometries of phosphorylation.[111] The monophosphorylated peptide obtained as one peak in reverse phase HPLC contains serines at positions 7 and 10 and is actually a mixture of mono-phosphorylated forms. Conversion of the serylphosphate to the S-ethylcysteine derivative and sequencing shows that phosphorylation occurs at both residues, but that approximately 80 percent of the phosphorylation occurred at position 7. From these results, it is concluded that levels of phosphorylation of serine 7 and serine 10, respectively, are 0.31 and 0.12 from animals injected with propanolol and 0.78 and 0.54 from the rabbits injected with epinephrine.[110]

Experiments *in vitro* are suggestive that major inactivation can occur by phosphorylation of serine 10, which is influenced by prior phosphorylation of serine 7. Casein kinase 1 can phosphorylate serine 10. Definition of enzymes involved in vivo require further attention, but the results are consistent with the view cyclic AMP-dependent protein kinase is not involved directly in the phosphorylation of glycogen synthase in vivo. Other work suggests the importance of glycogen synthase kinase 3 (Chapter 2).

Analogs of Natural Activators

One approach that may be taken to evaluate effects of activators in cells is to change an experimental condition to alter the intracellular concentration of the activator and determine how this change influences cellular modification reactions (in Chapter 7). A second approach is to use analogs of natural activators. Analogs of cyclic AMP and cyclic GMP have proven invaluable because many of these compounds can pass through cell membranes and activate specific enzymes acted upon normally by the natural materials. An important objective is to learn whether a correlation exists between the action of the compounds in vitro with the biological response. Such findings can help determine the site of action of the activator in the cell.

Cyclic GMP is a smooth muscle relaxant and a likely site for its action, although not proven, is cyclic GMP-dependent protein kinase. Vascular smooth muscle contains two isozymes for this enzyme, cGMP kinase Iα and cGMP kinase Iβ, in near equivalent amounts. Is one of these enzymes a site for the action of cylic GMP in smooth muscle? First, it was found that 8- substituted derivatives of cyclic GMP could activate cyclic GMP-dependent protein kinases differently. For example, 8-(2,4-dihydroxyphenylthio)cGMP is 22 times more effectve than cyclic GMP with isoform Iα but importantly it was much less effective with isoform Iβ. Second, by using a series of 8- substituted derivatives with pig coronary arteries, it was found that relaxation of smooth muscle occurred and that the results correlated with the in vitro effects of these compounds on isoform Iα. Relaxation could not be correlated with effects on Iβ. Hence, it was suggested that the cyclic GMP Iα was involved in mediating the effects of the cyclic nucleotide derivatives in smooth muscle.[112] Also, it was found that the relaxing effect of 8-phenylthioderivatives persisted for hours even after repeated buffer changes of the medium. The compounds once bound in their final form may show the same characteristics as described earlier for slow-acting inhibitors—entrappment and slow release. The long lasting effect of certain compounds may also be due to a resistance of hydrolysis by intracellular phosphodiesterases.[112]

REFERENCES

1. Frieden, C., Kurz, L.C., and Gilbert, H.R., *Biochemistry* 19, 5303–5309 (1980).
2. Morrison, J.F., and Walsh, C.T., *Advances in Enzymology* 61, 201–301 (1988).
3. Baker, B.R., Design of Active-Site-Directed Irreversible Enzyme Inhibitors, John Wiley and Sons Inc., New York, 1967.
4. Stone, S.R., and Hofstenge, J., *Biochemistry* 25, 4622–4628 (1986).
5. Naski, M.C., Fenton, J.W., II, Maraganore, J.M., Olson, S.T., and Shafer, J.A., *J. Biol. Chem.* 265, 13484–13489 (1990)
6. McPhalen, C.A., and James, M.N.G., *Biochemistry* 27, 261–269 (1987).
7. McPhalen, C.A., and James, M.N.G., *Biochemistry* 28, 6582–6598 (1988).
8. Leatherbarrow, R.J. and Salacinski, H.J., *Biochemistry* 30, 10717–10721 (1991).
9. Sealy, L., and Chalkley, R., *Cell* 14, 115–121 (1978).
10. Cousins, L.S., and Alberts, B.M., *J. Biol. Chem.* 257, 3945–3949 (1982).
11. Riggs, M.G., Whittaker, R.G., Neuman, J.R., and Ingram, V.M., *Nature* 268, 462–464 (1977).
12. Yoshida, M., Kijima, M., Akita, M., and Beppu, T., *J. Biol. Chem.* 265, 17174–17179 (1990).
13. Yoshida, M., Nomura, S., and Beppu, T., *Cancer Res.* 47, 3688–3691 (1987).
14. Suttie, J.W., *Ann. Rev. Biochem.* 54, 459–478 (1985).
15. Lowenthal, J., and MacFarlane, J.A., *J. Pharmacol. Exp. Ther.* 157, 672–680 (1967).

16. Grossman, C.P., and Suttie, J.W., *Biochem. Pharmacol.* 40, 1351–1355 (1990).
17. Duskin, D., and Mahoney, W.C., *J. Biol. Chem.* 257, 3105–3109 (1987).
18. Elbein, A.D., *Methods in Enzymol.* 138, 661–709 (1987).
19. Elbein, A.D., Tropea, J.E., Mitchell, M., and Kaushal, G.P., *J. Biol. Chem.* 265, 15599–15605 (1990).
20. Rhee, S.G., Chock, P.B., and Stadtman, E.R., in *Enzymology of Post-translational Modification Reactions* 2, (Freedman, R., and Hawkins, R.C., New York: Academic Press, 273–297 (1985).
21. Alberts, A.W., Chen, J., Kuron, G., Hunt, V., Hoffman, C., Rothrock, J., Lopez, M., Joshua, H., Harris, E., Patchett, A., Monaghan, R., Currie, S., Stapley, E., Albers-Schonberg, G., Hensens, O., Hirshfield, J., Koogsteen, K., Liesch, J., and Springer, J., *Proc. Natl. Acad. Sci.* 77, 3957–3961 (1980).
22. Crowell, P.L., Chang, R.R., Ren, Z., Elson, C.E., and Gould, M.N., *J. Biol. Chem.* 266, 17679–17685 (1991).
23. Steinman, B., Bruckener, P., and Superti-Furga, A , *J. Biol. Chem.* 266, 1299–1303 (1991).
24. Kivirikko, K.I., and Mylla, R., in *The Enzymology of Post-translational Modification of Proteins* 2, (Freedman, R. and Hawkins, R.C., Eds.), New York: Academic Press, 54–104 (1980).
25. Cardinale, G.J., and Udenfriend, S., *Adv. Enzymol.* 41, 245–300 (1974).
26. Rankin, P.W., Jacobson, E.L., Benjamin, R.C., Moss, J., and Jacobson, M.K., *J. Biol. Chem.* 264, 4312–4317 (1989).
27. Loesberg, C., Rooij, H.V., and Smets, L.A., *Biochim Biophys. Acta* 1037, 92–99 (1990).
28. Fanaka, H., Shinki, T., Takito, J., HeJin, C., and Suda, T., *Exp. Cell Res.* 192, 165–172 (1991).
29. Paige, L.A., Zheng, G., DeFrees, S.A., Cassady, J.M., and Geahlen, R.L., *Biochemistry* 29, 10566–10573 (1990).
30. Doering, T.L., Raper, J., Buxbaum, L.V., Adams, S.P., Gordon, J.I., Hart, G.W., and Englund, P., *Science* 252, 1851–1854 (1991).
31. Gilbert, B.A, Tan, E.W., Perez-Sala, D., and Rando, R.R., *J. Am. Chem. Soc.* 114, 3966–3973 (1992).
32. Shi, Yi-Qun and Rando, R.R., *J. Biol. Chem.* 267, 9547–9551 (1992).
33. Volker, C., Miller, R.A., McCleary, W.R., Rao, A., Poenie, M., Backer, J.M., and Stock, J.B., *J. Biol. Chem.* 266, 21515–21522 (1991).
34. Huzoor-Akbar, Wang, W., Kornhauser, R., Volker, R., and Stock, J.B., *Proc. Natl. Acad. Sci.,* 90, 868–872 (1993).
35. Clarke, S., *Ann. Rev. Biochem.* 61, 355–386 (1992).
36. Demaille, J.G., Peters, K.A., and Fischer, E.H., *Biochemistry* 16, 3080–3086 (1977).
37. Whitehouse, S., and Walsh, D.A., *J. Biol. Chem.* 258, 3682–3692 (1983).
38. Glass, D.B., Lundquist, L.J., Katz, B.M., and Walsh, D.A., *J. Biol. Chem.* 264, 14579–14584 (1989).
39. Reed, J., Kinzel, V., Kemp, B.E., Cheng, H-C., and Walsh, D.A., *Biochemistry* 24, 2967–2973 (1985).
40. Merlevede, W., and Riley, G.A., *J. Biol. Chem.* 241, 3517–3524 (1966).

41. Yang, S.-D., Vandenheede, J.R., and Merleved, W., *J. Biol. Chem.* 256, 10231–10234 (1981).
42. Hemmings, B.A., Resnik, T.J., and Cohen, P., *Eur. J. Biochem.* 138, 635–641 (1984).
43. DePaoli-Roach, A.A., *J. Biol. Chem.* 259, 12144–12151 (1984).
44. Price, D.J., and Li, H.C., *Biochem Biophys. Res. Commun.* 128, 1203–1210 (1985).
45. Jurgensen, S., Shacter, E., Huang, C.Y., Chock, P.B., Yang, S-D., Vandenheede, J.R., and Merleved, W., *J. Biol. Chem.* 259, 5864–5870 (1984).
46. Cohen, P., *Ann. Rev. Biochem.* 58, 453–508 (1989).
47. Holmes, C.F.B., Tonks, N.K., Major, H., and Cohen, P., *Biochim. Biophys. Acta* 929, 208–219 (1987).
48. Huchjon, D., Ozon, R., and Demaille, J.G., *Nature* 294, 358–359 (1981).
49. Foulkes, J.G., and Maller, J.L., *FEBS Lett.* 150, 155–160 (1982).
50. Maller, J.L., and Krebs, E.G., *J. Biol. Chem.* 252, 1712–1718 (1977).
51. Huchon, D., Ozon, R., Fischer, E.H., and Demaille, J.G., *Molecular and Cellular Endocrinology* 22, 211–222 (1981).
52. Brautigan, D.L., Sunwoo, J., Labbe, J-C., Fernandez, A., and Lamb, N.J.C., *Nature* 344, 74–78 (1990).
53. Cohen, P., Holmes, and Tsukitani, Y., *Trends in Biochemical Science* 98–102 (1990).
54. Bialojan, C., Ruegg, J.C., and Takai, A., *J. of Physiology* 398, 81–95 (1988).
55. Haystead, T.A.J., Sim, A.T.R., Carling, D., Hoonor, R.C., Tsukitani, Y., Cohen, P., and Hardie, D.G., *Nature* 337, 78–81 (1989).
56. Haystead, T.A.J., Weil, J.E., Litchfield, D.W., Tsukitani, Y., Fischer, E.H., and Krebs, E.G., *J. Biol. Chem.* 265, 16571–16580 (1990).
57. MacKintosh, C., Beattie, K.A., Klumpp, S., Cohen, P., and Codd, G.A., *FEBS Lett.* 264, 187–192 (1990).
58. Honkanen, R.E., Zwiller, J., Moore, R.E., Daily, S.L., Khatra, B.S., Dukelow, M., and Boynton, A.L., *J. Biol. Chem.* 265, 19401–19404 (1990).
59. Yanigihara, N., Tachikawa, E., Izumi, F., Yasugawa, S., Yamamoto, H., and Mujamoto, E., *J. of Neuroscience* 56, 294–298 (1990).
60. Yaish, P., Gazit, A., Gilon, C., Levitzki, A., *Science* 242, 933–935 (1988).
61. Geissler, J.E., Traxler, P., Regenass, U., Murray, B.J., Roesel, J.L., Meyer, T., McGlynn, E., Storni, A., and Lydon, N.B., *J. Biol. Chem.* 265, 22255–22261 (1990).
62. Toullec, D., Pianetti, P., Coste, H., Bellefergue, P., Perret-G., T., Ajakane, M., Baudet, V., Boissin, P., Boursier, E., Loriolle, F., Duhamel, L., Charon, D., and Kirilovsky, J., *J. Biol. Chem.* 266, 15771–15781 (1991).
63. Davis, P.D., Hill, C.H., Keech, E., Lawton, G., Nixon, J.S., Sedgwick, A.D., Wadsworth, J., Westmacott, D., and Wilkinson, S.E., *FEBS Lett.* 259, 61–63 (1989).
64. Takumitsu, H., Chijiwa, T., Hagiwara, M., Mizutani, A., Tereswa, M., and Hidaka, H., *J. Biol. Chem.* 265, 4315–4320 (1990).
65. Sumi, M., Kiuchi, K., Ishikawa, T., Ishii, A., Hagiwara, M., and Nagatsu, T., and Hidaka, H., *Biochem. Biophys. Res. Commun.* 181, 968–975 (1991).

66. Holzer, H., *Advances in Enzymol.* 32, 297–326 (1969).
67. Shapiro, B.M., Kingdon, H.S., and Stadtman, E.R., *Proc. Natl. Acad. Sci.* 58, 642–649 (1967).
68. Wulff, I., Mecke, D., and Holzer, H., *Biochem. Biophys. Res. Commun.* 28, 740–745 (1967).
69. Kingdon, M.S., and Stadtman, E.R., *Biochem. Biophys. Res. Commun.* 27, 470–473 (1967).
70. Rhee, S.G., Chock, P.B., and Stadtman, E.R., *Advances in Enzymol.* 62, 37–92 (1989).
71. Shui, R.P.C., Pouyssegur, J., and Pastan,I., *Proc. Natl. Acad. Sci.* 74, 3840–3844 (1977).
72. Briata, P., Briata, L., and Gherzi, R., *Biochem. Biophys. Res. Commun.* 169, 397–405 (1990).
73. Chang, S.C., Wooden, S.K., Nakaki, T., Kim, Y.K., Lin, A.Y., Kung, L., Attenello, J.W.,and Lee, A.S., *Proc. Natl. Acad. Sci.* 84, 680–684 (1987).
74. Lee, A.S., *Trends in Biochem. Sci.* 12, 2023 (1987).
75. Pike, M.C., Kredich, N.M., and Snyderman, R., *Proc. Natl. Acad. Sci.* 75, 3928–3932 (1978).
76. Najbauer, J., and Aswad, D.W., *J. Biol. Chem.* 265, 12717–12721 (1990).
77. Lognonne, J.L., and Wahrman, J.P., *Exp. Cell Res.* 166, 340–356 (1986).
78. Lognonne, J.L., and Wahrmann, J.P., *Exp. Cell Res.* 187, 90–97 (1990).
79. Kerwar, S.S., Marcel, R.J., and Salvador, R.A., *Biochem. Biophys. Res. Commun.* 66, 1275–1280 (1975).
80. Nolan, J.C., Ridge, S.,Ornonsky, A.L., and Kerwar, S.S., *Arch. Biochem. Biophys.* 189, 448–453 (1978).
81. Rosenbloom, J., and Prockop, D.J., *J. Biol. Chem.* 245, 3361–3368 (1971).
82. Cooper, J.B., and Varner, J.E., *Plant Physiol.* 73, 324–328 (1983).
83. Libby, P., and Goldberg, A.L., *Science* 199, 534–536 (1978)
84. Selgen, P.O., *Methods in Enzymol.* 96, 737–777 (1983).
85. Bernier,N., Laird,D.M., and Lane,N.D.,*Proc. Natl. Acad. Sci.* 84, 1844–1848 (1987).
86. Garcia-Morales, P., Minami, Y., Luong, E., Klausner, R.D., and Samelson, L.E., *Proc. Natl. Acad. Sci.* 87, 9255–9259 (1990).
87. Narayanan, A.S., Siegel, R.C., and Martin, G.R., *Biochem. Biophys. Res. Commun.* 46, 745–751 (1972).
88. Schmidt, A., Datta, K., and Marcus, A., *Plant Physiol.* 96, 656–659 (1991).
89. Karvonen, K., Kokko-A., L., Pihlajaniemi, T., Helaakoski, T., Henke, S., Gunzler, V., Kivirikko, K.I., and Savolainen, E-R.,*J. Biol. Chem.* 265, 8415–8419 (1990).
90. Sun, I.Y-C., and Allfrey, V.G., *J. Biol. Chem.* 257, 1347–1353 (1982).
91. Gratecos, D., and Fischer, E., *Biochem. Biophys. Res. Commun.* 58, 960 (1974).
92. Li, H-C., Simonelli, P.F., and Huan, L-J.,*Methods in Enzymol.* 159, 346–356 (1988).
93. Cowburn, J.D., Farrar, G., and Blair, J.A., *Chemistry in Britain* 1169–1172 (1990).

94. Buxbaum, J.D., Gandy, S.E., Cicchetti, P., Ehrlich, M.E., Czernik, A.J., Fracasso, R.P., Ramabhadran, V., Unterbeck, A.J., and Greengard, P.E., *Proc. Natl. Acad. Sci.* 87, 6003–6006 (1990).

95. Birchall, J.D., and Chappell, J.S., *Lancet ii*, 1008–1010 (1988).

96. Pernelle, J-J., Creuzet, C., Loeb, J., and Gacon, G., *FEBS Lett.* 281, 278–282 (1991).

97. Turner, J.M., Brodsky, M.H., Irving, B.A., Levin, S.D., Perlmutter, R.M., and Littman, D.R., *Proc. Natl. Acad. Sci.* 85, 5190–5194 (1988).

98. Csermely, P., Szamel, M., Resch, K., and Somogyi, J., *J. Biol. Chem.* 263, 6487–6490 (1988).

99. Forbes, I.J., Zalewski, P.D., and Giannakis, C., *Exp. Cell Res.* 195, 224–229 (1991).

100. Parker, P.J., and Ullrich, A., *J. Cell Physiol. (Suppl)* 5, 53–56 (1987).

101. Kikkawa, U., and Nishizuka, Y., in *The Enzymes* 17, 167–189 (1986).

102. Rando, R.R., and Young, N., *Biochem. Biophys. Res. Commun.* 122, 818–823 (1984).

103. Nakamura, H., Kishi, Y., Pauares, M.A., and Rando, R.R., *Proc. Natl. Acad. Sci.* 86, 9672–9676 (1989).

104. Kong, F., Kishi, Y., Perez-Sala, D., and Rando, R.R., *Proc. Natl. Acad. Sci.* 88, 1973–1976 (1991).

105. Burns, D.J., and Bell, R.M., *J Biol. Chem.* 266, 18330–18338 (1991).

106. Kraft, A.S., and Anderson, W.B., *Nature* 334, 661–665 (1983).

107. Mochley-Rosen, D., Khaner, H., and Lopez, J., *Proc. Natl. Acad. Sci.* 88, 3937–4000 (1991).

108. Mochley-Rosen, D., Khaner, H., Lopez, J., and Smith, B.L., *J. Biol. Chem.* 266, 14866–14868 (1991).

109. Haystead, T.A.J., Campbell, D.G., and Hardie, D.G., *Eur. J. Biochem.* 175, 347–354 (1988).

110. Nakielny, S., Campbell, D.G., and Cohen, P., *Eur. J. Biochem.* 199, 713–722 (1991).

111. Poulter, L., Ang., S.G., Williams, D.H., and Cohen, P., *Biochem. Biophys. Acta* 929, 296–301 (1987).

112. Sekhar, K.R., Hatchett, R.J., Shabb, J.B., Wolfe, L., Francis, S.H., Wells, J.N., Jastorff, B., Butt, E., Chankinala, M.M., and Corbin, J.D., *Molecular Pharmacology* 42, 103–108 (1992).

7

Biological Effects of Covalent Modification

REVERSIBLE AND IRREVERSIBLE MODIFICATIONS

The great repertoire of the protein modification reactions makes it almost impossible to generalize about the biological significance of cellular modifications. Each of the modifications apparently has its own unique biological functions. Together, they add tremendously to the diversity and versatility of protein properties and functions. In a broad sense, however, cellular protein modifications may be divided into reversible and irreversible categories. The two categories of modification reactions with their distinct characteristics appear to serve different general purposes in the cells.

A reversible protein modification requires at least two enzymes, or in rare cases an enzyme with two catalytic sites, for example, adenyl transferase, which separately catalyzes adenylation and deadenylation of *E. coli* glutamine synthetase.[1] The enzymes are often referred to as converting enzymes. The modification of the target protein is accompanied by a change in intrinsic activity of the protein such as an activation, an inactivation, or a change in other properties. Such a covalent protein modification has been compared to a light switch, since it turns a cellular process on and off by shuttling the target protein between its modified and unmodified state.[2] Just as frequently, however, both the modified and the unmodified forms of the target protein are present in the cells, and the activity of the target protein is adjusted to various levels by modulating the relative activities of the two converting enzymes. The switch analogue can still apply except that, in this case, it is a dimmer switch. The reversible protein modifications, as is apparent from their characteristics, are especially suited as cellular regulation devices.

While an irreversible protein modification may involve the action of one or more enzymes, the modification is unidirectional. Among these, there are two general categories. In the first category, the modification reaction may be considered an intermediate or final step in the biosynthesis of the protein molecule. Such protein modifications are closely coupled to protein synthesis to ensure that the modified protein becomes functional at the right time and at the right cellular compartment. Frequently, the modification reaction in fact plays a key role in directing the finished protein to the right cellular location. For example, protein fatty acylation[3] is suggested to localize the protein to the cell membranes. Proteolytic cleavage of the signal protein[4] represents an essential protein modification in the secretion of secretory proteins. These irreversible protein modifications are often referred to as protein processing and/or maturation reactions. A unique feature of these modifications is that the unmodified form of the target protein is usually present in transient and very low amounts during normal physiological conditions.

In the second category of irreversible protein modification, the target protein in its unmodified form accumulates under normal physiological conditions. The unmodified protein generally exhibits little or no biological activity, but becomes active (functional) upon modification. A good example is the complement fixation where a cascade of protease zymogen activation reactions is involved. All the zymogens are present in the normal plasma in optimum concentrations, poised to be triggered in response to the specific physiological stimulus[5]. Thus, the main function of this type of protein modification appears to be the induction of biological activity, so as to commit the cell to a specific biological state.

To establish whether a protein modification is reversible or irreversible depends ultimately on the demonstration of the presence of the relevant converting enzymes in the particular cell types. However, many reversible protein modifications were suggested prior to the discovery of the respective converting enzymes. Before protein kinases and protein phosphatases were demonstrated, it had been known that injection of ^{32}Pi into animals resulted in the rapid incorporation of radioactivity into proteins and other biochemical substances in many different tissues[6,7] and that the protein-bound phosphate turned over at a much greater rate than the protein molecules themselves.[8] Similarly, pulse label experiments using cultured cells revealed that acetyl group on histone molecules had a turnover rate orders of magnitude greater than the protein, thus providing the initial evidence for a reversible histone acylation.[9,10] With more recent examples, the demonstration of a more rapid turnover of palmitate moiety than that of the protein it attaches to has led to the suggestion that protein palmitoylation is reversible, whereas similarity in the turnover rates of the myristate and the protein moieties suggests that myristoylation is an irreversible protein modification.[11,12] Using protein synthesis inhibitors, a tight coupling between protein synthesis and protein modification may provide another criterion for an irreversible reaction and has been used to indicate that protein myristoylation is mostly irreversible.[13]

Although *in situ* labeling experiments are valuable in discriminating reversible and irreversible protein modifications, they do not provide conclusive results. Carboxylmethylation of proteins is a reversible reaction, that is, the modified protein undergoes spontaneous demodification. Since pulse-chase experiments showed that the turnover rate of the protein methyl ester was similar to that of the protein itself, it was suggested that protein carboxylmethylation was not reversed enzymatically.[14] However, evidence for the existence of an enzyme activity catalyzing the demethylation reaction in rod outer segment membranes has been reported.[15]

Protein modification such as protein glycosylation often involves a highly complex process consisting of many steps. Analysis of the turnover rates of the protein and the modifying group is not at all straightforward. On the other hand, judicial application of the pulse labeling procedure may provide insight into the detailed molecular mechanism of the protein modification. For example, when different labeled carbohydrates were used to trace the carbohydrates bound to dipeptide peptidase IV, a plasma membrane glycoprotein, it was found that the terminal carbohydrates turned over much faster than the protein molecule, whereas the core sugars displayed a similar turnover rate as the protein. Thus, it appeared that the terminal sugars of the enzyme underwent deglycosylation and reglycosylation.[16,17] As protein glycosylation involves the highly orchestrated action of many modifying enzymes, it may not be surprising that the overall protein modification consists of both reversible and irreversible components.

Although protein modification reactions are generally classified into reversible or irreversible categories, such a classification should not be considered as absolute. A good example is protein ubiquitination. Protein ubiquitination is often irreversible in its involvement in the process of intracellular protein degradation. However, ubiquitination of histone does not appear to commit the protein to degradation. This specific protein ubiquitination appears to be a reversible modification. The turnover of histone molecule in the cells is much slower than that of the conjugated ubiquitin.[18] Ubiquitination of histone occurs in a cell division cycle-dependent manner, disappearing during mitosis and reappearing thereafter.[19]

GENERAL CRITERIA FOR THE BIOLOGICAL ROLE OF PROTEIN MODIFICATION

Reversible Modifications

The recognition that reversible protein modifications are important cellular regulatory mechanisms may be traced back to the discovery of the interconversions between glycogen phosphorylase b and glycogen phosphorylase a which are catalyzed by a protein kinase, phosphorylase kinase, and a protein phosphatase, phosphorylase phosphatase.[20,21,22,23] Because the two forms of phosphorylase have different intrinsic activities and their relative concentrations in liver or muscle cells are influenced by hormones, the regulatory significance of the phosphorylase phosphorylation and phosphorylase dephosphorylation is apparent. Ensuing studies have uncovered numerous other protein kinases and protein phosphatase as well as cellular processes regulated by these kinases and phosphatase. A few of the multitudes of the enzymes and the processes are presented in Table 7.1 to show the wide range involvement of the protein phosphorylation and dephosphorylation reactions in cellular regulation. While the cellular processes diverge from metabolism to highly complex phenomena such as differentiation, the underlying regulation mechanisms for these processes appear to share a common theme in that they all implicate cellular signal transduction. Protein phosphorylation and dephosphorylation reactions have prominent roles in signal transduction systems.

The modern concept of signal transduction owes its origin to the studies of the biochemical mechanisms of hormonal regulation. The first breakthrough in this area came from the successful demonstration of the stimulation of phosphorylase b to phosphorylase a conversion in a cell free system.[20,21] This system made it possible to investigate specific biochemical substance and their action in such a hormonal effect, the study led to the discovery of cAMP and the formulation of the concept of the intracellular second messenger.[69] Detailed studies of the enzymology of phosphorylase b to a conversion led to the demonstration that phosphorylase kinase could also exist in an activated and phosphorylated and a nonactivated and nonphosphorylated state.[70] This was followed by detailed enzymological studies of the phosphorylase kinase activation reactions, resulting in the discovery by Walsh, Perkins, and Krebs[71] of cAMP-dependent protein kinase. For the first time, the complete series of biochemical reactions that mediate the effect of a hormone (epinephrine) on a specific cellular process (glycogenolysis) was elucidated (Figure 7.1).

Table 7.1. Protein phosphorylation in cellular regulation

Cellular function	Regulatory Target protein substrate and/ or cellular process	Kinase phosphatase
Contraction and motility	myosin regulatory light chain, muscle contraction	MLCK MLCPt[24,25,26]
	Caldesmon, muscle contraction and cell motility	PKC, CaMK II, CaM-PkII,[27,28,29,30,31]
	Myosin heavy chain, contraction and cell motility	MHCK[32]
Cell proliferation	Growth factor stimulation of cell proliferation	GFR-Ks[33,34]
	Neoplastic growth induced by oncogenes	ONC-PK[35,36,37,38]
	Tumor promoted-induced cell proliferation	PKC, PPt[39,40]
Neuronal function	Synapsin, synaptic vesicle exocytosis	CaM-PK I and II A-kinase[41]
	GAP 43, axonal growth and long term potentiation	PKC, CaM-PK II CaMPTK[42,43,44,45]
	Tyrosine hydroxylase	PKA, PKC[46,47]
	Neurotransmitter synthesis	CaM-PK-II[48,49] D-PK[50]
Differentiation and Development	Development of Drosphila compound eye	Sevenless PTK[51,52]
	Specification of Drosphila body patterns	Terse-PTK[53] c-raf[54]
	Development of mouse pigmentation	c-kit-PTK
Ion homeostasis	Phospholamban, regulation of cardiac SR Ca^{+2} pump	A-kinase,[55,56] CaM-PK[57] PKC
	Regulation of neuronal K^+ channel	A-kinase,[58,59] CaM-PK II,[60,61] PhK and PKE
	CFTR, chloride channel	A-kinase,[62,63] CaM-PK II,[64,65] and PKC
Metabolism	Pyruvate dehydrogenase	PDH-K,[66] PDH-Pt
	Hydroxmethyl glutaryl-CoA reductase	HGM-CoA red K[67]
	Glycolysis and gluconeogenesis	A-kinase,[68]

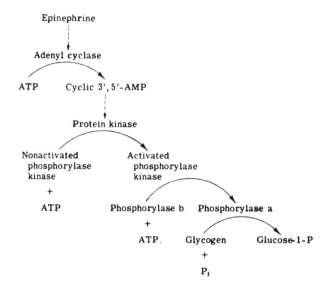

Figure 7.1. The mechanism of action of epinephrine in the regulation of glycogenolysis. Reprinted with permission from *The Enzymes*, 3rd edition, Vol. VIII, p. 561 (1973). Copyright 1973, Academic Press.

It had been well established that cAMP mediated the effect of a wide range of cell effectors on a multitude of cellular processes before cAMP-dependent protein kinase was discovered. Shortly afterward, cAMP-dependent protein kinase was implicated in many processes. Some of the biological processes were suggested to be regulated by the protein kinase because they were cAMP-dependent processes, whereas certain others were implicated by the identification of a specific *in vitro* substrate of the protein kinase. As more cAMP-mediated processes or more protein kinase substrates were discovered, more cellular functions were implicated as being regulated by cAMP-dependent protein kinase. It soon became clear that the suggestion of the involvement of cAMP-dependent protein kinase in a cellular process needed verification. Historically, Krebs[72] suggested five criteria that must be satisfied before an effect mediated by cAMP can be accepted to involve cAMP-dependent protein phosphorylation:

1. Cell type involved contains a cAMP-dependent protein kinase.
2. Protein substrate exists that bears a functional relationship to the process mediated by cAMP.
3. Phosphorylation of the substrate alters its function *in vitro*.
4. Protein substrate is phosphorylated *in vivo* in response to cAMP.

5. A phosphoprotein phosphatase exists to reverse the process.

Clearly, similar criteria may apply to other effector-regulated protein kinases, such as Ca^{+2}/CaM-dependent protein kinases, kinase C, EGF receptor kinase, and so on, as well as to protein kinases, which do not appear to be regulated by specific effectors. Since a cellular protein can often be phosphorylated by several protein kinases, it is also important to show that the site(s) phosphorylated *in vitro* by the protein kinase is the same as that phosphorylated *in vivo*.

With appropriate modification, a more general set of criteria may be formulated to judge the biological relevance of a reversible protein modification reaction. A protein modification can be accepted as biologically significant if the following criteria are satisfied:

1. The converting enzymes coexist with the target protein.
2. The protein modification can be demonstrated both *in vitro* and in the cells.
3. The sites of *in vitro* and *in situ* modification of the target protein are the same.
4. Functional activity of the target protein is altered similarly *in vitro* and *in situ*.

Although with the recent advances in cell biological and molecular biological techniques, more direct methods are available, the biological significance of a protein modification should still be examined against these criteria as part of its general characterization. The most important point contained in the set of criteria is that it takes both *in vitro* and *in vivo* experiments to establish biological significance of a protein modification reaction. As a protein modification reaction is usually detected either *in vitro* or *in vivo*, the respective complementary *in vivo,* or *in vitro* experiments, should be carried out.

Irreversible Modifications

Although it has not been possible to propose a set of formal criteria to evaluate the biological functions of irreversible protein modifications, general and unique biological functions have been suggested for some types of the modifications. For example, the various protein modifications involving conjugates of lipid moieties, such as palmitoylation, myristoylation, prenylation, and so on, appear to play the role of attaching the protein to cellular membranes; ubiquitination of proteins has been suggested to mark the proteins for degradation in the cells; and protein glycosylation has been implicated in many biological interactions, including cell-cell interactions, virus-host interactions, hormone-membrane receptor interaction, and cell-extracellular matrix interactions, and so on, and is therefore suggested to have the general function of biological recognition. On what basis are these suggestions about the general functions of the individual types of protein modifications made? An examination of a few selected examples may illustrate that complementary *in vivo* and *in vitro* experiments are also among the most important general approaches in the studies of biological functions of irreversible protein modifications.

Protein Ubiquitination

Cellular proteins undergo continuous turnover; both the protein synthesis and the protein degradation appear to be regulated. Among the early indications that protein degradation in cells are regulated processes was the observation that the release of amino acids from liver slices was strongly inhibited by agents known to deplete cell ATP such as cyanide and 2,4-dinitrophenol.[73] Later studies have shown that this ATP-dependent cellular protein degradation process is ubiquitously distributed in both prokaryotes and eukaryotes.[74] The rate of the protein degradation appears dependent of the nature of the protein as well as the physiological conditions of the cells. In particular, abnormal proteins are degraded much more rapidly than normal cell proteins. For example, incorporation of amino acid analogues in reticulocytes resulted in abnormal proteins, which were rapidly proteolyzed in an energy dependent manner.[75,76] The suggestion that protein ubiquitination is involved in cellular protein degradation has been derived, originally from the study of the mechanism of ATP-dependent protein degradation.

As is often the case, the major breakthrough in this study came from the successful demonstration of the ATP-dependent protein degradation in a cell-free system, that is, reticulocyte extract (Figure 7.2A).[77] The ATP-dependent protease system was soon fractionated into two fractions on DEAE-cellulose column. Fraction I displayed no protease activity, but could greatly stimulate the ATP-dependent protease activity of fraction II (Figure 7.2B).[78] The active principle in fraction I was purified and found to be a small and heat-stable protein.[79] The protein, ubiquitin, became covalently attached to reticulocyte proteins in an ATP-dependent manner, the ubiquitinated proteins were then proteolyzed.[80] Primarily on the basis of these observations, ubiquitination has been suggested to be an obligatory step in the ATP-dependent cell protein degradation. Strong support for this suggestion has subsequently come from both *in vitro* and *in vivo* studies.

From *in vitro* studies, ubiquitination of the protease target proteins has been shown to involve three consecutive reactions depending on three protein components: (1) formation of a complex between ubiquitin and ubiquitin activating enzyme, E_1, via a thioester bond in an ATP-

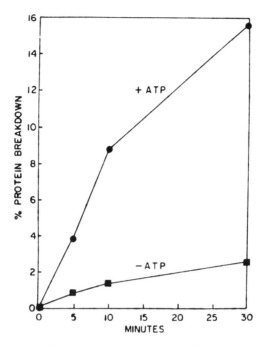

Figure 7.2A. Ubiquitin-mediated proteolysis—demonstration in cell free systems. Effect of ATP (1 mM) on the degradation of C1Abu-containing protein. Reprinted with permission from *Proceedings of the National Academy of Sciences, USA* 74:56 (1977).

dependent reaction; (2) transfer of ubiquitin moiety from the ubiquitin-E_1 adduct to a smaller carrier protein, E_2 via a transthiolation reaction; and (3) transfer of ubiquitin moiety from the E_2-ubiquitin adduct to an amino group of the target protein, a reaction catalyzed by ubiquitin protein ligase, E_3 (described more fully in Chapter 2). All these protein components of the ubiquitin pathway have been isolated and the ubiquitin pathway can be reconstituted from the isolated components.[81,82] Using the *in vitro* ubiquitination system, a complex ATP-dependent protease that specifically breaks down ubiquitinated proteins has been demonstrated.[83]

While *in vitro* studies have produced great insight into how the process of ubiquitination-mediated proteolysis may be carried out, *in vivo* studies are essential in establishing the occurrence of protein ubiquitination in the cells and to address the question of the relationship between protein ubiquitination and intracellular protein degradation. A number of studies have been carried out to show the correlation between protein ubiquitination and the degradation of cellular proteins. When

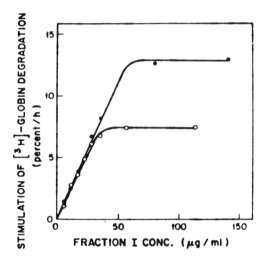

Figure 7.2B. Ubiquitin-mediated proteolysis—demonstration in cell free systems. Stimulation of DEAE Fraction I-catalyzed H³ globin degradation by DEAE Fraction 2 (Ubiquitin), Fraction I protein concentration: 2.9 (open circles) and 5.8 mg/ml (closed circles). Reprinted with permission from *Biochem. Biophys. Res. Commun.* 81:1104 (1978). Copyright 1978, Academic Press.

amino acid analogs were introduced into the cells, the increased rate of protein degradation was accompanied by increased intracellular protein ubiquitination.[84] When a mammalian cell division cycle mutant cell containing a temperature sensitive ubiquitin activating enzyme, E_1 was placed at the restrictive temperature, the degradation of normally short-lived protein was greatly retarded.[85] When a series of b-galactosidase derivatives with modified NH_2-terminal residues was introduced into yeast, the rates of protein ubiquitination and the extents of protein degradation were found to be similarly dependent of the N-terminal residue of the protein derivative.[86] In *S. cerevisiae*, the form of E_3 was recently shown to be required for the integrity of this N-end rule dependent protein turnover.[87]

From these early studies it was clear that the ubiquitin-mediated protein degradation plays a role in the removal of abnormal proteins from the cells. The biological significance for such a role is apparent to cells under a variety of pathological conditions. For example, upon a sudden elevation of the temperature, cells undergo heat shock response, which involves the production of a set of specific proteins — heat shock proteins. One of the heat shock proteins is ubiquitin. Most of the ubiquitin in cells undergoing heat shock exists as protein conjugates,

Table 7.2. Biological role of protein ubiquitination

Protein target	Cellular function affected
Phytochrome	Red light-mediatedmorphogenesis in plants[97]
MAT α 2 repressor	Cell type specific transcription[98]
Cyclin	Cell cycle progression[47]
p53	Tumor suppression[99]
Myc, Fos	Nuclear oncogene function[100]
Histones H2A, H2B	Transcription control[101]

presumably to channel the partially denatured proteins into the ubiquitin proteolysis pathway.[88] Similarly, ubiquitin has been implicated in a number of degenerative diseases; many of these diseases are associated with the appearance of cytoplasmic inclusion bodies, which can be stained by ubiquitin antibodies.[89]

More importantly, ubiquitin proteolysis pathway carries out protein degradation with specific physiological functions in normal cells. Its involvement in cell cycle has been suggested from studies of cell mutants. The primary defect of a temperature sensitive cell cycle mouse cell mutant, Ts65, is a ubiquitin-activating enzyme that is thermally labile.[90] At a nonpermissive temperature, the degradation of many short-lived proteins as well as protein ubiquitin reactions was severely curtailed in the mutant cell.[85] Similarly, a yeast cell division cycle gene, CDC34, implicated in Gs to S phase transition, has been shown to be a member of the ubiquitination enzyme family.[91] Direct evidence for the vital role of ubiquitin-mediated proteolysis in cell cycle has come from the observation that cyclin B in Xenopus egg extracts is degraded via the ubiquitin proteolysis pathway.[92] The involvement of ubiquitination in cyclic degradation and the observation that degradation of cyclin B alone is sufficient and essential for the cell to exit mitosis,[93] have clearly shown that protein ubiquitination plays vital physiological roles. Table 7.2 lists a number of other cell proteins that are degraded by the ubiquitin proteolysis pathway and the physiological functions of these proteins. Excellent reviews[94,95,96] on protein ubiquitination are available.

Covalent Attachment of Long Chain Fatty Acids

A wide range of eukaryotic cellular proteins as well as proteins of viral origins have been demonstrated to contain covalently bound long chain fatty acids. The mode of the protein modifications by fatty acids may be categorized into four groups, which differ from each other in the structure of the lipid moiety, the chemical linkage between the protein and the lipid, and the biochemical mechanism of the protein modification. These are: (1) attachment of palmitate to the proteins via a cysteine or a serine/

threonine residue by a thioether or ester bond; (2) attachment of myristate to an aminoterminal glycine of the protein via an amide bond; (3) polyprenylation of the protein at a carboxyterminal cysteine via a thioester bond to introduce either a farnesyl or a geranylgeranyl residue to the protein; and (4) linking a glycosylated phosphatidylinositol via an amide bond between the carboxyl group of the protein carboxyterminal residue and ethanolamine of the glycosylated phosphatidylinositol moiety (Chapter 4). There has never been any doubt that these protein modifications occur in living cells. Cellular proteins belonging to each of the categories have been isolated from tissues and cells under physiological conditions. In addition, these protein modifications have all been demonstrated to occur in living cells using labeled fatty acids or fatty acid precursors.

An early suggestion for the primary function of these fatty acid adducts is the association of the modified proteins to cellular membranes. It is generally believed that a stable interaction between a membrane protein and the membrane depends at least in part on the hydrophobic interaction between the protein and the hydrophobic core of the lipid bilayer. Therefore, it seems intuitively correct to suggest that the long chain fatty acid adducts play the role of anchoring the modifiedproteins to the membrane structure. Many cell proteins that contain covalently attached long chain fatty acids have indeed been found to be associated with membrane structures.

More than forty proteins containing the glycosylated form of phosphatidylinositol have now been identified; all of these are membrane proteins.[102,103] Some of these proteins are presented in Table 7.3. These proteins are from various organisms spanning a wide range of the evolutionary scale, and possess diverse biological functions. However, they share the common property of being at cell surfaces. The membrane anchoring function of the glycosylated phosphatidylinositol is most strongly supported by the observation that these proteins could be released from membrane surfaces by treating the cells with bacterial phospholipase C, which is specific for phosphatidyl inositol.[104,105] The released protein could not be rebound to lipid vesicles, thus suggesting that the effect of the bacterial phospholipase C was on the protein molecule rather than on phosphatidyl inositol of membrane lipids.[106] These studies provide the basis for the notion that these proteins are held to the cell surface solely by their glycolipid anchors.

Table 7.4 lists selected proteins modified by other lipid modifications, palmitoylation, myristoylation, and isoprenylation. Like proteins with glycolipid anchors, almost all palmitoylated proteins display high affinity association with membrane structures, whereas a few among the known myristoylated proteins are cytosolic. The first two proteins

Table 7.3. Selected examples of proteins with a glycosyl-
phosphatidylinositol membrane anchor[a]

Hydrolytic enzymes:
    Trehalose (rabbit tissues)
    Acetylcholinesterase (human erythrocytes, Torpedo)
    Dipeptidylpeptidase (pig kidney)
    Alkaline phosphatase (human placenta, liver, intestine, etc.)
    Lipoprotein lipase (mammalian tissues)
    Merozoite protease (plasmodium falciparum)
Antigens:
    CD16-II (natural killer cells, neutraphils)
    Thy-1 (rodent)
    Ly-6.1 (ELT-cells)
    Variant surface glycoprotein (VSG; Trypanosoma)
Cell adhesion proteins:
    Heparan sulfate proteoglycan (rat liver)
    Dictyostelium contact site A (Didiscoideum)
    Neural cell adhesion molecule (mouse, chicken)
    LFA-3 (lymphoid cells, erythrocytes)
Miscellaneous
    Hepatoma glycoprotein (130 kDa)
    Folate binding protein
    Oligodendrocyte-myelin protein
    Scrapie prion protein (infectious particle, hamster brain)
    Decay-accelerating factor (blood cells, other tissues)
    Zymogen granule GP-2 (pancreatic secretory granules)

---

[a]Selected from review articles.[102,103]

discovered to be myristoylated, cAMP-dependent protein kinase (catalytic subunit)[117] and calcineurin (b subunit)[118] are mostly present in the cytosol. In several studies[13,119,120] cultured cells were labeled by [³H]myristate or [³H] palmitate, the labeled proteins were then analyzed by SDS-PAGE and autoradiograph following subcellular fractionation. Palmitoylated proteins were found exclusively in the particulate fraction. On the other hand, a small fraction of myristoylated protein was found in the cytosolic fraction. The cytosolic and particular myristoylated proteins represented distinct sets of proteins, rather than the same proteins bimodally distributed. Isoprenylation,[121] the most recently discovered protein lipid modification, affects many G proteins. Essentially all the prenylated proteins are membrane associated.

The best support for a direct role of the lipid moiety in the protein membrane association has come from protein mutagenesis studies. The

Table 7.4. Representative proteins with covalent fatty acid modifications

Palmitoylated proteins[107,108]
Sodium channel
Rhodopsin
Vinculin
Nicotinic receptor
Insulin receptor
Vesicular stomatitis virus glycoproteins
p21[ras]
Simian virus 40 large T antigen
Myristolylated proteins[107,108]
Catalytic subunit of cAMP-dependent protein kinase
Calcineurin β-subunit
P60src
a-subunit guanine nucleotide binding protein
NADH cytochrome b5 reductase
gag proteins of various viruses
Polyomer virus capsid protein VP2
Prenylated proteins
Rap protein[109,110]
lamins A and lamins B[111,112,113]
Photoreceptor G protein[114]
Xenopus Ko kDa protein[115]
ras proteins[116]

[a]From the review articles of Hu et al.[107] and Towler et al.[108]

oncogene G protein, p21[v-H-ras] is a palmitoylated protein.[122] When the cysteine residue of the acylation site was changed by mutagenesis, the protein membrane interaction was completely blocked.[123] Similarly, mutating the site of isoprenylation of ras protein abolished the membrane association and transforming ability of the protein.[124] Likewise, the b_2-adrenergic receptor was reported to be modulated by palmitoylation.[125] Prevention of palmitoylation by site directed mutagenesis yielded a protein which was uncoupled to G_s protein in the membrane. Incorporation of other fatty acids also may influence the function of the target protein. The protein tyrosine kinase oncogene product, pp60[v-src] is normally a myristoylated protein.[126] When the amino terminal glycine residue was mutated by site-directed mutagenesis, the protein no longer showed a membrane localization.[127] For the oncogene proteins, the intrinsic activities of these proteins did not seem to be affected by the protein mutations, but the cell transforming activities of the proteins were lost.[123,128]

In addition to playing a role in membrane association, the fatty acid moiety of the protein conjugates may have other functions. Vesicular stomatitis virus glycoprotein (VSV) is palmitoylated. The palmitoylation of this protein in the infected animals could be blocked by the antibiotic cerulenin.[129] While the protein synthesis and glycosylation was not significantly affected by the antibiotics, the virus particle formation was markedly inhibited, suggesting a role of the fatty acid moiety on virus assembly. Hemagglutinin of influenza viruses contains palmitate. When the virus was treated by hydroxylamine to remove the fatty acid, it lost the ability to mediate erythrocyte hemolysis.[130]

The observation that a few myristoylated proteins are cytosolic suggests that the myristate moiety has functions in addition to membrane association. Although the exact functions of this fatty acid moiety are not known, one of these may be related to protein-protein interactions. Indeed, even for some of the membrane associated myristoylated proteins, the association with membrane may depend on the presence of specific protein receptors in the membrane. Thus, newly synthesized v-src protein was shown to bind specifically to plasma membranes in a saturable manner. This binding could be eliminated by protease treatment of the membrane or competitively inhibited by amino acid myristated peptide corresponding to the amino terminal sequence of v-src protein.[131] Similarly, the trimeric G_o protein, $\alpha_{Go}\beta\gamma$, contains the myristoylated α_{Go} subunit. Upon dissociation, the $\beta\gamma$ subunit but not α_{Go} subunit could bind to phospholipid vesicles, suggesting that the membrane localization of this protein was due to $\beta\gamma$ subunits.[132,133]

Glycosylation

The sugar moieties of glycoproteins have been suggested to play the roles of specificity determinants in many vital cellular processes, which involve specific interactions of whole cells. There have been increasing examples of the complex carbohydrates of glycoproteins serving recognition determinants in host-pathogen interaction, cell-cell interaction, cell-extracellular matrix interaction, and cell membrane-protein targeting process. The enormous structural diversity of the protein-conjugated carbohydrates, the complexity of the biosynthetic pathways of glycoproteins, the tissue-specific and the developmentally controlled expressions of glycoproteins, and the plasticity of the structure of the complex sugars of the glycoproteins in response to physiological and pathological conditions provide further support to the importance of the carbohydrate of glycoproteins in specific cell interactions.

Virus infection. The earliest indication that cell surface sugars might be important biological recognition determinants was derived from studies of the interaction between human erythrocytes and influenza virus. Human erythrocytes had been known to be induced to undergo agglutination upon incubation with influenza viruses. This virus-induced erythrocyte agglutination was found to be eliminated if the cells had been pretreated with neuraminidase,[134] thus implicating cell surface carbohydrates in the virus-induced cell agglutination. Subsequently, a glycoprotein enriched fraction was isolated from human erythrocyte membranes and shown to inhibit the virus-induced erythrocyte agglutination. The inhibitory activity of this protein sample could be eliminated if the sample was treated by neuraminidase.[135]

The most convincing evidence for the specificity role of the membrane sugars in virus-host cell interaction came from studies using specific sialyltransferases. In one study[136] three isolated sialyltransferases, each catalyzing the formation of a distinct sialic acid linkage, were tested for their ability to restore cell agglutination of the neuraminidase-treated erythrocytes by each of 4 distinct viruses. Only β galactoside α 2- 3 sialyltransferase was capable of restoring the agglutination of the neuraminidase-treated erythrocytes to all four viruses. The other two sialyltransferase could restore the agglutination of the erythrocytes to influenza A viruses only. In addition to demonstrating the involvement of carbohydrates in the virus-host cell interactions, the results showed the importance of carbohydrate in species-specific cellular interactions. The crystallographic elucidation of the structure of influenza HA complexed with the trisaccharide sialyllactose provided the final proof for the role of sialylgalactose as the specificity determinant in the influenza virus and host cell interaction.[137]

For virus-host cell interactions, the sugar moiety that determines the interactions specificity may also be from a viral glycoprotein. For the infection of T-cells by human imuno-deficiency virus (HIV), the carbohydrate specificity determinant is provided by the virus. The retrovirus HIV infects specifically a subset of human T-lymphocytes. This specific infection is mediated by the interaction between a viral envelope protein gp120 and a T-cell surface glycoprotein CD4.[138] Like many other lymphocyte-specific surface proteins, CD4 was identified as antigens of specific antibodies. Some of these monoclonal antibodies are also instrumental in identifying CD4 as the T-lymphocyte receptor of HIV, as they could inhibit the virus infection. Similarly, specific antibodies of gp120 are capable of blocking virus-T lymphocyte interactions, thus providing initial evidence for the involvement of the viral envelope proteins in targeting the virus to T-lymphocytes.[139]

Although both gp120 and CD4 are glycoproteins, only the carbohydrate moieties of the viral protein appears to participate in the function of recognition determinant of this interaction. While the isolated viral envelope protein blocks the interaction between the virus and T-lymphocytes, a recombinant gp120 analog expressed in *E. coli* thus devoid of carbohydrate moieties was found to be inactive in blocking the virus-cell interaction.[140] Similarly, treatment of the isolated viral envelope protein by endo F to remove its conjugated sugar resulted in drastic reduction in the protein's effectiveness in inhibiting virus infection.[141] In contrast, various recombinant CD4 analogs, including those deleted of all potential glycosylation sites, have been found to retain the inhibitory activity toward virus infectivity.[142,143]

Cell-cell interaction. The specific cell-cell interactions and cell-extracellular matrix interactions are among the most important and versatile cellular interactions, as they play key roles in such fundamental biological processes as egg fertilization, embryonic development, tissue differentiation, neuronal functions, immune responses, and so on. They have also been implicated in various pathological processes including tumor transformation and metastases. There are two general types of cell-cell interactions—those between cells of the same type, and those involving cells of distinct types. The specific cell-cell interaction of the first type is one of the underlying principles for the organization of multicellular organisms into tissues. The fertilization that is initiated by the specific interaction between a sperm and an egg is probably one of the simplest cell-cell interactions involving distinct cell types.

Marine sponges, the simplest multicellular organisms containing only a few cell types, have been used to demonstrate specific cell-cell interactions. Sponges can be dissociated into single cells under relatively gentle conditions, and the dissociated cells are capable of reaggregating and rearranging to form a normal sponge. In a classical study by H. V. Wilson, different species of sponges that had different colors were dispersed into single cells and then mixed to allow reaggregation. The cells reaggregated in a species-specific manner—that is, cells aggregated into sponges of distinct colors.[144] If the sponge cell dissociation was carried out in sea water depleted of Ca^{+2} and Mg^{+2}, a factor called the aggregation factor was separated from the cells; and the cells could not associate into sponges unless the factor was readded to the aggregation reaction.[145] The aggregation factor has since been extensively purified and found to be a proteoglycan.[146] The dissociated sponge cells cannot be induced to associate by the aggregation factor in the presence of glucuronic acid.[147] No other sugars are capable of substituting for glucuronic acid in preventing the aggregation, thus implicating the sugar moiety of the aggregation factor as the specificity determinant of the cell-cell interaction.

Cell-cell interaction and morphogenesis. Studies using dissociated embryonic cells have provided strong support to the notion that specific cell-cell interaction is among the most important principles underlying the morphogenetic development of the embryos. Embryonic tissues can be dissociated into cell suspension by mild protease-treatment, and then reassociated into cell aggregates. If cells of two different embryonic tissues are mixed to allow reassociation, they form segregated regions of cell assemblies of homologous cells in a time-dependent manner.[148] This tissue-specific cell-cell interaction may be vividly illustrated by using two types of chick embryo cells, those of neural retina and pigmented retina, which are nonpigmented or darkly pigmented, respectively (Figure 7.3). When cells dissociated from neural retina and pigmented retina of a 7–day-old chick embryo were maintained on a suspension culture, they rapidly associated into aggregates of mixed cells. Gradually the pigmented and nonpigmented cells sorted themselves out from each other to form segregated assemblies of homologous cells. After 2 days, pigmented retinal cells formed a single mass, which was surrounded by the nonpigmented neural retina cells.[148]

There were numerous clues in literature suggesting that cell-cell interaction depended on specific cell membrane proteins. Plasma membranes isolated from neural retina of chick embryo bound preferentially to neural retinal cells, resulting in specific inhibition of assembly of these cells.[149] The specific association of embryonic cells is developmentally regulated. Cells dissociated from the same embryonic tissues associate preferentially with those from the embryos of the same age. In accordance with this observation, plasma membranes of neural retina from chick embryos of different ages showed preferential inhibition of the assembly of neural-retinal cells from the same age chick embryos.[150] Glycoproteins were isolated from various cells and tissues that could block cell assembly. However, definitive identification of membrane proteins that mediate the specific cell-cell interaction had to wait for the methodological advances in the assay of cell-cell interaction, protein and immunological chemistry, and molecular and cell biology.

The introduction of the immunological approach may represent the single most important innovation in the study of the molecular basis of cell adhesion. For example, Edelman and coworkers[151] initially developed antibodies against surface proteins of chick embryonic neural retinal cells. Some of these antibodies were capable of blocking the *in vitro* assembly of the dissociated neural retinal cells. The availability of

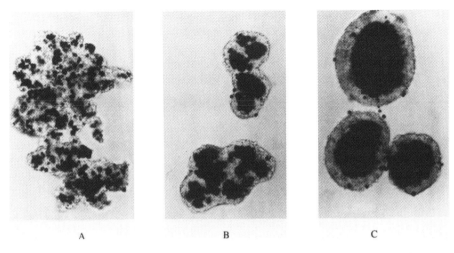

Figure 7.3. Specificity of cell interaction as demonstrated by mixing 7-day old chick embryo neural retinal (unpigmented) and pigmented retinal cells (darkly pigmented) in suspension culture: (a) after 5 hrs., (b) after 19 hrs., and (c) after 2 days. Reprinted with permission from *CRC Critical Reviews in Biochem. & Molec. Biol.* 24:120 (1991). Copyright 1991, CRC Press, Inc., Boca Raton, FL.

these antibodies made it possible to identify putative membrane proteins involved in cell-cell interaction, thus leading to the purification of a unique adhesion protein, neural cell adhesion molecule (N-CAM).[152] Of equal importance, these antibodies could be used to quantify the presence of specific cell adhesion molecules.[153] The strategy has been used for the identification of other cell adhesion molecules.

Carbohydrate moieties of cell surface glycoproteins appear to play essential as well as modulatory roles in cell adhesion. Several stage-specific embryonic antigens of mouse have been identified and their respective antibodies shown to be directed toward specific carbohydrate structures of surface glycoproteins. For example, one stage-specific antigen appearing at 8-16 cell stage of the mouse embryo has the epitope structure of [Gal]β 1,4(Fucα1,3) GlcNAc-R]. The appearance of this antigen during embryogenesis coincides with the packing of loosely-associated embryonic cells into a smooth ball, a process known as compaction.[154] A soluble analog of the antigenic carbohydrate is capable of inhibiting the cell compaction,[155] thus directly implicating the carbohydrate moiety in the cell assembly process.

Carbohydrate moieties of N-CAM are strikingly variable; this variation in structure appears to play important roles in morphogenesis. For example, during embryonic development, the carbohydrate moiety of N-

CAM has the unique structure of containing polysialic acid chains with the sequence NeuAcα 2,8-(NeuAcα2,8)$_n$ NeuAcV8,1 2,3 Gal-R where n = 20-200.[156,157] While this polysialic acid structure is not found in N-CAM of adult animals, it is present in N-CAM of highly metastatic tumor cells such as Wilm's tumor cells.[158] Nerve fibers in an animal usually grow as assembled bundles. N-CAM has been suggested to play an important role in modulating this nerve fiber bundling. There is good evidence to suggest that the polysialic acid structure prevents tight association between cells, so as to provide dynamic interactions among cells.[159]

Fertilization. Fertilization is among the most studied systems involving interactions between distinct cell types. The interaction is carried out between two well-defined cells, the sperm and the egg, and it is of obvious biological importance. Since the process of fertilization is species specific, molecular interactions derived from the fertilization reaction can be assessed for their biological relevance on the basis of their species specificity. The reaction of sperm and egg is also relatively easy to manipulate and to monitor. This is especially true for that of simple marine organisms such as sea urchins. Fertilization of sea urchin eggs takes place in sea water. A typical female and male sea urchin produces millions of eggs and billions of sperms, respectively. Sea urchin eggs are surrounded by a cellular structure, an outmost envelope called the jelly coat, and a layer of glycoproteins, the vitelline layer. During egg fertilization, the sperm first reacts and penetrates the jelly coat, and then interacts with specific glycoproteins of the vitelline layer.

The species-specific interaction between the sperm and the egg is mediated by a sperm-specific protein called bindin[160] and a glycoprotein from the vitelline layer of sea urchin eggs. The egg glycoprotein and the sperm-derived Bindin have been purified from several sea urchin species.[161,162] The purified proteins are capable of blocking the interaction between the sperm and egg in a species-specific manner. An early event of sea urchin egg fertilization involves the proteolytic removal of a glycopeptide from the vitelline glycoprotein. This glycopeptide appears to contain the specificity determinants of the sperm-egg interaction, since its removal from sea urchin eggs is accompanied with loss of the egg's ability to undergo specific interactions with additional homologous sperm cells.[163] In addition, the isolated glycopeptide is capable of binding to homologous sperms and to block the fertilization of fresh eggs.[164]

The carbohydrate moieties of the sea urchin egg glycoprotein play an important part in the specificity determination of the sperm-egg interaction. The glycopeptide of certain species of sea urchin eggs is capable of binding to concanavalin A, the fertilization of these eggs is blocked by this lectin.[165] Bindin is capable of inducing agglutination of its homologous sea urchin eggs. This ability of the egg to undergo agglutination by bindin is inhibited by low concentrations of galactose, lactose, and fucose, or abolished when the eggs have been subjected to periodate oxidation.[166]

Mammalian eggs are also surrounded by a layer of glycoprotein coat called the Zona Pellucida. This layer is similar to the vitelline layer of sea urchin eggs in that it is the site of species-specific recognition of the interacting sperm cells. The mammalian glycoprotein layer contains relatively few glycoproteins, one of these, ZP3, has been suggested to be the species-specific sperm receptor protein on the basis of various observations.[167] For example, ZP3 purified from mammalian eggs and oocytes has been shown to bind to the plasma membrane of the sperm head of homologous species, whereas other glycoproteins from Zona Pellucida do not bind to the sperm.[168]

Both N- and O-linked oligosaccharides are present in ZP3, however, only the O-linked sugar appears to participate as a species-specific sperm receptor determinant.[169,170] Treatment of ZP3 with mild alkaline conditions (β-elimination reaction) to remove the O-linked sugar destroys the sperm-binding activity of the protein. On the other hand, removal of the N-linked sugar from ZP3 by endoglucoseminidase F does not significantly affect the sperm-binding ability of the protein (Figure 7.4). Furthermore, the O-linked carbohydrate derived from ZP3 by alkaline-borohydride hydrolysis can potently inhibit the fertilization of the eggs of homologous animal species.

Although a mammalian egg receptor protein similar to Bindin of sea urchin sperms has not been unequivocally identified, galactosyl transferase, an enzyme catalyzing the transfer of galactase from CCDPGal to terminal N-acetylglucosamine residue to form N-acetyllactocalinine, has been implicated as the egg receptor of mouse sperm,[171] in consistence with the notion that the O-linked carbohydrate of ZP3 plays the role of specificity determinant.

Protein Targeting. The study of Ashwell and co-workers[172] on the mechanisms of protein clearance from the bloodstream of mammals has been among the most influential in bringing into focus the role of the carbohydrates of glycoproteins in biological recognition. Purified ceruloplasmin, treated with neuraminidase to remove the terminal sialicacid of its conjugated sugar and then reinjected into rabbits, was found to be rapidly cleared from serum and taken up by the liver, whereas the untreated ceruloplasmin has a long-circulating life span.[173] This rapid clearance of asialated ceruloplasmin can be attributed to the nonreducing end galactose residue of the carbohydrates that is exposed by the neuraminidase treatment. The neuraminidase-treated ceruloplasmin can be restored to close to the original circulating life span

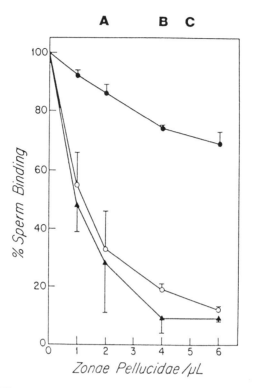

Figure 7.4. Effect of NaOH treatment on 2P3 inhibition of sperm-egg interaction. The inhibitory activity of the NaOH treated protein (closed circles) is compared with the untreated (closed triangles) and distilled water-treated (open circles) samples. Reprinted with permission from *Cell* 41:315 (1985). Copyright 1985, Cell Press.

if the protein has been treated with galactose oxidase, β-galactosidase, or asialyltransferase.[172,174] From these observations, it was deduced that rabbit liver has a specific asialoprotein receptor. A long list of other plasma glycoproteins (Table 7.5) have subsequently been shown to be cleared rapidly from circulation if they are bound to the asialoforms.[172]

Asialoprotein binding factor has been purified from Triton X-100-solubilized liver membranes of a number of mammals, and the purified membrane factor is often referred to as mammalian hepatic receptor.[175,176,177] Studies of the interaction between the purified receptor and its ligands have provided strong support to the pivotal roles of the carbohydrate moieties in the specificity determination of the interaction. The receptor-ligand interaction can be inhibited by free galactose, methylglycosides of galactose, and methylglycosides of N-

Table 7.5. Selected circulating glycoproteins undergoing clearance by
asialoglycoprotein receptor[a]

Erythropoietin
Interferon
Ceruloplasmin
Thyroxin-binding globulin
Haptoglobulin
Immunoglobulin CT
Carcino-embryonic antigen
Prothrombin

[a]From a review article by Ashwell and Harford.[172]
Reprinted with permission from Annual Review of Biochemistry 51:531 (1982). Copyright
1982 Annual Review Inc.

acetylgalactosamine. The sugar binding site of the receptor appears to be
relatively small, capable of accommodating the terminal galactose and
portion of the adjacent monosaccharide unit.[178] Methylglycosides of N-
acetylgalactosamine are much more potent inhibitors for the receptor-
ligand interaction than the corresponding methylglycosides of galactose.[179]

Proteins that normally do not contain carbohydrates can be chemically linked to specific sugars. These synthetic glycoproteins are especially useful in the study of biological roles and the mechanism of action
of carbohydrate moieties of glycoproteins. For example, while a galacto-
side-conjugated bovine serum albumin could undergo high affinity
binding to the hepatic receptor, mannoside-conjugated or N-
acetylglucosaminide-conjugated bovine serum albumin could not.[180] Oligosaccharides analogous to the sugar moieties of the reacting glycoproteins displayed similar affinities towards the receptor as the glycoproteins themselves.[181,182] It appears, therefore, that the sugar moieties of
the protein ligands are solely responsible for the specific interaction.

What is the biological role of such interaction between circulating
glycoproteins and hepatic receptor; or more specifically, the biological
role of the sugar moieties of these glycoproteins? Several observations
suggest that this is a mechanism for the regulation of the life span of
circulating protein. The binding of the glycoproteins to the membrane
receptor occurs partly in coated pits, a specialized membrane structure
involved in endocytosis. The internalized glycoproteins are degraded in
lysosomes.[183,184] Normal mammalian blood contains very low levels of
asialoproteins that are elevated in patients of various liver diseases.[185]
Presumably, most circulating glycoproteins contain N-linked oligosaccharides with terminal sialic acid residues. They are, therefore, stable in
the bloodstream of healthy individuals until attacked by a neuramini-

dase to expose a terminal galactose residue. On the other hand, patients with liver diseases are defective in cleaving the asialoproteins. Together, these observations suggest that the neuraminase digestion and the hepatic receptor uptake reaction are both parts of the general process for the modulating of the life spans of circulating proteins. Such a suggestion is supported by the observation that avian liver contains a glycoprotein receptor specific for N-acetylglucosamine instead of galactose;[186] and correspondingly, avian conculation has relatively high levels of glycoproteins with terminal galactose on the sugar moieties.[187]

The notion that the carbohydrate moieties of circulating glycoproteins play the general role of determining the life span of the proteins has been discussed in a recent review.[188] Hepatic receptor is a protein of multiple subunits, the multiple subunit may result in selective binding of the sugar moieties. For example, the receptor binding of a cluster of galactose termini in a branched oligosaccharide has a much higher affinity than that for a single galactose residue.[189,190] Thus, glycoproteins with multiple N-linked saccharides may have a shorter circulating life span than those with single N-linked oligosaccharides. In addition, there are other membrane glycoprotein receptors with different sugar specificities. These receptors may be involved in the clearance of circulating glycoproteins that contain sugars of distinct structure. For example, liver also contains a mannose-specific receptor.[191] Very recently, a receptor specific for sulfated N-acetylgalactosamine[192] has been identified. Such a sugar signal is found in glycoprotein hormones such as pituitary hormones,[193] thus rendering the hormones short-lived in circulation. Figure 7.5 shows a schematic representation of the glycoproteins clearance systems, which may illustrate how the sugar moieties of glycoproteins are processed to destine the glycoproteins to specific clearance pathways.

One of the best characterized targeting functions of a protein-conjugated carbohydrate moiety is the targeting role of mannose-6-P residue of lysosomal enzymes. Neufeld and co-workers.[194,195] originally observed the uptake of lysosomal enzymes by cultured cells and showed that this protein uptake had the characteristics of a receptor-mediated process. Some catalytically active lysosomal enzymes could be shown to exist in either a low uptake or a high uptake state; periodate oxidation of the high uptake state form could convert the enzyme to the low uptake state. The observation provided the initial indication that the carbohydrate moieties play a role in the cell uptake of lysosomal enzymes.[195]

The suggestion that mannose-6-P residue is the structural determinant of the process has come mainly from the studies of competitive inhibition of the cell uptake of lysosomal enzymes by glycoproteins and

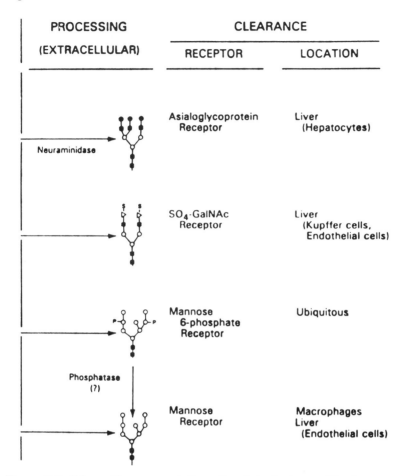

Figure 7.5. Schematic representation of carbohydrate-specific glycoprotein clearance systems. Reprinted with permission from *Cell* 67:1029
(1991). Copyright 1991, Cell Press.

carbohydrates. Among the various glycoproteins and carbohydrates
examined, those containing mannose-6-P residue all displayed potent
inhibition of the cell uptake of the lysosomal enzymes. Many phosphates
of monosaccharides were tested, and only mannose-6-P and fructose-1-
P were found to be potent inhibitors (Table 7.6).[196] The existence of a
mannose-6-P specific membrane receptor was soon established by the
demonstration of the specific binding activity with isolated cell membranes and with the solubilized membranes.[197]

Two mannose-6-P specific receptors of different molecular weights
and showing different dependence of metal ions for ligand binding

activities have been purified.[198,199,200] The high molecular weight receptor, ~Mr 215,000, binds ligand independent of metal ions; the low molecular weight receptor, ~Mr 46,000, displays a metal ion-stimulated binding activity.[201,202] Both receptors are involved in transporting newly synthesized lysosomal enzymes to lysosome, whereas the high molecular weights receptor has the additional function of carrying out endocytosis of lysosomal proteins from outside of the cells. Cultured cells deleted of the high molecular weight receptor secrete large amounts of lysosomal enzymes, and they do not carry out endocytosis of extracellular lysosomal enzymes.[203]

A number of examples have been given above to document the roles of the carbohydrates of glycoproteins as recognition determinants in various biological interactions. Although the experimental systems varied greatly from dissociated cells of a whole organism (a marine sponge), to a defined glycoprotein interacting with defined cells, proteins (usually glycoproteins) that mediate the interaction could almost always be identified and isolated. Common approaches and criteria have often been used to establish the role of carbohydrates in these reactions. They include competitive inhibition of the interaction by specific glycoproteins and specific carbohydrates, blocking the interaction by carbohydrate-specific antibodies, modification of the interaction by carbohydrate enzymes, for example, sialyltransferases and endo-β-N-acetyl-glucosaminidase H, and so on. With the improved protein purification techniques and the advances in molecular biology approaches, a large number of glycoproteins and/or specific carbohydrate binding proteins

Table 7.6. Inhibition of pinocytosis of the lysosomal enzyme β glucuronidase by sugar phosphates[a]

Sugar phosphate 0.5 mM	Pinocytosis Rate (units/mg,hr)	Inhibition (%)
None	132 + 5	No
D-mannose-6-P	25 + 1	81
D-Fru-1-P	39 + 2	70
D-Gal-1-P	108 + 4	18
D-Glc-6-P	97 + 10	27
2-Deoxy-D-Glc-6-P	127 + 4	4
D-Fru-6-P	113 + 3	14
D-Fru-1,6-bisP	96 + 3	27
D-Sedoheptulose 1,7–P	109 + 1	17
2-Deoxy-D-Ribose-5-P	121 + 1	8

[a]From Kaplan et al.[196] Reprinted with permission from *Proceedings of the National Academy of Sciences, USA* 74:2026 (1977).

have now been purified and characterized, and their cDNA clones isolated and sequenced. Many of these are cell membrane proteins and extracellular matrix proteins, studies of these proteins have brought about a new level of the understanding of the role of carbohydrates of glycoproteins in a wide variety of cellular interactions. At the same time, it has become increasingly clear that much more about the biological functions of glycoproteins are still to be learned.

APPROACHES FOR THE STUDY OF BIOLOGICAL ROLE

The biological role of a covalent protein modification reaction can be considered at several levels. At its simplest, if the protein modification is originally discovered in a cell-free system, it addresses the question of whether the reaction occurs in living cells and how the intrinsic activity and/or property of the target protein is effected by the modification. At a more complex level, it addresses what cellular processes are effected by the protein modification and how they are effected. At yet another level, it deals with the change in cell physiology that is brought about by the protein modification and regulatory factors and physiological conditions that induce and modulate the protein modification. No single approach is adequate to address the question at all its varied levels. Even at its simplest level, the question may have to be dealt with by using complementary approaches.

Biochemical Approaches

Biochemical Characterization of Protein Modification Reactions.

Knowledge of how the intrinsic activity of the modified protein is affected by the protein modification reaction is basic to the understanding of the biological function of the protein modification. A biochemical approach where the protein modification reaction is studied in a cell-free system is especially useful for the elucidation of the effect of the protein modification. The approach usually entails isolation of various cellular components so that the protein modification reaction can be reconstituted. If both the modified and the unmodified potein forms are isolated, comparison of intrinsic activities of the two forms will show how the protein activity may be affected by the protein modification. Alternatively, the target protein may be isolated in one form, which is then converted to the other form by using the appropriate converting enzyme. Change in intrinsic activity of the substrate protein that accompanies the protein modification reaction will elucidate the effect of the modification.

A change in intrinsic activity is but one of many effects on a protein that may result from covalent modifications. Frequently, more subtle effects than a simple change in intrinsic activity of the protein are observed in a biochemically reconstituted system of protein modification reaction. For example, certain calmodulin-dependent enzymes such as smooth muscle myosin light chain kinase[204] and a brain CaM-dependent cyclic nucleotide phosphodiesterase[205] can be phosphorylated by cAMP-dependent protein kinase to result in a marked decrease in their affinities towards calmodulin. When the enzyme activity is assayed at a constant concentration of calmodulin, the phosphorylated forms of these CaM-dependent enzymes require much higher concentrations of Ca^{+2} for activity than the respective nonphosphorylated forms. Thus, the biological function of the cAMP-dependent protein phosphorylation has been suggested to be the modulation of the Ca^{+2} sensitivity of these CaM-dependent enzymes.

The demonstration of the effect of protein modification may sometimes require cellular components in addition to the target protein and the converting enzymes. The biochemical approaches are especially useful for identifying such components. Phospholipase C catalyzes the hydrolysis of phosphatidylinositol-4, 5-bisphosphate to give rise to the two general second messenger molecules diacylglycerol and inositide 1,4,5-trisphosphate, thus playing important roles in cellular signal transduction.[206] One isozyme of the enzyme, phospholipase C-γ, is believed to mediate the regulatory action of certain receptor tyrosine kinases such as epidermal growth factor (EGF) receptor. The enzyme is phosphorylated on tyrosine residues in EGF-stimulated cells, and the phosphorylation correlates with the increase in hydrolysis of phosphatidyl inositol-4,5-bisphosphate.[207] Although phospholipase C-γ couldn be readily phosphorylated *in vitro* by EGF receptor, the phosphorylation had no effect on the activity of the phospholipase C.[208] However, *in vitro* activation of the enzyme by EGF-receptor could be achieved when a less pure phospholipase sample, a sample immunoprecipitated from cell extracts, was used, suggesting the requirement of additional factors for the enzyme activation.[209] It is now established that phospholipase C-γ activity can be stimulated by the EGF-receptor catalyzed phosphorylation only when the enzyme is assayed in the presence of the cytoskeletal protein profilin.[210]

Muscle contraction is generally considered to proceed by a sliding motion between the thick (myosin) and the thin (actin) filaments of myofibril. The energy for this sliding movement is derived from ATP hydrolysis.[211] A biochemical reaction thought to simulate muscle contraction is actin-activated myosin ATPase activity. Myosin isolated from various smooth muscles were found to undergo phosphorylation on one of the light chains[212,213,214] by a specific protein kinase, myosin light chain kinase (MLCK);[215] only the phosphorylated myosins possess actin-activated ATPase, thus implicating the smooth muscle light chain phosphorylation reaction as an essential reaction in smooth muscle contraction. Enzymological characterization of smooth muscle MLCK revealed that the activity of MLCK was dependent on Ca^{+2} and calmodulin.[216] Since Ca^{+2} is believed to be a general cellular messenger for muscle contraction, the biological function of the Ca^{+2}-dependent smooth muscle light chain phosphorylation may be the mediation of stimulus-contraction coupling of smooth muscle contraction (Figure 7.6). Thus, as the biochemical characterization of a protein modification becomes more detailed, the biological role of the reaction becomes better defined.

In addition to elucidating the biological roles of the protein modification, detailed characterization of the modification reactions creates a fertile ground for the discovery of new protein modifications and relevant cellular factors, as already shown in many earlier examples. In the protein phosphorylation field, biochemical studies have lead to important discoveries that form some of the main bases of the current view of the mechanism of signal transduction. Shortly after the discovery of cAMP-dependent protein kinase, many cellular proteins were found to be substrates of the kinase in biochemical studies, thus leading to the notion that the protein kinase is a general mediator of cAMP effect.[217] With the special insights that cGMP and Ca^{+2} were general second messengers and that calmodulin was a multifunctional Ca^{+2} mediatory protein, cGMP-dependent and CaM-dependent protein kinases were sought for and discovered in the 1970s.[218,219] Similarly, the discovery of protein kinase C in the early 1980s was made at a time when the notion that phospholipid turnover may be a cell signaling device was gaining acceptance.[220]

Multi-step Protein Modification. The targeting and processing of a cell protein through the ubiquitin-mediated proteolysis pathway consists of at least three enzymatic reactions for protein ubiquitination as a proteolytic step in the degradation of the ubiquitin-protein conjugates.[93,94] The conversion of fibrinogen to fibrin, an essential agent in blood clotting involves the cascade of more than half a dozen protein modification reactions.[221] The processing of N-linked carbohydrates of a glycoprotein is a highly complex process resulting from more than 10 reaction steps, which take place in various cell compartments.[222] The regulation of E. coli glutamine synthetase and rabbit muscle glycogen phosphorylase by adenylation and phosphorylation reaction, respectively, each involves several steps of protein modification reactions.[217] The common feature of these varied protein modification processes is that they occur as series

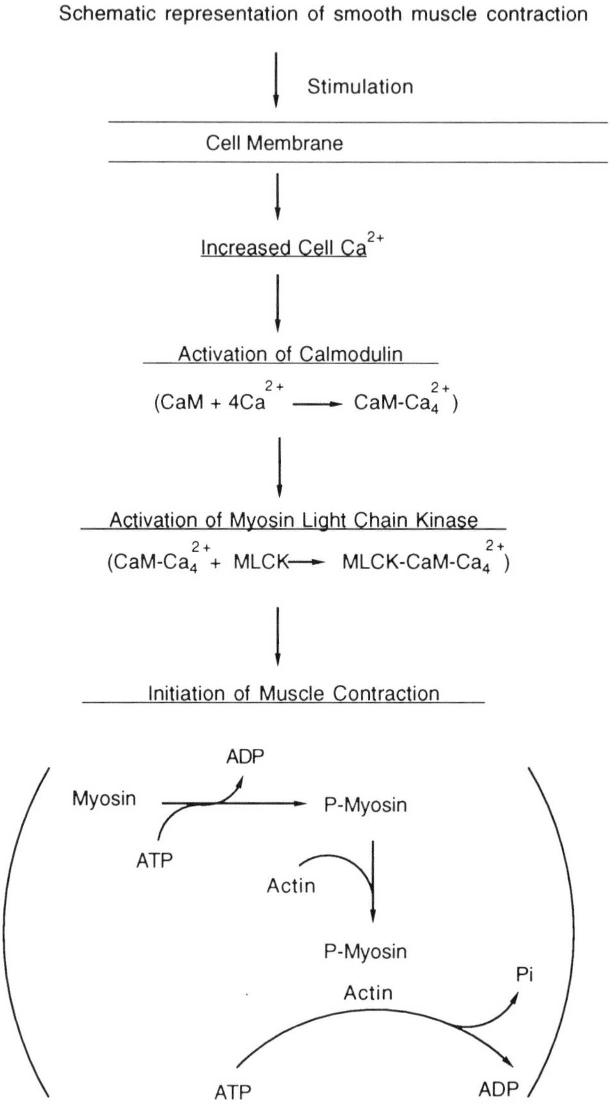

Schematic representation of smooth muscle contraction

$$(CaM + 4Ca^{2+} \longrightarrow CaM\text{-}Ca_4^{2+})$$

$$(CaM\text{-}Ca_4^{2+} + MLCK \longrightarrow MLCK\text{-}CaM\text{-}Ca_4^{2+})$$

Figure 7.6. Schematic representation of the mechanism of stimulus-contraction couple of smooth muscle. Equations in parenthesis are biochemical reactions simulating the biological phenomena.

of modification reactions. Both the protein modification reactions and the organization of the reaction sequence have distinct biological significance. Biochemical approaches are especially valuable for the elucidation of such higher orders of the protein modification reactions.

As an example, recent studies of the mechanism of action of receptor protein tyrosine kinases demonstrate how biochemical approaches contribute to the elucidation of the signaling cascades involving these kinases. Several membrane receptors, including those of insulin,[223] epidermal growth factor,[224] and nerve growth factor[225,226] possess ligand-stimulated protein tyrosine kinase activity. When cells are stimulated by a ligand of such a receptor, there is, in addition to an increased protein tyrosine phosphorylation in the cells, a rapid increase in phosphorylation on serine/threonine residues of cellular proteins.[227,228,229,230] One of the cellular proteins undergoing increased serine/threonine phosphorylation in response to membrane receptor tyrosine kinases is a 40 S ribosomal protein, S6 protein.[231,232,233] Typically, such receptor tyrosine kinase mediated protein serine/threonine phosphorylations are difficult to be reproducibly demonstrated in cell-free systems. On the other hand, extracts of ligand-treated cells can be shown to contain elevated serine/threonine phosphorylation activity towards S6 protein. The observations have led to the suggestion that these receptor protein tyrosine kinases activate the S6 ribosomal protein phosphorylation through a kinase cascade.

One of the biochemical approaches commonly used for the elucidation of the multi-step protein modifications involves the purification and characterization of the enzyme, catalyzing the individual steps so that the series of the protein modification reactions can be reconstituted *in vitro*. Using this approach, several steps of the enzyme cascade involved in mitogen-stimulated S6 ribosomal protein phosphorylation have been elucidated and schematically represented in Figure 7.7. The growth factor-stimulated S6 protein kinases have been purified from different sources; several clones have also been obtained. Structural characterization of the purified S6 protein kinases and the cDNA clones have shown that there are two classes of S6 protein kinases with distinct molecular masses, about 83 kDa and 59 kDa, with apparent molecular weights on SDS-PAGE of 90k and 70k, respectively.[234] The activated forms of the enzyme can be deactivated by treatment with phosphoserine/phosphothreonine phosphatases such as phosphatase 1 or phosphatase 2A.[235,236] Using $^{32}P_i$-labeled cells, the phosphorylation of S6 kinases during cell stimulation has been directly demonstrated. Analysis of phosphoamino acids of the labeled protein has revealed that only serine and threonine residues are phosphorylated.[237,238]

The observation that S6 protein kinases are phosphorylated exclusively on serine/threonine residues suggests that the growth factor receptor protein tyrosine kinase signaling pathway consists of intermediary protein kinases. A widely distributed protein serine/threonine kinase variously known as MAP (microtubules associated protein) kinase, MBP (myelin basic protein) kinase or S6 kinase kinase has been shown to catalyze the phosphorylation and activation of the 90 kDa S6 protein kinases isolated from unstimulated cells or inactivated by phosphatase treatment, thus suggesting that it is immediately upstream of S6 kinase in the kinase cascade.[236,239] The MAP kinase isolated from stimulated cells can be inactivated by either a protein serine phosphatase, phosphatase 2A, or a protein tyrosine phosphatase, CD45,[240,241] suggesting that the enzyme itself depends on the phosphorylation on both tyrosine and serine residue for enzyme activity. However, MAP kinase does not apear to be a direct substrate of the receptor kinases, because either inactivated forms of the enzyme cannot be reactivated by isolated receptor kinases.

Using inactivated MAP kinase, a factor capable of reactivating the enzyme has been demonstrated in extracts of growth factor stimulated cells.[242,243,244] The activation of MAP kinase by the factor depends on ATP, suggesting that the factor may be a kinase, MAP kinase kinase. The possibility that the factor acts allosterically to stimulate the autophosphorylation of MAP kinase[243,244] rather than to catalyze MAP kinase phosphorylation has also been considered[243,244] and then discarded. MAP kinase mutants which are devoid of catalytic activity have been found to be phosphorylated by the activating factor at regulatory sites.[245,246,247] Thus, the activating factor is appropriately named the MAP kinase kinase (Figure 7.7). The MAP kinase kinase has been extensively purified from EGF-stimulated A431 cells; there appears to be at least two different forms of MAP kinase kinase in this cell line.[247] This enzyme catalyzes the phosphorylation of both serine/threonine and tyrosine residues. MAP kinase kinase is inactivated upon treatment by a protein serine phosphatase, phosphatase 2A, but not by a protein tyrosine kinase. The inactivated MAP kinase kinase is a poor substrate of NGF receptor-linked tyrosine kinase. These observations suggest that there is (are) additional steps in the receptor kinase mediated S6 ribosomal protein phosphorylation pathway.

Raf oncogenes which posses acute cell transforming activity encode a family of serine/threonine kinase, v-raf. The cellular homologs of v-raf, c-rafs, have been shown to undergo phosphorylation on both serine and tyrosine residues in response to a variety of extracellular signals, including platelet derived growth factor,[248] insulin,[249,250] epidermal growth factor,[251] and nerve growth factor,[252] to mention a few (reviewed by Rapp[253]). Very recently, a member of the raf family kinase c-Raf-1 has been implicated as the activator of MAP kinase kinase.[254] NIH 3T3 cells stably transformed with v-raf have been found to contain constitutively

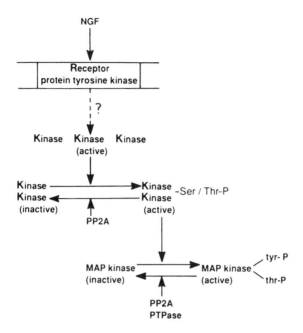

Figure 7.7. Schematic representation of receptor protein tyrosine kinase-regulated protein kinase cascades. Reprinted by permission from *Nature* Vol. 353:173 (1991). Copyright 1991 Macmillan Magazines Limited.

activated MAP kinases as well as MAP kinase kinase. MAP kinase kinase can be partially inactivated by treatment with protein phosphatase 2A. The partially inactivated MAP kinase kinase can be reactivated by an activated c-raf preparation isolated from insulin and vanadate-stimulated hepatocyte cell line. While MAP kinase kinase appears to be activated by raf kinase, the mechanism of raf activation has not been elucidated. Cell activation by a number of extracellular stimuli directed at tyrosine kinase receptors appears to be accompanied by extensive serine phosphorylation of the raf kinase. Whereas in a few cases where the stimulus action is mediated by a nonprotein kinase receptor, the cell activation is accompanied by tyrosine phosphorylation of raf.[252] Thus, additional protein phosphorylation steps have to be elucidated before the complete enzyme cascade mediating the S6 ribosomal phosphorylation is established. In addition, there may also be key steps that do not involve protein phosphorylation. For example, the low molecular weight of guanine nucleotide binding protein, ras, has been suggested as playing an essential role in this enzyme cascade.[255,256]

Biochemical Tools

techniques to the research in the protein modification field have been illustrated in previous chapters. There are also a number of techniques specifically derived from or developed for the studies of biological systems. From these techniques, useful biochemical tools have been produced; some of them have found general and widespread applications in studies of protein modification reactions. One such biochemical tool is synthetic peptide. Certain peptide derivatives from a protein may perform, in a limited sense, the function of the protein in a biochemical system, and therefore can be considered as a protein model. Such a peptide as well as its analogs with appropriate amino acid substitutions and/or deletions may be readily prepared in large quantities to define the structural determinants for the specific protein function.

Protein kinases phosphorylate their respective substrates at discrete sites. For example, of the more than 40 serine and threonine residues in glycogen phosphorylase, only one, ser-14, is phosphorylated in the phosphorylase kinase-catalyzed reaction.[257] The observation suggests the existence of unique structural features in phosphorylase that are essential to the ability of the protein to serve as the kinase substrate. The substrate structure determinants of phosphorylase appear to be present, at least partly, proximal to the phosphorylation site, since a chymotryptic peptide of the protein is readily phosphorylated by phosphorylase kinase.[258] The suggestion has been substantiated by the successful phosphorylation of the short synthetic peptide corresponding to the sequence of the phosphorylated-serine site, KRKQISVR.[259] By using synthetic peptides with systematic amino acid substitutions, structural determinants for the substrate activity have been determined for a large number of protein kinases, a selected few are listed in Table 7.7 to illustrate this point. More extensive listing may be found in recent reviews.[260,261] A few examples for the use of synthetic peptides in elucidating substrate determinants of other protein modifications such as acetylation and myristoylation are presented in Chapter 3.

Table 7.7. Selected protein phosphorylation site motifs[a]

Protein kinase	Site motif
Myosin light chain kinase	XKKRXXRXXSX
CaM-protein kinase II	XRXXSX
Cyclic AMP-dependent protein kinase	XRRXSX
Cyclic GMP-dependent protein kinase	XSRX
Protein kinase C	XRXXSRX
Casein kinase II	XSXXEX

[a]From a review article by Kemp and Pearson.[260]

The knowledge of consensus sequences of protein kinase substrates has many practical applications. One of these is the design of specific and readily available protein kinase substrates. For example, p34^{cdc2} catalyzes the phosphorylation of pp60^{c-src} at three sites, all three sites conform to the pattern:S/T-P-X-R/K.[262] A synthetic peptide constructed on the basis of amino acid sequences of the three sites consisting of this motif and with the serine in the motif as the only amino acid containing hydroxy group is a good substrate of p34^{cdc2} kinase, but not phosphorylated by many other protein kinases including cAMP-dependent kinase, CaM-dependent protein kinase II, protein kinase C, and casein kinases I and II. Furthermore, substitution of proline in the motif by glycine or alanine completely abolished the p34^{cdc2} kinase substrate activity of the peptide. This peptide, along with its inactive analogs have been used to assay for p34^{cdc2} kinase or related kinases in crude biological samples.[263] Similarly, a synthetic peptide derived from the amino terminal region of p34^{cdc2} containing the tyrosine residue phosphorylated in dividing cells has been shown to be specifically phosphorylated by src family protein tyrosine kinases,[264] and it may be used as a specific reagent for quantifying src family kinase activity in cell extracts.

Antibodies and immunochemical techniques are powerful biochemical tools with widespread applications; they are especially versatile, in some cases, indispensable in protein modification studies. Antibodies specific for protein modifying groups or the modified residues, such as anti-phosphotyrosine or anti-ubiquitin antibodies, have been developed. They are important analytical tools and are used widely in the identification and quantification of specifically modified proteins. The high degree of diversity in the structure of carbohydrates of glycoproteins has made the use of antibodies especially valuable in glycoprotein studies. A large number of monoclonal antibodies with diverse specificities toward carbohydrate moieties of glycoproteins have been produced and characterized;[265] some of these are listed in Table 7.8 to illustrate the versatility of antibodies. More extensive lists may be found in recent reviews.[266,267] These antibodies may be used to identify and assay for glycoproteins containing specific carbohydrates. In spite of the advances in physicochemical techniques for carbohydrate analysis, it is unlikely the sensitivity, versatility, and specificity offered by the immunological tools in the sugar analysis can be achieved by physicochemical means in the near future.

Perhaps the most common application of antibodies in biochemical studies is as the reagent for specific cell proteins. Antibodies against specific proteins can be produced by either using the purified protein or synthetic peptides based on the protein sequence. Monoclonal antibodies against a cellular protein can also be obtained by using specific screening

Table 7.8. Selected carbohydrate specific monoclonalantibody[a]

Oligosaccharide structure	Name	Antibody name
Galβ-1-3GlaNAcβ1-3Galβ$_1$-R	LNT	FC10.2
Fucα1-2Galβ1-4GlcNAcβ$_1$-R	H type 2	H-11
Galβ1-3GlcNAcβ$_1$-R	Le[a], LNF11	CF6-C4
Galβ1-4GlcNAcβ$_1$-R	Le[x], LNF111,	SSEA-1
NeuAcα2-3Galβ1-3GlcNAcβ$_1$-R	Disiaylated Le	FH7
NeuAcα2-3Galβ1-3GalNAcβ$_1$-R	Sialylated Globe Somes	SSEA-4
NeuAcα2-3Galβ1-3GlcNAcβ1-3-Galβ$_1$-R	Disialyl LNT	FH9

[a]From the compilation by Magnani.[266] Reprinted with permission. Copyrioght 1987 Academic Press, Inc.

procedures. By using antibodies against a specific protein in combination with antibodies specific to a modification group, questions such as whether or not a cellular protein is modified by a unique protein modification reaction may be addressed. For example, the cell cycle regulated protein kinase, p34[cdc2] is phosphorylated on a tyrosine residue in dividing cells.[268,269] This can be demonstrated by first immuno-precipitating a cell extract using an antiphosphotyrosine antibody, followed by Western immunoblot analysis of the precipitated sample, using a p34cdc2 specific antibody.[267]

A specific protein modification may also result in a well-defined mobility change of the modified protein on SDS-PAGE gel. Under such a condition, the protein modification may be monitored by using the protein specific antibody alone. For example, a cardiac sarcoplasmic reticulum (SR) Ca^{+2} pump regulatory protein, phospholamban can be phosphorylated by either cAMP-dependent or CaM-dependent protein kinases at distinct sites.[270,271] The protein exists in SR vesicles as a pentamer; the monomeric unit contains one phosphorylation site for each of the kinases. Thus, the pentameric phospholamban can exist in multiple phosphorylated forms of distinct phosphorylation stoichiometry.[272] Figure 7.8 shows that under well-defined conditions, the differently phosphorylated phospholamban can be separated into discrete bands on SDS-PAGE. These bands are well stained by an antiphospholamban antibody, although only about 2 percent of the total protein of the crude SR vesicles used in this experiment is represented by phospholamban.

Physiological Approaches

Although they are powerful in the analysis of complex cellular processes, biochemical approaches have at least two serious shortcomings. The reconstituted biochemical reaction is, in most cases, a crude approxima

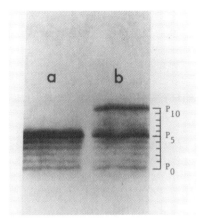

Figure 7.8. Analysis of phosphorylated phospholamban. Phospholamban in SR (0.3 mg/ml) was phosphorylated by cAMP-dependent protein kinase (a) and subsequently by endogenous CaM-dependent kinase (b) was analyzed by SDS-PAGE followed by Western blotting. Reprinted with permission from *Journal of Biological Chemistry* 266:17488 (1991). Copyright 1991, The American Society for Biochemistry & Molecular Biology.

tion of what is occurring in the cells; as a result, certain aspects of the reaction are misrepresented or not represented at all in the biochemical system. The more serious problem is that the reaction demonstrated *in vitro* may not occur in the living cells. Thus, for the study of biological significance of covalent protein modification, it is often necessary to complement biochemical studies with studies using living cells, tissues or the whole organisms under physiological conditions. Over the years, various procedures have been established to study specific cellular in this section whereas many others that rely on more sophisticated cell processes in living specimens. Some of the common ones will be discussed in this section whereas many others that rely on more sophisticated cell biological, molecular biological, or genetic methods will be discussed separately.

Demonstration and Characterization of in Situ Protein Modification

The basic method for the demonstration of specific protein modification in living cells often involves the isolation of the protein from the cell extract to determine its state of the specific modification. Conditions have to be established for the tissue and/or cell extraction and sample preparation so that the state of the protein modification is not altered. For example, for protein phosphorylation studies, the cell extraction and sample preparation are usually carried out with buffers containing chelators and phosphatase inhibitors. If the state of the protein modification under a specific physiological condition is to be determined, procedures to rapidly terminate cellular activities during cell or tissue sampling will have to be established. General procedures include rapid sample freezing or extracting the sample in boiling SDS-containing buffers.

Specific and sensitive analytical methods have been developed for various protein modifications, such as immunological methods and radiolabeling approaches. Antibodies such as those specific for phosphotyrosine,[273] ubiquitin,[274] and carbohydrate moieties of glycoproteins[265] have been used to detect and assay for specific protein modifications. Alternatively, radioactivity labels can be introduced into proteins by specific modifications, such as fatty acylation or lycosylation. In particular, *in situ* ^{32}Pi labeling is widely used to identify and to assay for phosphoproteins. To determine the identities of the proteins modified, a procedure that is often used consists of SDS-PAGE, or two-dimensional isoelectric focusing and SDS-PAGE separation of the extracted proteins, followed by the detection of modified proteins using appropriate techniques, for example, immunotransblot or autoradiograph.

Cell Perturbation to Probe the Biological Role

Once tools for the analysis of *in situ* protein modification are available, the question about the biological function of the specific protein modification can be addressed. As the biological roles of the protein modifications are diverse, experimental designs dealing with the questions are varied. One of the common features for many of these experiments involves a perturbation of the normal state of the experimental specimen (cells, tissues, and so on), and then correlate the change in the protein modification with specific cellular processes suspected of involving the modification reaction in the cells. Depending on the modification reaction and the specific cellular processes to be examined, the perturbation condition applied may be a hormone, a toxin, an electric charge, or a change of medium pH.

In an interesting study designed to test the notion that ubiquitination plays the role of targeting specific cellular proteins for proteolytic degradation, the condition of the perturbation was the introduction of abnormal proteins.[274] In earlier studies, it had been shown that reticulocytes could use amino acid analogs such as the lysine analog, 4-thialysine (S-2-aminoethyl-L-cysteine), to make abnormal proteins, and these proteins were degraded much faster than normal protein.[275] To test whether the enhanced degradation of the abnormal proteins was channeled through a ubiquitin-mediated pathway, reticulocytes fed with the amino acid analog and control cells were both pulse labeled by [³H] leucine. Extracts from the cells were immunoprecipitated by a ubiquitin antibody and analyzed. The precipitate of the lysine analog-treated cells was found to contain markedly increased radioactivity, suggesting that more of the newly synthesized abnormal proteins were ubiquitinated. The total amounts of ubiquitin in the two cell populations, however, were about the same. Thus, the increased degradation of the abnormal proteins was accompanied by an increased protein ubiquitination.

This Ca^{+2}-CaM-dependent myosin light chain phosphorylation has been suggested to be the molecular basis of the Ca^{+2}-mediated smooth muscle contraction.[211] Physiological studies using different smooth muscle preparations have invariably shown that both muscle contraction and light chain phosphorylation occurred under conditions of increased cell Ca^{+2}.[276] However, a more complex relationship between cell Ca^{+2} myosin light chain phosphorylation, muscle contraction, and muscle relaxation has emerged from more rigorous physiological studies of the smooth muscle contraction. When the time-courses of the stimulation induced smooth muscle contraction, force development, Ca^{+2} flux, and light chain phosphorylation were compared; a complex relationship between the various time-courses were observed, as schematically represented in Figure 7.9. Upon stimulating the smooth muscle preparation, the cell Ca^{+2} concentration increases rapidly and then declines to a relatively stable level slightly higher than that of the resting muscle. In approximate correlation with the increase in Ca^{+2} concentration, the muscle contracts and the light chain is phosphorylated. However, while the light chain undergoes dephosphorylation, as the elevated cell Ca^{+2} subsides, the contraction state of the muscle is maintained. The relaxation of the muscle can be induced only by using chelating agents to return the cell Ca^{+2} to the level existing in resting muscle. Thus, the maintenance of the contraction, designated latch state by Murphy,[277] who made the initial observation of this complex relationship in smooth muscle contraction, requires Ca^{+2} but is independent of the light chain phosphorylation. This example illustrates the point made earlier that biochemical reconstituted systems are often crude approximations of the cellular system. In this case, although the actin-activated myosin ATPase reaction has been used successfully to elucidate the role of myosin light chain phosphorylation in smooth muscle contraction (see above, Figure 7.9), important aspects of the smooth muscle regulation are not represented in the reconstituted system.

Protein Modification in Signal Transduction

Many of the physiological studies of protein phosphorylation reactions deal with specific signal transduction systems. One of the general

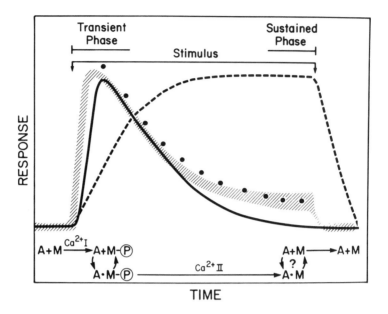

Figure 7.9. Schematic representation of the relationships between cell Ca2+, myosin phosphorylation and smooth muscle contractions. Shaded area represents cell Ca^{+2} concentrations, (--) for force development, (closed circles) for maximal shortening velocity, and (—) for the level of myosin phosphorylation A and or stand for actin and myosin respectively. Reproduced, with permission, from the *Annual Review of Pharmacology & Toxicology* 25:681 (1985). Copyright 1985 by Annual Reviews, Inc.

questions raised in such a study is: What signaling pathway mediates this process? To delineate the signaling pathway employed in a particular cellular regulatory function, three general types of physiological experiments may be carried out: (1) determine whether cell concentration of a specific second messenger is increased during the cell stimulation, (2) learn whether artificial increases in the concentration of a second messenger mimic the effect of the cell stimulation, and (3) evaluate the effect of inhibitors specific for a second messenger-dependent protein kinase on the cell stimulation.

Sensitive and specific procedures are available for the assay of cellular second messengers. Cyclic AMP and cGMP are capable of high affinity and relatively specific binding to cAMP and cGMP-dependent protein kinases, respectively. These proteins can be used in a competitive binding procedure in quantifying the respective cyclic nucleotides.[278] Antibodies specific to cAMP or cGMP have been produced and used in place of the protein kinases in the binding assay with greatly increased sensitivity and specificity.[279,280] Many procedures have been developed for the assay of cellular diacylglycerol. Some of the procedures involve the purification of diacylglycerol from the cell extract, and the pure sample is then assayed by a sensitive chemical or enzymological method. In other procedures, cells are labeled by the radioactive precursor molecule, arachidonic acid, the neutral lipid fraction was then isolated and diacylglycerol is quantified by the radioactive contents of the fraction. A procedure that is gaining popularity consists of isolating the neutral lipid fraction, phosphorylating the fraction with [γ-32P]ATP by *E. coli* diacylglycerol kinase, and determining the amount of radioactive phosphatidic acid produced.[281,282] The advantage of this procedure is that diacylglycerol does not have to be purified and that there is no need for cell labeling.

The second messenger, inositol 1,4,5-trisphosphate (IP_3), is one of several inositol phosphates produced in a phospholipase C-mediated cell activation. The assay of cellular IP_3 depends on the separation of the various inositol phosphates. In a commonly used procedure, the cells are labeled with [3H] inositol. The aqueous cell extract that contains the inositol phosphates are then separated by one of several methods. These include HPLC, anion exchange, electrophoresis, and so on. The separated inositol phosphate species can then be quantified by radioactivity analysis.[283]

The assay procedures described above all require cell extraction and isolation of the second messenger. The determination of cell Ca^{+2} concentration as a second messenger presents a different challenge. Cell Ca^{+2} is highly compartmentalized, and it concentrates in cell organelles such as mitochondria and endoplasmic reticulum.[284] The second messenger action of Ca^{+2}, however, is exerted in the cytosol, where Ca^{+2} concentrations typically fluctuate between 10^{-7} to 10^{-6} M. Therefore, the free cytosolic Ca^{+2} concentration rather than the total cell Ca^{+2} is of interest. Cytosolic Ca^{+2} concentration has been assayed by using a Ca^{+2} specific microelectrode[285] or Ca^{+2} indicator.[286] These procedures are designed to measure the free cytosolic Ca^{+2} in living cells. A Ca^{+2} indicator can be a Ca^{+2} binding photoprotein or a Ca^{+2} binding fluorescence dye. A good Ca^{+2} indicator shows high affinity and high selectivity for Ca^{+2} under the cytosolic solution condition, which is, typically 10^{-7} - 10^{-6} M Ca^{+2}, 150 mM K^+ and millimolar range Mg^{+2}. A frequently used Ca^{+2} indicator, Fura 2,[287] meets these criteria. Another advantage of Fura 2 is that it can be easily introduced into small living cells. The carboxylate groups of Fura 2 that are involved in Ca^{+2} binding can be esterified, so as to allow easy entry of the indicator into the cells. Once inside the cell, the esterified Fura 2 is hydrolyzed by the cytosolic esterase. The carboxylate groups of the compound are regenerated, and the indicator is trapped in the cytosol to monitor cytosolic Ca^{+2}.

The use of Ca^{+2} indicators has made it possible to develop the fluorescence imaging techniques to determine the cytosolic Ca^{+2} dynamics in single cells. The technique can provide a means to examine cytosolic Ca^{+2} fluctuation with a temporal resolution of fractions of a second, and spatial resolution of micrometers.[288] Fluorescence imaging has also been applied to the studies of cellular cAMP.[289] The sensor for cAMP is a fluorescent-labeled cAMP-dependent protein kinase. Cyclic-dependent protein kinase is a tetramer of two catalytic and two regulatory subunits. The activation of the enzyme by cAMP involves the binding of cAMP to the regulatory subunit, resulting in the dissociation of the catalytic subunits and the regulatory subunits. To prepare a useful sensor for cAMP, the two types of the subunits are labeled by two distinct fluorescence dyes capable of fluorescent energy transfer. While the fluorescent holoenzyme can carry out the energy transfer, the cyclic AMP-dissociated enzyme cannot, thus possessing a different fluorescent property.

Another general procedure often used to implicate a signaling pathway in a cellular regulatory process is the examination of the effect of artificially raised cellular second messenger concentration on the cellular process. There are various procedures to increase cellular concentrations of specific second messengers. For example, cAMP or cGMP analogs, which are more hydrophobic than the parent nucleotides and therefore more cell permeable, may be used to increase cellular cAMP or cGMP concentrations. Calcium ionophore is often used to increase cell Ca^{+2} concentration. Protein kinase C can be specifically activated in the cells by tumor-promoting phorbol esters,[290] an analog of diacylglycerol. Cellular concentration of the second messenger can also be raised by activating or inhibiting the second messenger metabolizing enzymes. Forskolin, an adenylate cyclase activator, can be used to artificially increase cellular cAMP concentration.[291] Similarly, isobutyl methyl xanthine, an inhibitor of cyclic nucleotide phosphodiesterase, has been used to raise the cellular cyclic nucleotide concen-trations.[292]

Inhibitors as Probes for Covalent Modifications

Inhibitors of protein modification enzymes with varying degrees of specificities are available. They are valuable tools, among other applications, for the elucidation of the biological effects of covalent modifications. A comprehensive discussion of the categorization of the inhibitors, general procedures, and necessary precaution of their various applications has been presented in Chapter 6.

The complex pathways of the processing of N-linked carbohydrates and their roles in biological function and cellular routing of the various glycoproteins have been extensively studied using combined biochemical, cell biological, and molecular biological approaches. A schematic representation of the carbohydrate processing pathways for glycoproteins destined for secretion and for lysosome are shown in Figure 7.10. The N-linked oligosaccharides are initially assembled in endoplasmic reticulum by the stepwise addition of various sugar units to the lipid

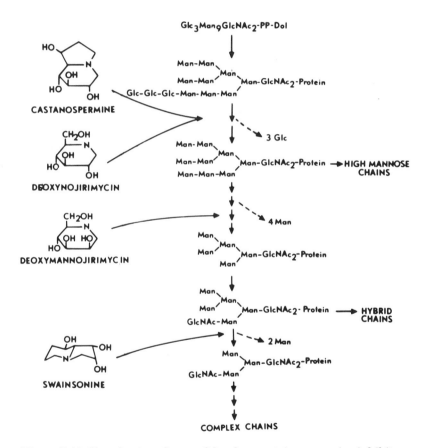

Figure 7.10. Site of action of some of the glycoprotein-processing inhibitors. Castanospermine and deoxynojirimycin are inhibitors of glucosidase I, deoxymannojirimycin inhibits mannosidase I, and swainsonine inhibits mannosidase II. Reprinted with permission from *Federation of American Societies for Experimental Biology Journal* 5:3057 (1991). Copyright 1991, FASEB.

carried dolichyl-P and then transferred to specific asparagine residues of the apoprotein. The processing pathways involve both the addition and the removal of sugar units by specific enzymes. As the reactions proceed, the protein intermediates move by vesicular transport through the ER and Golgi, some of them are transported by specific receptors to lysosome.

Tunicamycin, an inhibitor blocking the transfer of the complex oligosaccharide from the lipid carrier to the protein, has been used widely to address the question of the biological function of N-linked sugar moieties of glycoprotein.[293] For example, lipoprotein lipase, an enzyme secreted from adipocytes, has been studied in terms of its N-linked carbohydrate function by using this inhibitor. In one study, treatment of 3T3-L1 adipocytes by 1 μg tunicamycin for 18 hours resulted in close to complete inhibition of the lipase activity, but the amount of the lipase protein produced, as determined by immunoreactivity, was 75 percent that of the control cell. The protein was retained in the cells and it had an SDS-PAGE mobility indicative of the unglycosylated form of the protein. Thus, in this case, the N-glycosylation appears to be required for both the enzyme activity and the enzyme secretion.[294]

Potent inhibitors specific for various oligosaccharide processing enzymes are available and have been used effectively in elucidating the structure-function relationship of the N-linked oligosaccharides of specific glycoproteins. Many of these inhibitors are glycosidase inhibitors; their application to the study of glycoprotein function and processing has recently been reviewed.[295] The sites of action of some of these inhibitors in the glycoprotein processing pathways are indicated in Figure 7.10. By using different inhibitors, different forms of incompletely processed glycoproteins may be produced. Characterization of these incompletely processed glycoproteins may reveal the function of the specific oligosaccharide structural element of the protein. For example, treatment of guinea pig adipocytes by deoxynojirimycin resulted in the production of the high mannose form of the lipoprotein lipase. The incompletely processed protein displayed full enzyme activity and was secreted.[296] Thus, while the N-linked oligosaccharide is essential for the lipoprotein lipase activity and secretion, most of the processing steps do not appear to play functional roles in these respects.

Molecular Biology Contributions

Advances in molecular biology have drastically changed the landscape of biological research in many areas, including that of protein covalent modification studies. There are basically two types of molecular biology contributions: (1) those resulting from increased protein structure information due to the high efficiency of DNA sequencing and the successful application of complementary screening, and (2) those derived from exploring new research avenues uniquely accessible to molecular biology approaches. A few specific advances made from molecular biology approaches will be used to exemplify how these approaches may contribute to the elucidation of biological roles of protein modifications.

Insights from Amino Acid Sequence Information

The large volume of protein sequence information has made it possible to analyze the structure-function relationship of proteins in a more systematic fashion. For example, a large number of protein kinases have been cloned and sequenced. Although members of this family appear at first glance to be highly diverse in structure with widely different molecular weights and subunit structures, sequence analysis has shown that they all contain a homologous region of about 30,000 to 35,000 dalton molecular mass.[297] A data base of the catalytic domain sequences has been established by Hanks and co-workers.[298] and used to deduce phylogeny of the protein kinase family (Figure 7.11). From the analysis of 65 sequences of the protein kinase family, the catalytic domain of the family has been subdivided into eleven subdomains. Each of the domains contain highly conserved or essentially invariable sequences and/or amino acid residues. The organization of the subdomains of the protein kinase family is schematically shown in Figure 7.12. The highly conserved sequences or residues of each domain are summarized in Table 7.9.

Table 7.9. Primary structural features of protein kinase catalytic domain

Sub-domains	Amino acid or sequence	Residue in A-Kinase
I	LGXGXXGXV	49-57
II	AXKXL	70-74
III	E	91
IV	I	103
VI	TXXYL	153-157
VI	YXDLXXXNL	167–172
VII	VXDFG	182-186
VIII	TXXYXAXE	201-208
IX	DXWAXGV	220-226
XI	L and R	269 and 280

Derived from Hanks et al.[298]

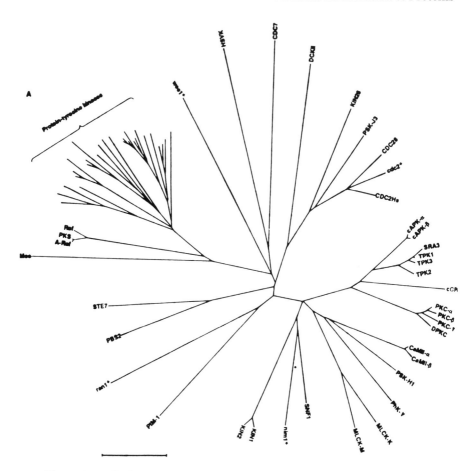

Figure 7.11. Deduced phylogeny of protein kinase catalytic domains. The
phylogenetic tree was constructed from the multiple alignment. The
protein-tyrosine kinases are not labeled in (A), but are shown in the cluster
enlargement in (B). The tree is shown "unrooted" in (A) as the branches are
all measured relative to one another with no outside reference point.
Reprinted with permission from *Science* 241:50 (1988). Copyright 1988 by
the AAAS.

The sequence analysis has many useful applications. One of these is
the identification of additional members of the family. With the advances
in molecular cloning, it is not uncommon to obtain the sequence of a
cDNA without any knowledge of the protein product. On the basis of the
nucleotide sequence, and the deduced amino acid sequence of the protein

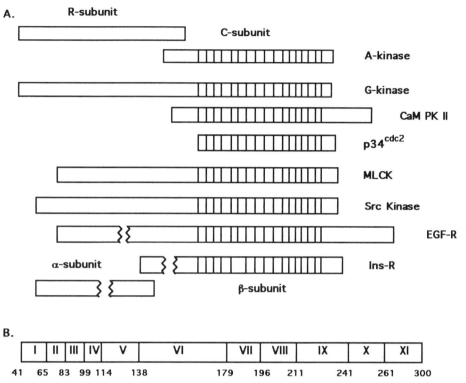

Figure 7.12. Structure domain organizations of selected protein kinases.

product, it is possible to suggest whether the protein may be a protein kinase, a protein phosphatase, or a protein of other general functions.

The protein product of the cell division cycle regulatory gene cdc2, p34[cdc2], known to be essential for the yeast cell to traverse G_1/S as well as G_2/M boundaries,[299] was identified as a protein kinase on the basis of its gene sequence before the protein product was obtained.[300] Similarly, the p197[wee1] protein, a negative regulator of cdc2 gene product[301] was identified as a protein kinase on the basis of gene sequence without the benefit of any information about the protein.[302] In addition, sequence information may provide some insights into the protein function. Known protein kinases have been divided into two subfamilies on the basis of their amino acid specificities, protein serine/threonine kinase, and protein tyrosine kinase subfamilies. Comparison of catalytic domain amino acid sequences have shown that there are regions of the sequence that may distinguish between members of the two subfamilies. For example, there is, in the subdomain VI of the kinase catalytic domain (Figure 7.12), a concensus sequence for protein serine/ threonine kinase subfamily, DLKPEN, whereas for protein tyrosine kinase subfamily, the concensus sequence is DL R/A A A/R N[298].

The protein sequence information, however, has to be used with some reservation and caution. For example, the negative regulatory action of weel protein kinase is countered by the protein product of cdc25 gene.[303] It was therefore suggested that the protein product of cdc25 was a protein phosphatase when weel was identified as protein kinase gene. However, upon sequencing of cdc25 gene, it was suggested that the sequence showed no homology to those of protein phosphatase family members.[304] Only after further biochemical studies and more careful analysis of the protein sequence, it was established that the product of cdc25 is indeed a protein phosphatase.[305] Similarly, amino acid sequence of weel gene suggested that the protein product is a protein kinase of the protein serine kinase group.[306] Biochemical evidence, however, suggests that the kinase has dual specificities, that is, capable of phosphorylating both protein serine/threonine and protein tyrosine residues.[307]

Amino acid sequences outside of the catalytic domain of the modification enzymes also contain important information about the protein function. Recent advances in protein structure analysis have resulted in the identification of many structural motifs or structural features in proteins, which may be the basis for specific protein functions. Several protein kinases and protein phosphatases display in their primary structure a segment having amino acid sequence characteristic of transmembrane domain structure. All of these enzymes are transmembrane proteins. Second messenger-regulated protein kinases and phosphatases all have specific regulatory domains housing the binding sites of their regulatory ligands. These regulatory domains exist either on a separate subunit, a reversibly associated regulatory protein, or on the same peptide that contains the catalytic domains, such as in the case of cAMP-dependent protein kinase and phosphorylase kinase, CaM-dependent kinase II and CaM-dependent phosphatase, or cGMP-dependent protein kinase and protein kinase C, respectively.

Protein sequence analysis sometimes reveals important insights into the regulatory and functional properties of the modification enzymes. Protein kinase C was initially shown to exist as different isozymes by protein separation and immunological differentiation.[308] Subsequent molecular cloning studies have detected and characterized at least eight mammalian protein kinase C clones.[309,310] The primary structure of protein kinase C may be divided into regulatory and catalytic domains. Comparison of the primary structures of the kinase C isozymes suggests that there are four conserved and five variable regions (Figure 7.13).[311]

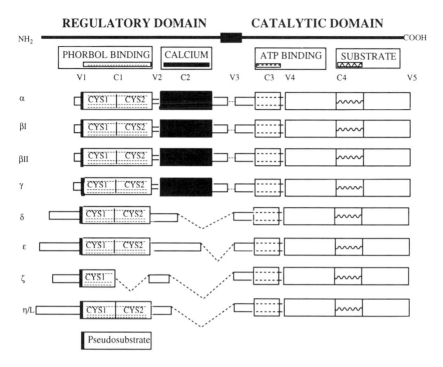

Figure 7.13. The domain structures of protein kinase C family members. V, variable regions of the PKC molecule; c, conserved regions of the PKC molecule. Reprinted with permission from *Journal of Biological Chemistry* 266:4662 (1991). Copyright 1991, The American Society for Biochemistry & Molecular Biology.

One conserved region, C_2, appears to be the Ca^{+2} regulatory domain. This domain is not present in some of the isozymes, PKC δ, ε, σ, and η, suggesting that these PKC isozymes are not Ca^{+2}-dependent[247, 250,251]. Such Ca^{+2}-independent protein kinase C has since been demonstrated in certain tissues. To cite another example, the first protein tyrosine phosphatase purified, characterized, and cloned was a cytosolic protein tyrosine kinase from human placenta.[312] Gene bank searches for homologous protein sequences has led to the discovery of a number of membrane proteins, such as the lymphocyte antigen, CD 45,[313] which contains on its cytosolic domain protein tyrosine phosphatase homologous sequences.[253,314] Since functions of the cytosolic domain of CD45 were not known before, the discovery has generated tremendous excitement in both protein phosphorylation and the immunology fields.[315] For review on phosphotyrosine phosphatases, see Fischer et al.[316]

Protein amino acid sequence analysis also helps in the identification of target proteins. Amino acid determinants around the phosphorylation sites of substrate proteins have been elucidated for many protein kinases (see above, Table 7.7). It is possible to inspect the sequence of a protein to identify potential phosphorylation sites for a number of protein kinases. Similarly, amino acid sequence motif has been determined for N-linked glycosylation. While such a prediction may be useful, the existence of the structural motif is by no means a proof of the occurrence of the modification reaction in cells.

The protein sequence analysis may extend beyond those involving the modification enzymes and protein targets. A case in point is its application to predict potential lectins, proteins displaying specific carbohydrate binding activities. The complex and diverse carbohydrate processing mechanisms result in an enormously wide spectrum of complexed carbohydrate structures in glycoproteins. This large amount of cellular information appears to be decoded by an equally wide variety of sugar-binding proteins known as lectins. A large number of animal lectins have now been purified, molecularly cloned, and sequenced. Based on their sugar-binding properties, these lectins have been classified into a Ca^{+2}-dependent group and a Ca^{+2}-independent group.[317] Comparison of amino acid sequences of these lectins has shown that within a group, the carbohydrate recognition domains of these proteins display common structural features. As illustrated schematically in Figure 7.14, members of the Ca^{+2}-dependent lectin family contain in their binding domains a set of invariable amino acid residues. Many cell adhesion molecules have been found to be carbohydrate-binding proteins containing carbohydrate-binding domain structures characteristic of the Ca^{+2}-dependent type lectins.[318,319]

Nucleotide-directed Mutagenesis

Probably the most commonly used and having the widest impact of molecular biological approaches in expanding the understanding of protein structure and function is nucleotide-directed mutagenesis. This is due to the relative ease and specificity in using this technique to target almost any specific residue or regions of the protein for substitution or deletion. The ATP binding site of protein kinases is highly conserved, consisting of a nucleotide binding fold and a downstream lysine believed to interact with the β phosphate of ATP.[297] As shown by many investigators using several protein tyrosine kinases, when the lysine residue is substituted, the resulting enzyme has no activity.[320,321] The effect of protein modification on cell physiology can also be explored. For example, the lysine substituted EGF-receptor has been found to express normally as a glycoprotein on the cell surface and shows normal EGF

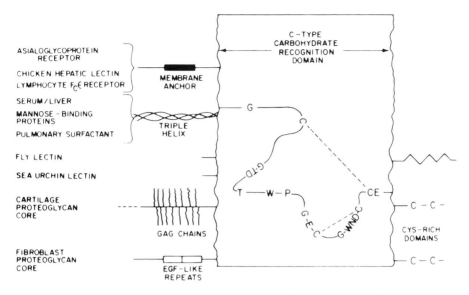

Figure 7.14. Summary of structural features of C-type animal lectins. The (nearly) invariant residues found in the common carbohydrate-recognition domain of the C-type lectins are shown, flanked by schematic diagrams of the special effector domains (if any) found in individual members of the family. GAG, glycosaminoglycan; EGF, epidermal growth factor. The disulfide bond linkages have been assigned for the sea urchin lectin and are assumed for the other C-type lectins based on the presence of invariant cysteine residues at these position. Reprinted with permission from *Journal of Biological Chemistry* 263:8558 (1988). Copyright 1988, The American Society for Biochemistry & Molecular Biology.

binding activity. However, the receptor does not mediate cellular modulation of the ligand EGF.[322] As another example, all src family protein tyrosine kinase contain a myristoylated NH_2-terminal glycine. When the glycine at the NH_2 terminus of the viral src is substituted or deleted, the protein product is translocated from the membrane to the cytosol. Although the kinase extract is still active, it loses the ability to transform cultured cells.[322] Thus, cell transformation by the src family oncogene appears to require both the activity and the proper cell localization of the protein tyrosine kinase.

There are also interesting examples where a specific amino acid of a mediating enzyme was targeted for mutation to address the question of the residue's involvement in the regulation of the enzyme. Viral src product, pp60[v-src] and the cellular counterpart, pp60[c-src], are highly conserved over their amino terminal 514 amino acids. The remaining 19

Table 7.10. Sequence homology of cyclic nucleotide binding sites

Protein, site	Sequence
CAP	GELGLFEGQERSAWVRA
A-Kinase	
RI, A site	GELALIYGTPRAATVKA
RII, A site	GELALMYNTPRAATIYA
RI, B site	GEIALLMNRPRAATVVA
RII, B site	GEALVTNKPRAASAYA
G Kinase	
A site	GELAILYNCTRTATVKT
B site	GEKALQGEDVRTANVIA

Note: E and R represent residues substituted to result in the abolition of cAMP binding. A is the residue substituted by threonine to produce high affinity cGMP binding site.

amino acid carboxyterminal segments of the chicken c-src, however, is totally different from the 12 amino acid carboxyterminal sequence of the v-src proteins.[323] A tyr residue, Tyr_{527} in c-src protein is conserved in all the protooncogene (cellular oncogene) products of the src family, and it exists as a phosphorylated tyrosine in the cells. Substitution of Tyr_{527} of the c-src kinase by phenylalanine results in a high activity mutant of the c-src kinase.[324,325] The result suggests that c-src protein kinase, in contrast to the v-src enzyme, contains a negative regulatory domain; the carboxyterminal 12 amino acid sequence in which tyrosine527 plays an important regulatory role.

There are two major forms of cAMP-dependent protein kinases called A-kinase I and A-kinase II. Regulatory subunits of both isozymes, RI and RII, contain two cAMP binding sites, cAMP binding sites A and B. Sequence analysis of the R subunits has revealed strong homologous regions between the sequences of R subunits and that of the E. coli catabolite activating protein, a cAMP-dependent transcription activator for which crystallographic structure is available.[326] A model on the basis of this structure homology has been built for the cAMP binding structure on the R subunit of cAMP kinase and key residues for cAMP binding have been suggested.[327] Substitutions of these residues in RI, including Gly_{200}, Arg_{209} of the A site, and Gly_{324} and Arg_{332} of the B site by using recombinant DNA either singularly or in one case, dual substitution of Gly_{324} and Arg_{332} (see Table 7.10), resulted in mutant RI subunits with only one cAMP binding site. The result supports the prediction that these residues play essential roles of cAMP binding. All mutants could still interact with and mediate the cAMP effect on the catalytic unit of cAMP-dependent protein kinase, suggesting that either one of the two cAMP binding sites is sufficient for *in vitro* regulation of the kinase.[328,329]

Comparative analysis of the two cAMP binding site sequences of cAMP protein kinase with the other cyclic nucleotide binding sites along with a model building of these sites have led to the suggestion that a threonine residue in each of the two cGMP binding sites, Thr_{177} or Thr_{301}, participates in the cGMP binding.[330] The ala residue at the corresponding position in cyclic AMP binding sites of the A-kinases does not appear to participate in the cAMP binding (Figure 7.15). Mutating ala residue in one of the R units of A kinase to thr greatly increased the affinity of this site towards cGMP with no effect on cAMP binding.[331]

For many protein covalent modifications, the reaction cannot be readily reconstituted in a simple biochemical system. In other cases, the effect of protein modification may not be readily studied in a cell-free system. Site-directed mutagenesis is especially effective in addressing specific questions under such difficult conditions. The glycoprotein processing is a highly complex process, in addition to glycosylation reactions, the protein often undergoes other types of modification. Furthermore, a single protein may be glycosylated at multiple sites with carbohydrates of different structures in the mature protein. Site-directed mutagenesis has been used to address the question of the role of individual glycosylation site in the protein. The cation-dependent mannose-6-P receptor, one of the two receptors mediating the targeting of lysosomal enzymes, is a 46,000 dalton glycoprotein. It is N-glycosylated at four sites, enzymological deglycosylation, or blocking glycosylation at the receptor by inhibitors results in loss of ligand binding activity. Deduced amino acid of the receptor indicated that there are five potential sites for N-glycosylation; asparagines at positions 57, 83, 94, 107, and 113. These asparagines were individually substituted by threonine using site directed mutagenesis, and expressed in cultured cells. Analysis of the various mutant proteins for carbohydrate and comparing with the wild type protein showed that residues 57, 83, 107, and 113 but not residue 94 were glycosylated with high mannose on residues 57 and 113, the complex carbohydrate type on residues 83 and 107.[332] All the mutant proteins had decreased stability, but only the thr-113 mutant also showed decreased binding affinity.

Mechanism of Thrombin Receptor Activation

The general mechanism of receptor activation involves the binding of the receptor agonist to induce a change in receptor conformation, leading to the stimulation of specific intracellular signaling pathways. The initial reaction may involve certain protein modifications, such as the insulin-induced autophosphorylation of insulin receptor; the modification usually occurs at the cellular domain of the receptor. The activation of

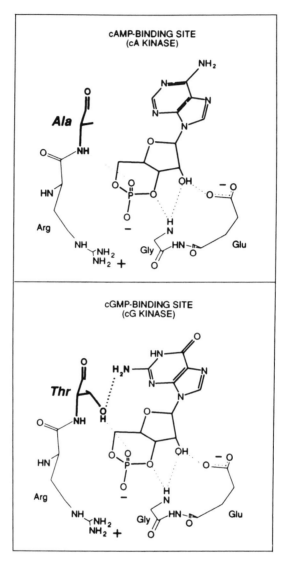

Figure 7.15. Predicted protein-cyclic nucleotide contacts for cAMP (cA) and cGMP (cG) kinases. The models are valid for both sites. Proposed hydrogen bonds are indicated by dotted lines. The proposed interaction that defines cyclic nucleotide binding specificity is highlighted by the shaded area. Reprinted with permission from *Journal of Biological Chemistry* 265:16032 (1990). Copyright 1990, The American Society for Biochemistry & Molecular Biology, and with permission from *Biochemistry* 29:4537 (1990). Copyright 1990, American Chemical Society.

thrombin receptor represents a unique case, where the receptor is covalently modified externally as part of the activation mechanism. Thrombin, a protease, is the best activator of platelets aggregation, and it also involves cell responses during vascular injury and inflammation process.[333] Although thrombin action has been known to be receptor mediated, attempts to identify the receptor protein have not been successful until very recently. Conventional methods using agonist or antagonist affinity labeling have led to the finding of several thrombin-binding proteins. However, in the absence of a demonstrable signal transduction activity, the relationship between these thrombin-binding proteins and thrombin receptor cannot be established. The recent success in the identification of thrombin receptor and the discovery of a novel function of proteolytic protein modification as a mechanism of receptor activation provides a good illustration of how molecular biology contributes to the study of important biological questions.[334,335,336]

The elucidation of the mechanisms of thrombin receptor activation is achieved by the application of novel cloning procedure, insightful sequence analysis, and recombinant DNA techniques. In the absence of an identified protein as the receptor, the cloning of the receptor was carried out using a functional assay for the receptor mRNA. One of the early events of the thrombin activation is the mobilization of cell Ca^{+2}. The mRNAs from a thrombin-responsive cell were microinjected into a Xenopus oocyte, a thrombin nonresponsive cell, and the oocyte was found to become responsive to thrombin as judged by the thrombin-inducible Ca^{+2} mobilization assay. This method was therefore used to screen for functional receptor clones. A cDNA clone was obtained and its sequence established.

The deduced amino acid sequence of the protein contains a region of seven transmembrane domains and a number of N-linked glycosylation motifs in the aminoterminal portion, characteristic of the membrane receptor proteins. Analysis of the deduced amino acid sequence suggested that the protein contains a number of unique features, especially in the cell surface aminoterminal portion of the protein. As schematically presented in Figure 7.16, a region about 40 amino acid downstream from the start methionine contains a five amino acid sequence, which is almost identical to that of the thrombin cleavage site of protein C, a physiological substrate of thrombin. This observation suggests that thrombin may hydrolyze the receptor at this site, Arg_{41}-Ser_{42}, resulting in the exposure of a new aminoterminal sequence. Thirteen amino acid further down from the putative thrombin cleavage site is a region of abundancy of acidic amino acid residues. This region bears similarities with two potent peptide thrombin inhibitors, desulfato-hiruden and hirudin, which block thrombin action by binding at specific sites of the protein.

Based on the sequence analysis, a working hypothesis has been postulated for the mechanism of thrombin receptor activation, as schematically represented in Figure 7.16. It is proposed that the highly acidic region provide anchoring site for thrombin binding. Once bound, the protease cleaves the peptide bond Arg_{41}-Ser_{42}, resulting in a new aminoterminal sequence. In the well-studied trypsinogen activation process, the newly exposed amino terminal sequence binds to a specific site on the enzyme during the activation process. This specific intramolecular interaction is essential for trypsin activity.[337] By analogy, it is proposed that the newly created aminoterminal sequence of the thrombin receptor binds to a specific site on the receptor protein to induce receptor activation.

A number of experiments have been carried out to test this model. A synthetic peptide corresponding to the sequence of the first 14 amino acids of the new aminoterminal sequence has been tested and found to be a potent activator of thrombin receptor, suggesting that the new aminoterminal sequence does bond to the receptor at specific site to induced receptor activation. Site-directed mutagenesis of the receptor has been carried out to test the postulate that receptor activation involves the cleavage of peptide bond Arg_{41}-Ser_{42}. Three receptor mutants with Arg_{41}, Arg_{46}, or Arg_{70} substituted by alanine have been constructed and expressed in Xenopus oocyte, only the Arg_{41} mutant is thrombin nonresponsive. Furthermore, a mutant with the thrombin cleavage sequence, LDPR, replaced by a enterokinase recognition sequence, DDDDK, renders the oocyte responsive to enterokinase but nonresponsive to thrombin. To test the notion that the highly acidic region of the receptor functioning as a thrombin bonding site, a receptor mutant deleted of this region has been constructed and functionally tested. This deletion mutant retains the maximal response to thrombin, but requires almost 1,000-fold higher thrombin concentrations.

Cell Biology Approaches

Like advances in molecular biology that have had a great impact on the biochemical studies of covalent protein modifications, advances in cell biology have significantly expanded the horizon of cell physiology studies. Many of the cell biology advances have provided badly needed tools for the study of the biological functions of protein modification reactions. Many covalent protein phosphorylation reactions have specific effects on certain physiological processes or physiological phenomena of the cells. Since these processes or phenomena are complex and not well under

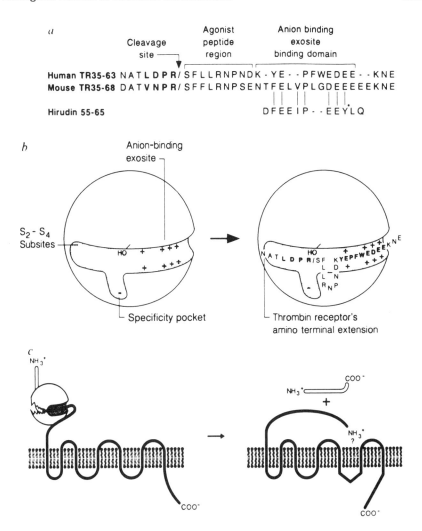

Figure 7.16. Working model for thrombin-receptor interaction. (a) Functional domains in human and murine thrombin receptor's N-terminal extension and comparison of hirudin's C tail with thrombin receptor's anion-binding exosite binding domain. Y* represents the sulphated tyrosine normally found in this position in hirudin. (b) The receptor's LDPR sequence interacts with thrombin's S1-S4 subsites and active site. The receptor's YEPFWEDEE sequence binds thrombin's anion-binding exosite region. Thrombin-receptor interaction through the latter sites is a major determinant of thrombin's potency at the receptor. (c) Proteolysis at the position 41/42 arginine-serine bond is sufficient for receptor activation. Reprinted with permission from *Nature* 353: 674 (1991). Copyright 1991 Macmillan Magazines Limited.

stood in terms of their biochemical basis, biochemical reactions simulating these processes cannot be reconstituted *in vitro*. Many cell biological techniques have been applied to the elucidation of the mechanism and function of various types of protein covalent modifications. For example, immunohistochemical approaches can be used to elucidate the cellular distribution of both the protein modification enzymes and the modified proteins, patch clamp techniques used to study covalent modification on single channel activity, microinjection of antibodies, and specific inhibitors or specific activators of modification enzymes used to perturb the cell physiology, just to mention a few. The application of some commonly used cell biological approaches in the study of protein covalent modification will be presented in this section as examples.

DNA Transfection

If the transfer of a DNA molecule encoding for a specific protein modification enzyme induces a cellular change, it may imply that protein modification reactions catalyzed by this modification enzyme are involved in this cellular activity. Similarly, if the cDNA of a covalent modification target protein is used to transfect the cell to induce the cellular change, this protein may be implicated in the cellular process. The procedure of DNA transfection is used extensively in many areas of cell research. For example, it represents the most direct method for the identification and testing of oncogenes. Thus, the cDNA may transfect a cell culture; that the transfected cells undergo oncogenic transformation will indicate that it is an oncogene.[338]

One area where DNA transfection has made especially important contributions is the study of cell adhesion phenomena. The involvement of cell adhesion molecules in specific cell-cell interactions had come mostly from the use of cell adhesion-specific antibodies, which were shown to block cell-cell interaction. Direct evidence for the involvement of individual cell adhesion molecules in specific cell-cell interaction can be obtained by DNA transfection experiment when a mouse fibroblast cell line that does not express cell adhesion molecules was transfected by chicken L-CAM[339] or mouse E-cadherin cDNA using a eukaryote expression system.[340] The transfected cell expressed the adhesion molecule on the cell surface and at the same time acquired Ca^{+2}-dependent aggregation activity, which was proportional to the adhesion protein expression. When these cells were plated, they, unlike the untransfected cells, formed cell aggregates. Such aggregate formation was inhibited by antibodies against the expressed adhesion proteins. These observations, in addition to providing direct evidence of the cell-cell interaction involvement of the cell adhesion molecule, suggest that DNA transfection may be used to study the functional characteristics of these molecules.

The specificity of the cell-cell interaction mediated by cell adhesion molecules has been addressed by using the DNA trans-fection approach. Mouse fibroblasts were transfected by either E-cadherin or P-cadherin; cells expressing these adhesion molecules were obtained. When the two types of cells were dispersed by trypsin and EDTA treatment, then mixed in a 1:1 ratio and plated, they formed segregated colonies with respect to cell types.[338] The specificity studies were extended to the interaction of the transfected cells with normal embryonic cells. Mouse embryonic lung is comprised of epithelial and mesenchymal cells; only epithelial cells contain E-CAM. When mouse embryonic lung cells were dispersed by trypsin-collagenase treatment and mixed with the dispersed E-CAM transfected cells, the E-CAM transfected cells were found to associate with epithelial cells, whereas the untransfected cells associated mainly with mesenchymal cells.[341] Although the general reactions of processing N-linked carbohydrate moieties of glycoproteins are elucidated, many details remain to be delineated. Glycoproteins often contain multiple potential glycosylation sites, but not all of them are glycosylated. Is the selectionof protein glycosylation site determined by the structure of the glycoproteins, or is this an attribute of the glycoprotein producing cells? Similarly, sugar moieties at different sites of the same glycoprotein or at the same site on different molecules of the protein may be processed differently to result in different structures. The question about the relative contributions of the glycoprotein structure and cellular glycosylation enzyme pattern to such diverse sugar processing mechanisms may be raised. Chicken ovalbumin, a glycoprotein normally produced in chicken ovary, contains in its sequence two potential N-glycosylation sites; only one of these asparagines is glycosylated in the mature protein. The isolated protein is comprised of multiple forms representing the protein molecule covalently attached to different sugar moieties. The sugar moieties fall into two general types—the high mannose type and a type of hybrid structure.[342]

The question about relative contributions of the protein structure and the cell characteristics to the specific features of ovalbumin-conjugated sugars was examined by transfecting the cDNA of chicken ovalbumin into a heterologous cell, mouse L-353.[342] The protein expressed in the mouse cell was then isolated and its carbohydrate structure analyzed. Like the normal ovalbumin, the protein from the mouse cells is glycosylated on only one and the same site, and the protein is heterogeneous in terms of the con-jugated carbohydrates. Carbohydrates of the mouse cell-expressed ovalbumin are also represented by a high percentage of the hybrid structure. On the other hand, the sugars of the mouse cell-expressed ovalbumin differ from those of the normal protein in being completely sialylated, and in containing some complex oligo-saccharide structure. Thus, some of the glycosylation characteris-tics of a protein appear to arise from the intrinsic properties of the glycoprotein, whereas cells producing the glycoprotein contribute to certain other characteris-tics. The different contributions from the protein and the cells may be probed by a system described here. The system may have practical applications; they may be exploited to produce glycoproteins with specific sugar structure by using selected cell lines.

Transfection by Recombinant DNA

Specific questions can be probed by transfecting recombinant DNA into cells. The recombinant DNAs, which are designed to investigate structure-function relationship of a protein modification enzyme, may be used for transfections to examine how the protein product affects cellular activities. For example, when a recombinant DNA of EGF receptor where the codon for the essential lysine was changed, was transfected and expressed in a cultured cell, the expressed protein was localized on the membrane and displays high affinity EGF binding, but failed to modulate cellular activities in response to EGF.[343]

Ras oncogenes are a group of low molecular weight GTP-binding proteins of highly conserved sequences. All of these proteins are localized at the inner surface of plasma membrane.[344] The mature proteins of Ras oncogenes are modified at the carboxy-terminal region by a complex process that can be separated into two discrete steps. The first consists of proteolytic removal of a carboxyterminal three residue sequence to expose a new carboxy-terminal cysteine residue, followed by carboxymethylation and polyprenylation of the new carboxyterminal cysteine. The second step involves palmitoylation of specific cysteine residues proximal to the carboxyterminus. By using recombinant DNA specifically modified at the carboxyterminal region, the question about the role of palmitoylation of the protein has been addressed.

Various DNAs with specific deletion at regions carboxy-terminal to the polyprenylation signal CAAX were expressed in COS cells. The expressed proteins were analyzed for the state of palmitoylation and subcellular distribution. As can be seen in Table 7.11, a minimum of nine amino acids of the carboxyterminal squence are required for full level of palmitoylation, whereas a minimum of six amino acids of the carboxyterminal are required for partial palmitoylation. Only the cells containing the palmitoylated proteins showed transformation, and only the palmitoylated proteins were found in the membrane fraction. Since

Table 7.11. Carboxyterminal modification and the processing of ras protein

Modification	Amino acid Sequence	Processing Step 1	Step 2
w.t.	QHKLRKLNPPDESGPGEMSCKCVLS	N	++++
a.m.	Q ————————E————————K—S	N	0
d.m.	Q—K————————P————————S	N	++++
d.m.	Q—K————————M————————S	N	++
d.m.	Q—K————————C—S	N	++
d.m.	Q—K————————C—S	N	++++
d.m.	Q————————————K 0	0	
s.m.	Q————————S————————S	0	0

Abbreviations are as follows: w.t., wild type; d.m., deletion mutant; s.m., substitution mutant; N, normal processing. Numbers indicate residue number. —— indicates identical sequence as the wild type and —— indicates deletion. Nomenclature modified from *Cell* 57: 1167 (1989). Reprinted with permission. Copyright 1989 Cell Press.

all of the mutant proteins could undergo first-step modification, the result of this study specifically implicated palmitoylation as essential for the membrane localization and transforming activity of the protein.[345,346]

Many proteins contain multiple domains; the function of a specific domain of the protein can be addressed by using recom-binant DNAs. For example, the question whether or not the sequence contained in the carboxyterminal nine amino acids of the Ras protein contains all the structure information for protein palmitoylation has been approached by the expression of specifically engineered proteins. A nucleotide sequence encoding the last nine amino acids of Ras protein was ligated to the cDNA of protein A, and the recombinant DNA transfected the COS cells; the expressed mutant protein was palmitoylated and distributed mainly in the membrane fraction. The normal protein A, or protein A expressed by a cDNA cloned with nucleotides coding for the last four amino acids of Ras protein added, were not palmitoylated and were fractionated with the soluble proteins.[346] Similar experiments have been carried out to show that the aminoterminal 14 amino acids of p60[v-src] contain the information for protein myristoylation and membrane localization.[347]

One type of experiment involves deleting part or the whole of a protein domain and then transfecting the deletion DNA to test for the cellular function of the protein domain. The cDNA of cell adhesion molecular from chicken liver, L-CAM, can be transfected to a mouse cell line to impart the cell with cell adhesion properties characteristic of that exhibited by L-CAM containing cells. When a truncated cDNA of L-CAM cDNA deleted of nucleotide sequence encoding for the last 50 amino acid residues of the cytosolic domain was used to transfect the mouse cells, the

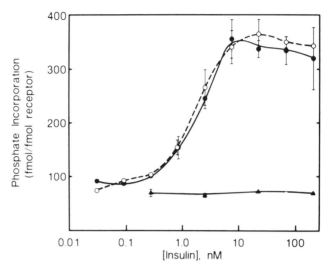

Figure 7.17A. Phosphorylation of src peptide by HIRc, HIRDCT, and Rat 1 insulin receptors. 22 fmol of HIRc insulin receptors 25 fmol of HIRDCT receptors, and an equivalent amount of Rat 1 protein (1 mg) eluted from WGA columns were assayed as described. Results are the means of three determinations (+1 S.D.) and are normalized for the fmol of phosphate incorporated per fmol of HIRc or HIRDCT receptor. The data for Rat 1 receptors are not normalized for femtomoles of receptor but are included to demonstrate that the amount of incorporation caused by the HIRc and HIRDCT proteins cannot be accounted for by endogenous rat receptors in the amounts of WGA-eluates used. Reprinted from *Journal of Biological Chemistry* 263:8914 (1988). Copyright 1988, The American Society for Biochemistry and Molecular Biology.

transfected cells, though expressing the protein on the cell surface, did not display cell aggregation. The results suggest that cytosolic domain of CAM participates in the cell adhesion activity.[348]

A chimeric cDNA encoding a protein containing domains from different protein molecules may be used to examine how functional domains interact in a protein, and the effect of such interactions on cellular processes. Both insulin receptor and EGF receptor are receptor protein tyrosine kinases. However, there are little structure similarities between these two protein kinases other than the kinase catalytic domains. When a chimeric cDNA encoding for extracellular portion of insulin receptor and cytosolic domain of EGF receptor was expressed in simian cells, the protein kinase activity derived from the EGF receptor could be activated by insulin, but not by EGF. The results suggest that the two receptor kinases use a common mechanism to transmit signals across the membrane.[349]

In a separate study,[350] chimeric protein composed of extracellular domain of insulin receptor linked to p68[gag-ros], an oncogene containing src kinase catalytic domain, was expressed in cultured cells. The protein was properly expressed on the cell surfac. Insulin could stimulate autophosphorylation of the chimeric receptor on the β subunit, similar to normal insulin receptor — again, suggesting that the insulin can transmit the signal to the cytosolic domain contributed by p68[gag-ros]. However, the receptor did not mediate cellular effects of insulin, such as the stimulation of glucose uptake or thymidine incorporation of the cells. One possible interpretation for this observation is that insulin receptor and the src kinases have different protein substrate specificities. Alternatively, it may be suggested that the kinase activity alone is not sufficient for the receptor to express all its cellular effects. This latter suggestion is supported by a study using a carboxyterminal truncated insulin receptor.[351] A human insulin receptor mutant with 43 amino acid

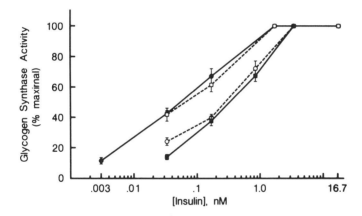

Figure 7.17B. Effect of insulin on glycogen synthase I activity in control and transfected cells. Confluent monolayers of Rat 1 (closed square), HIRc-A (open square), HIRc-B (closed circle), and HIRDCT (open circle) in 100-mm dishes were incubated with various concentrations of insulin at 37 C for 60 min, and then cells were harvested, homogenized, and centrifuged. Glycogen synthase activities were measured in the supernatant fractions at 0.3 and 6 mM glucose 6-phosphate concentrtions. Each value is plotted as the percentage of maximal insulin effect as a function of insulin concentration. Reprinted with permission from *Journal of Biological Chemistry* 263:8915 (1988). Copyright 1988, The American Society for Biochemistry & Molecular Biology.

326	Co- and Post-Translational Modification of Proteins

deletion at the carboxyterminus was expressed in a mouse fibroblast. While the expressed receptor mutant shows essentially identical insulin-stimulated kinase activity as the wild type receptor (Figure 7.17A), it does not have the ability to mediate the insulin stimulation of glycogen synthase activity of the cells (Figure 7.17B).

A genetic approach for examining protein-protein interactions has been developed in *S. cerevisiae*.[352,353] The technique, designated the two hybrid system, exploits the properties of the *GAL4* activating protein in yeast. This protein has two functional domains, a DNA binding domain and an activating domain resulting in galactosidase activation. The two functional domains are separately fused to different DNA sequences of known proteins. If the target proteins are capable of binding, the two functional domains of the *GAL4* activating protein are brought into proximity resulting in activation of the galactosidase gene. Activation can be monitored by formation of blue colonies.[352,353] A variation of this technique has been developed in which the GAL4 activating domain is fused to multiple genes in a cDNA library.[353] This modification provides a means to identify new proteins which bind to a specific target protein under study. A novel protein binding to a yeast protein kinase has been identified with this variation of the two-hybrid system.[354] This genetic approach has the potential for identifying many novel binding proteins and substrates for previously known proteins.

REFERENCES

1. Stadtman, E.R., and Ginsburg, A., in "The Enzymes" (P.D. Boyer, Ed.) 3rd. Ed., Vol. 10, 755–807, Academic Press, New York (1974).
2. Chock, P.B., and Stadtman, E.R., Fundamentals of Medical and Cell Biology, V3B, Chemistry of Living Cell, pp. 55–75 (1991).
3. Towler, D.A., Gordon, J.I., Adam, S.P., and Glaser, L., *Annu. Rev. Biochem.* 57, 69–99 (1988).
4. Wichner, W.T., and Lodish, H.F., *Science* 230, 400–404 (1985).
5. Porter, R.R., *CRC Crit. Rev. Biochem.*, 16–51 (1984).
6. Johnson, R.M., and Albert, S., *J. Biol. Chem.* 200, 451–455 (1952).
7. Kennedy, E.P., and Smith, S.W., *J. Biol. Chem.* 207, 153–163 (1954).
8. Findley, M., Strickland, M.P., and Rossiter, R.J., *Can. J. Biochem.* 32, 504–514 (1954).
9. Byvoet, P., *Biochim. Biophys. Acta* 160, 217–223 (1968).
10. Inoue, A., and Fjuimoto, D., *Biochim. Biophys. Acta* 220, 307–316 (1970).
11. Omary, M.B., and Trowbridge, I.S., *J. Biol. Chem.* 256, 12888–17892 (1981).
12. Olson, E.N., and Spizy, G., *J. Biol. Chem.* 261, 2548–2466 (1986).
13. Magee, A.I., and Coutueidge, S.A., *EMBO J.* 4, 1137–1144 (1985).
14. Clarke, S., *Annu. Rev. Biochem.* 54, 479–506 (1985).

15. Perez-Sala, D., Tam, E.W., Canada, F.J., and Rando, R.R., *Proc. Natl. Acad. Sci., USA* 88, 3043–3046 (1990).
16. Kreisel, W., Volk, B.A., Buchsel, R., and Reutter, W., *Proc. Natl. Acad. Sci., USA* 77, 1828–1831 (1980).
17. Kreisel, W., Hanski, C., Tran-Thi, T.-A., Katz, N., Decher, K., Reutter, W., and Gerok, W., *J. Biol. Chem.* 263, 11736–11742 (1988).
18. Grove, G.W., and Zweidler, A., *Biochemistry* 23, 4436–4443 (1984).
19. Mueller, R.D., Yasuda, H., Hatch, L.L., Bonner, W.M., and Bradbusy, E., *J. Biol. Chem.* 260, 5147–5153 (1985).
20. Fischer, E.H., and Krebs, E.G., *J. Biol. Chem.* 216, 121–132 (1955).
21. Sutherland, E.W., and Wosilait, W.D., *Nature* 175, 169–171 (1955).
22. Cori, G.T., and Green, A.A., *J. Biol. Chem.* 151, 31–38 (1943).
23. Cori, G.T., and Cori, C.F., *J. Biol. Chem.* 158, 321–332 (1945).
24. Dabrowski, R., and Hartshorne, D.J., *Biochem. Biophys. Res. Commun.* 85, 1352–1357 (1978).
25. Yaki, K., Yazawa, M., Kakiuchi, S., Oshima, M., and Uenishi, K., *J. Biol. Chem.* 253, 1338–1341 (1978).
26. Pato, M.D., and Adelstein, R.S., *J. Biol. Chem.* 256, 7047–7054 (1983).
27. Umekawa, H., and Hidaka, H., *Biochem. Biophys. Res. Commun.* 132, 56–62 (1985).
28. Vorotnikov, A.V., Shirinsky, V.P., and Gusev, N.B., *F.E.B.S. Lett.* 236, 321–324 (1988).
29. Ikebe, M., and Reardon, S., *J. Biol. Chem.* 263, 3055–3058 (1990).
30. Yamushiro, S., Yamakita, Y., Ishikawa, R., and Matsuwara, F., *Nature* 344, 675–678 (1990).
31. Mak, A.S., Watson, M.H., Litwin, C.M., and Wang, J.H., *J. Biol. Chem.* 266, 6678–6681 (1991).
32. Korn, E.D., and Hammer, J.A., *Annu. Rev. Biophys. Chem.* 17, 23–45 (1988).
33. Carpenter, G., *Annu. Rev. Biochem.* 56, 881–914 (1987).
34. Gill, G.N., in "Oncogenes and Molecular Origins of Cancer," R.A. Weinberg, ed., pp. 67–96, Cold Sring Harbor Lab. Press (1989).
35. Collett, M.S., and Erickson, R.L., *Proc. Natl. Acad. Sci., USA* 75, 2021–2024 (1978).
36. Levinson, A.D., Oppermann, H., Levintow, L., Varmus, H.E., and Bishop, J.M., *Cell* 15, 561–572 (1978).
37. Hunter, T., in "Oncogene and Molecular Origins of Cancer," R.A. Weinberg, ed., pp. 147–173, Cold Spring Harbor Lab. Press (1989).
38. Rapp, W.R., Cleveland, J. L., and Bonner, T.I., in "The Oncogene Handbook," E.P. Reddy et al. eds., pp. 213–253, Elsevier Pub. (1988), Amsterdam, New York, Oxford..
39. Castagna, M., Takai, Y., Kaibuchi, K., Sano, K., Kikkawa, U., and Nishizuka, Y., *J. Biol. Chem.* 257, 7847–7851 (1982).
40. Takai, A., Bialejan, C., Troschka, M., and Ruegg, J.C., *F.E.B.S. Lett.* 217, 81–84 (1987).
41. DeCamille, P., and Greengard, P., *Biochem. Pharmacol.* 35, 4349–4357 (1986).

42. Zwier, H., Schotman, P., and Gispan, W.H., *J. Neurochem.* 34, 1689–1699 (1980).
43. Liu, Y., and Storns, D.R., *J. Biol. Chem.* 264, 12800–12804 (1989).
44. Madison, D.V., Malenka, R.C., and Nicoll, R.A., *Annu. Rev. Neurosci.* 14, 379–398 (1991).
45. O'Dell, T.J., Kandel, E.R., and Grant, S.G.N., *Nature* 353, 558–560 (1991).
46. Ledendre, C.H., MacDonnell, P.C., and Guroff, G., *Biochem. Biophys. Res. Commun.* 74, 891–897 (1971).
47. Joh, T.H., Park, D.H., and Reis, D.J., *Proc. Natl. Acad. Sci., USA* 75, 4744–4748 (1978).
48. Yamauchi, T., and Fujisawa, H., *Biochem. Int.* 1, 98–104 (1980).
49. Albert, K.A., Helmer-Matyjek, E., Nairn, A.C., Muller, T.H., Haycock, J.W., Green, L.A., Golstein, M., and Greengard, P., *Proc. Natl. Acad. Sci., USA* 81, 7713–7717 (1984).
50. Vulliet, P.R., Hall, F.L., Mitchell, J.P., and Hardie, D.G., *J. Biol. Chem.* 264, 16293–16298 (1989).
51. Hafen, E., Baster, K., Edstrom, J.E., and Rubin, G.M., *Science* 236, 55–63 (1987).
52. Ambrosio, W., Mahowald, A.P., and Pertimon, N., *Nature* 342, 288–291 (1989).
53. Springer, F., Stevens, L.M., and Nusslein-Volhard, C., *Nature* 338, 478–483 (1989).
54. Besmer, P., *Current Opinion in Cell Biology* 3, 939–959 (1991).
55. Tada, M., Kirchberger, M.A., and Katz, A.M., *J. Biol. Chem.* 250, 2640–2647 (1975).
56. LePeuch, C.J., Haiech, J., and Demaille, J.G., *Biochemistry* 18, 5150–5157 (1979).
57. Movesessian, M.A., Nishikawa, M., and Adelstein, R.S., *J. Biol. Chem.* 259, 8029–8032 (1984).
58. Acosta-Urquidi, J., Alkorn, D.L., and Neary, J.T., *Science* 224, 1254–1257 (1984).
59. Sakahibara, M., Alkon, D.L., DeLorenzo, R., Goldenring, J.R., Neary, J.T., and Heldman, E., *Biophys. J.* 50, 319–327 (1986).
60. Alkon, D.L., Acosta-Urquidi, J., Olds, J., Kuzma, G., and Neary, J.T., *Science* 219, 303–306 (1983).
61. Alkon, D.L., Kubota, M., Neary, J.T., Naito, S., Coulter, D., and Rasmussen, H., *Biochem. Biophys. Res. Commun.* 134, 1245–1253 (1986)
62. Cheng, S.H., Rich, D.P., Marshall, J., Gregory, R.J., Welsh, M.J., and Smith, A.E., *Cell* 66, 1027–1036 (1991).
63. Wagner, J.A., Cozen, A.L., Schulman, H., Gruemert, D.C., Stryer, L., and Gardner, L., *Nature* 349, 793–796 (1991).
64. Hwang, T-C, Lu, L., Zeitlin, P.L., Grunert, D.C., Huganir, R., and Guggino, W.B., *Science* 244, 1351–1353 (1988).
65. Li, M., McCaan, J.D., Anderson, M.P., Clancy, J.P., Liedtke, C.M., Nairn, A.C., Greengard, P., and Welsh, M.J., *Science* 244, 1353–1356 (1988).
66. Reed, L.J., and Yeoman, S.J., *The Enzymes* 18, 77–96 (1987).

67. Gibson, D.M., and Parker, R.A., *The Enzymes* 18, 180–217 (1987).
68. Pilkis, S.J., Claus, T.H., and El-Maghrabi, M.R., *Adv. in 2nd Messenger and Phosphoprotein Res.* 22, 175–191 (1991).
69. Sutherland, E.W., *Science* 177, 401–408 (1972).
70. Krebs, E.G., Graves, D.J., and Fischer, E.H., *J. Biol. Chem.* 234, 2867–2873 (1959).
71. Walsh, D.A., Perkins, J.P., and Krebs, E.G., *J. Biol. Chem.* 243, 3763–3765 (1968).
72. Krebs, E.G., *Int. Cong. Ser.-Excerpta. Med.* 273, 17–24 (1973).
73. Simpson, M.V., *J. Biol. Chem.* 201, 143–154 (1953).
74. Goldberg, A.L., and St. John, A.C., *Annu. Rev. Biochem.* 45, 747–803 (1976).
75. Rabinowitz, M., and Fisher, J.M., *Biochim. Biophys. Acta* 91, 313–322 (1964).
76. Bradley, M.O., Hayflick, L., and Schimke, R.T., *J. Biol. Chem.* 251, 3521–3529 (1976).
77. Etlinger, J.D., and Goldberg, A.L., *Proc. Natl. Acad. Sci., USA* 74, 54–58 (1977).
78. Ciechanover, A., Hod, Y., and Hershko, A., *Biochem. Biophys. Res. Commun.* 81, 1100–1105 (1978).
79. Ciechanover, A., Heller, H., Elias, S., Haas, A.L., and Herskho, A., *Proc. Natl. Acad. Sci., USA* 77, 1365–1368 (1980).
80. Hershko, A., Ciechanover, A., Heller, H., Haas, A.L., and Rose, I.A., *Proc. Natl. Acad. Sci., USA* 77, 1365–1786 (1980).
81. Hershko, A., Heller, S., Elias, S., and Ciechanover, A., *J. Biol.Chem.* 258, 8206–8214 (1983).
82. Hershko, A., Leshinsky, E., Ganoth, D., and Heller, H., *Proc. Natl. Acad. Sci.* 81, 1619–1623 (1984).
83. Hough, R., Pratt, G., and Rechsteiner, M., *J. Biol. Chem.* 262, 8303–8313 (1987).
84. Hershko, A., Eytan, E., Ciechanover, A., and Haas, A.L., *J. Biol. Chem.* 257, 13964–13970 (1982).
85. Ciechanover, A., Finley, D., and Varshavsky, A., *Cell* 37, 57–66 (1984).
86. Backmair, A., Finley, D., and Varshavsky, A., *Science* 234, 179–186 (1986).
87. Watkins, J.F., Sung, P., Prakash, S., and Prakash, L., *Genes Develop.* 7, 250–262 (1993)
88. Parug, H.A., Rabey, B., and Kulka, R.G., *EMBO J.* 6, 55–61 (1987).
89. Lowe, S., and Mayer, R.J., *Appl. Neuropath.* 16, 281–291 (1990).
90. Finley, D., Ciechanover, A., and Varshasky, A., *Cell* 37, 43–55 (1984).
91. Goebl, M.G., Yochem, J., Jentsch, S., McGrath, J.P., Varshavsky, A., and Byers, B., *Science* 241, 1331–1335 (1988).
92. Glotzer, M., Murray, A.W., and Kirschner, M.W., *Nature* 349, 132–138 (1991).
93. Murray, A.W., Solomon, M.J., and Krishner, M.J., *Nature* 339, 280–286 (1989).
94. Hershko, A., *J. Biol. Chem.* 263, 15237–15240 (1988).

95. Finley, D., and Chau, V., *Annu. Rev. Cell. Biol.* 1, 25–69 (1991).

96. Rechsteiner, M., *Cell* 66, 615–618 (1991).

97. Shanklin, J., Jabben, M., and Vierstra, R.D., *Proc. Natl. Acad. Sci., USA* 84, 359–363 (1987).

98. Hochstrasser, M., Ellison, M.J., Chau, V., and Varshavsky, A., *Proc. Natl. Acad. Sci* 88, 4606–4610 (1991).

99. Scheffner, M., Werness, B.A., Huibregtse, J.M., Levine, A.J., and Howley, P.M., *Cell* 63, 1129–1136 (1990).

100. Ciechanover, A., DiGiuseppe, J.A., Bercovich, B., Orian, A., Richter, J.D., Schwartz, A.L., and Brodeur, G.M., *Proc. Natl. Acad. Sci., USA* 88, 139–143 (1991).

101. Bonner, W.M., Hatch, C.L., and Wu, R.S., in *Ubiquitin*, M. Rechsteiner, ed., 157–172, Plenum Publ. (1988).

102. Cross, A.M., *Annu. Rev. Cell Biol.* 6, 1–40 (1990).

103. Ferguson, M.A.J., and Williams, A.F., *Annu. Rev. Biochem.* 57, 285–320 (1988).

104. Ikezawa, H., Yamenego, M., Takuchi, K., Miyashita, T., and Ohyabu, T., *Biochim. Biophys. Acta* 450, 154–164 (1976).

105. Low, M.G., and Fineau, J.B., *F.E.B.S. Lett.* 82, 143–146 (1977).

106. Low, M.G., and Ziversmit, D.B., *Biochemistry* 19, 3913–3918 (1980).

107. Hu, J.S., James, G., and Olson, E.N., *Biofactors* 1, 219-226 (1988).

108. Towler, D.A., Gordon, J.I., Adams, S.P., and Glaser, L., *Annu. Rev. Biochem.* 57, 69–99 (1988).

109. Casey, P.J., Solski, P.A., Der, C.J., and Buss, J.E., *Proc. Natl. Acad. Sci., USA* 86, 8323–8327 (1989).

110. Maltese, W.A., Sheridan, K.M., Repko, E.M., and Erdman, R.A., *J. Biol. Chem.* 265, 2148–2155 (1990).

111. Weber, K., Plessmann, U., and Traub, P., *F.E.B.S. Lett.* 257, 411–414 (1989).

112. Wolda, S., and Glomset, J.A., *J. Biol. Chem.* 263, 5497–6006 (1988).

113. Vorburger, K., Kitten, G.T., and Nigg, E.A., *EMBO J.* 8, 4007–4013 (1989).

114. Fukada, Y., Takao, T., Ohguro, H., Yoshizawa, T., Akino, T., and Shimonsihi, Y., *Nature* 346, 658–660 (1990).

115. Kloc, M., Reddy, B., Crawford, S., and Etkin, L.D., *J. Biol. Chem.* 266, 8206–8212 (1991).

116. Schafer, W.R., Kim, R., Sterne, R., Thorner, J., Kim, S.H., and Rine, J., *Science* 245, 379–385 (1989).

117. Carr, S.A., Biemann, K., Shozo, S., Parmalee, D.C., and Titani, K., *Proc. Natl. Acad. Sci., USA* 79, 6128–6131 (1982).

118. Aiken, A., Cohen, P., Santikaru, S., William, D.H., Calder, A.G., Smith, A., and Klee, C.B., *F.E.B.S. Lett.* 150, 314–318 (1983).

119. McIlinney, R.A., Pelly, S.J., Chadwick, J.K., and Cowleyl, G.P., *EMBO J* 4, 1145–1152 (1985).

120. Olson, E.N., Towler, D.A., and Glazer, L., *J. Biol. Chem.* 260, 3784–3790 (1985).

121. Schmidt, R.A., Schneider, C.J., and Glomset, J., *J. Biol. Chem.* 259, 10175–10180 (1984).

122. Buss, J.E., and Sefton, B.M., *Mol. Cell Biol.* 6, 116–122 (1986).
123. Williamsen, B.M., Christensen, A., Hubert, N.L., Papageorge, A.G., and Lowy, D.R., *Nature* 310, 583–586 (1984).
124. Hancock, J.F., Magee, A.I., Childs, J.E., and Marshall, C.J., *Cell* 57, 1167–1177 (1989).
125. Moffett, S., Mouillac, B., Bonin, H., and Bouvier, M., *EMBO J.* 12, 349-356. (1993)
126. Buss, J.E., Kamp, M.P., and Sefton, B., *Molec. Cell Biol.* 4, 2697–2704 (1984).
127. Buss, J.E., Kamp, M.P., Goul, K., and Sefton, B.M., *J. Virol.* 58, 468–474 (1986).
128. Kamp, M.P., Buss, J.E., and Sefton, B.M., *Cell* 45, 105–112 (1986).
129. Schlesinger, M.J., and Malfer, C., *J. Biol. Chem.* 257, 9887–9890 (1982).
130. Lambrecht, B., and Schmidt, M.F., *F.E.B.S. Lett.* 202, 127–132 (1986).
131. Resh, M.D., *Cell* 58, 281–286 (1989).
132. Linder, M.E., Ewald, D.A., Miller, R.J., and Gilman, A.G., *J. Biol. Chem.* 265, 8243–8251(1990).
133. Sternweis, P.C., *J. Biol. Chem.* 261, 631–637 (1986).
134. Burnet, F.M., *Physiol. Rev.* 31, 131–150 (1951).
135. Tiffany, J.M., and Blough, H.A., *Virology* 44, 18–28 (1971).
136. Paulson, J.C., Sadler, J.F., and Hill, R.L., *J. Biol. Chem.* 254, 2120–2124 (1979).
137. Weiss, W., Brown, J.H., Cusak, S., Paulson, J.C., Skelel, J.J., and Wiley, D.C., *Nature* 333, 426–431 (1988).
138. McDougal, J.S., Kennedy, M.S., Sligh, J.M., Cort, S.P., Mawle, A., and Nicholson, J.K.A., *Science* 231, 382–385 (1986).
139. Lifson, J.F., Reyes, G.R., McGrath, M.S., Stein, B.S., and Englen, E.G., *Science* 232, 1123–1127 (1986).
140. Putney, S.D., Mathew, T.J., Robey, W.G., Lynn, D.L., Robert-Gruroff, M., Mueller, W.T., Langlois, A.J., Ghrayeb, J., Petteway, S.R., Weinhold, K. J., Fischinger, P.J., Wong-Staal, F., Gallo, R.C., and Boloaues, D.P., *Science* 234, 1392–1395 (1986).
141. Mathew, T.J., Weinhold, K.J., Lyerly, K.H., Langlois, A.J., Wigzell, H., and Bologues, *Proc. Natl. Acad. Sci., USA* 84, 5424–5428 (1987).
142. Trannecher, A., Wolfgang, L., and Karjalainen, K., *Nature* 331, 84–86 (1988).
143. Deen, K.C., McDougal, J.S., Inacker, R., Folena-Wasserman, G., Arthos, J., Rosenberg, J., Madden, P.J., Axel, R., and Sweet, R.W., *Nature* 331, 82–84 (1988).
144. Frazier, W., and Glaser, L., *Annu. Rev. Biochem.* 48, 491–523 (1979).
145. Humphrey, T., *Devel. Biol.* 8, 27–47 (1963).
146. Humphrey, S., Humphrey, T., and Sano, J., *J. Suppra Mol. Struc.* 7, 339–351 (1977).
147. Turner, R.S., and Burger, M.M., *Nature* 244, 509–510 (1973).
148. Armstrong, P.B., *CRC Crit. Rev. Biochem. Molec. Biol.* 24, 119–149 (1989).
149. Merrell, R., and Glaser, L., *Proc. Natl. Acad. Sci., USA* 70, 7294–7298 (1973).

150. Gottlieb, D.I., Merrell, R., and Glaser, L., *Proc. Natl. Acad. Sci., USA* 71, 1801–1802 (1974).
151. Rutishauser, U., Thiery, J.P., Brackenbury, R., Sela, B.A., and Edelman, G.M., *Proc. Natl. Acad. Sci., USA* 73, 577–581 (1976).
152. Thiery, J.P., Brackenbury, R., Rutishauser, U., and Edelman, G.M., *J. Biol. Chem.* 252, 6341–6345 (1977).
153. Brackenbury, R., Thiery, J.P., Rutishauser, U., and Edelman, G.M., *J. Biol. Chem.* 252, 6335–6340 (1977).
154. Solter, D., and Knowles, B.B., *Proc. Natl. Acad. Sci., USA* 75, 5565–5571 (1978).
155. Fenderson, B.A., Zehavi, U., and Hakamori, S., *J. Exp. Med.* 160, 1591–1596 (1985).
156. Finne, J., Finne, U., Deagostini-Bazin, H., and Goridis, C., *Biochem. Biophys. Res. Commun.* 112, 482–487 (1983).
157. Edelman, G.M., and Chuong, C-M., *Proc. Natl. Acad. Sci., USA* 79, 7036–7040 (1982).
158. Roth, J., Zuber, C., Wagner, P., Taatjes, D.J., Weisgerber, C., Heitz, P.U., Goridis, C., and Bitter-Suermann, D., *Proc. Natl. Acad. Sci., USA* 85, 2999–3003 (1988).
159. Rutishauser, U., Acheson, A., Hall, A.K., and Sunshine, J., *Science* 240, 53–57 (1988).
160. Vacquier, J.D., and May, G.W., *Proc. Natl. Acad. Sci., USA* 74, 2456–2460 (1977).
161. Tsuzuki, H., Hoshida, H., Onitake, K., and Aketa, K., *Biochem. Biophys. Res. Commun.* 26, 502–511 (1977).
162. Schnell, E., Earle, B.J., Breaux, C., and Lennary, W.J., *J. Cell Biol.* 72, 25–46 (1977).
163. Vacquier, V.D., and Payne, J.E., *Exp. Cell Res.* 82, 227–235 (1973).
164. Summers, R.G., and Hylander, B.C., *Exp. Cell. Res.* 96, 63–68 (1975).
165. Aketa, K., *Exp. Cell Res.* 90, 56–62 (1975).
166. Glabe, C.G., and Vancquier, V.D., *Nature* 267, 836–838 (1977).
167. Wassarman, P.M., *Annu. Rev. Biochem.* 57, 415–442 (1988).
168. Bleil, J.D., and Wassarman, P.M., *J. Cell Biol.* 104, 1363–1371 (1986).
169. Florman, H.M., and Wassarman, P.M., *Cell* 41, 313–324 (1985).
170. Florman, H.M., Bechtol, K.B., and Wassarman, P.M., *Dev. Biol.* 106, 243–255 (1984).
171. Shur, B.D., and Hull, N.G., *J. Cell. Biol.* 95, 574–579 (1982).
172. Ashwell, G., and Harford, J., *Ann Rev. Biochem.* 51, 531–554 (1982).
173. Morell, A.G., Irvine, R.A., Sternlieb, I., Scheinberg, I.H., and Ashwell, G., *J. Biol. Chem.* 243, 155–159 (1968).
174. Hickman, J., Ashwell, G., Morell, A.G., van den Hauser, C.J.A., Scheinberg, I.H., *J. Biol. Chem.* 245, 759–766 (1970).
175. Sawyer, S.T., Sunfer, J.P., and Doyle, D., *J. Biol. Chem.* 263, 10534–10538 (1988).
176. Tanabe, T., Pricer, W.E., and Ashwell, G., *J. Biol. Chem.,* 254, 1038–1043 (1979).

177. Baenziger, J.W., and Maynard, Y., *J. Biol. Chem.* 255, 4607–4613 (1980).
178. Sarker, M., Liao, J., Kabat, E.A., Tanabe, T., and Ashwell, G., *J. Biol. Chem.* 254, 3170–3174 (1979).
179. Steer, C.J., and Ashwell, G., *J. Biol. Chem.* 255, 4607–4613 (1980).
180. Van Tamelen, E.E., and Heys, J.R., *J. Am. Chem. Soc.* 97, 1252–1253 (1975).
181. Regoeczi, E., Taylor, P., Debanne, M.P., Marz, F., and Halton, M.W.C., *Biochem. J.* 184, 399–407 (1979).
182. Lee, Y.C., Townsend, R.R., Hardy, M.R., Conngren, J., Arnarp, J., Haraldsson, M., and Lonn, H., *J. Biol. Chem.* 258, 199–202 (1983).
183. Dunn, W.A., Hubbard, A.L., and Aronson, N.N., Jr., *J. Biol. Chem.* 255, 5971–5978 (1980).
184. Gregoriadis, G., Morell, A.G., Steinlieb, I., and Scheongberg, I.H., *J. Biol. Chem.* 245, 5833–5837 (1970).
185. Sebue, G., and Kosaka, A., *Hepato-Gastroenterology* 27, 200–203 (1980).
186. Regoeczi, E., Hatton, M.W.C., and Charlwood, P.A., *Nature* 254, 699–701 (1975).
187. Lunney, J.K., and Ashwell, G., *Proc. Natl. Acad. Sci., USA* 73, 341–343 (1976).
188. Drickamer, K., *Cell* 67, 1029–1032 (1991).
189. Lee, R.T., Lin, P., and Lee, Y.C., *Biochemistry* 23, 4255–4261 (1984).
190. Rice, K.G., Weisg, O.A., Barthel, T., Lee, R.T., and Lee, Y.C., *J. Biol. Chem.* 265, 18429–18434 (1990).
191. Steer, C.J., and Clarenberg, R., *J. Biol. Chem.* 254, 4456–4461 (1979).
192. Fiete, D., Srivaslavan, V., Hindsgaul, O., and Baenziger, J.V., *Cell* 87, 1105–1110 (1991).
193. Smith, P.L., and Baenziger, J.V., *Science* 42, 930–933 (1988).
194. Neufeld, E.F., Lim, T.W., and Shapiro, L.J., *Annu. Rev. Biochem.* 44, 357–376 (1975).
195. Hickman, J., Shapiro, L.J., and Neufeld, E.F., *Biochem. Biophys. Res. Commun.* 57, 5561 (1974).
196. Kaplan, A., Achord, D.T., and Sly, W.S., *Proc. Natl. Acad. Sci., USA* 74, 2026–2030 (1977).
197. Fischer, H.D., Gonzalez-Noriego, A., Sly, W.S., and Morre, D.J., *J. Biol. Chem.* 255, 9608–9615 (1980).
198. Dahmis, N.M., Lobel, P., Breitmeyer, J., Chirgwin, J.M., and Kornfeld, S., *Cell* 50, 181–192 (1987).
199. Morgan, D.O., Edman, J.C., Standring, D.N., Fried, V.A., Smith, M.C., Roth, R.A., and Rutter, W.J., *Nature* 329, 301–307 (1987).
200. Oshima, A., Nolan, C.M., Kyle, J.W., Grubb, J.H., and Sly, W.S., *J. Biol. Chem.* 263, 2553–2562 (1988).
201. Hoflack, B., and Kornfeld, S., *J. Biol. Chem.* 260, 12008–12014 (1985).
202. Tong, P.Y., Tollefsen, S.E., and Kornfeld, S., *J. Biol. Chem.* 263, 2585–2588 (1988).
203. Stein, M., Zijderhand-Bleekemolen, J.E., Geuze, H., Hasilik, A., and von Figura, K., *EMBO J.* 6, 2677–2681 (1987).
204. Conti, M.A., and Adelstein, R.S., *J. Biol. Chem.* 256, 3178–3181 (1981).

205. Sharma, R.K., and Wang, J.H., *Proc. Natl. Acad. Sci., USA* 82, 2603–2607 (1985).
206. Majerus, P.W., Connolly, T.W., Deckmyn, H., Ross, T.S., Bross, T.E., Ishii, H., Bansal, V.S., and Wilson, D.H., *Science* 234, 1519–1526 (1986).
207. Rhee, S.G., Sab, P.G., Ryu, S.H., and Lee, S.Y., *Science* 244, 546–550 (1989).
208. Meisanholder, J., Suh, P.G., Rhee, S.G., and Hunter, T., *Cell* 57, 1109–1122 (1989).
209. Nishibe, M.I., Wahl, M.I., Hernandez-Sotomayer, S.M.T., Tonks, N.K., Rhee, S.G., and Carpenter, G., *Science* 250, 1253–1256 (1990).
210. Goldschmidt, C.P., Kim, J.W., Mochesky, L.M., Rhee, S.G., and Pollard, T.D., *Science* 251, 1231–1233 (1991).
211. Adelstein, R.S., and Eisenberg, F., *Annu. Rev. Biochem.* 49, 921–956 (1980).
212. Sobieszek, A., and Small, J.V., *J. Mol. Biol.* 112, 559–576 (1977).
213. Lebowitz, E.A., and Cook, R., *J. Biol. Chem.* 85, 1489–1494 (1979).
214. Chacko, S., Conti, M.A., and Adelstein, R.S., *Proc. Natl. Acad. Sci., USA* 74, 129–133 (1977).
215. Pires, E.M.V., and Perry, S.V., *Biochem. J.* 167, 137–143 (1977).
216. Dabrowska, R., and Hartshorne, D.J., *Biochem. Biophys. Res. Commun.* 85, 1352–1359 (1978).
217. Walsh, D.A., and Krebs, E.G., in "The Enzymes," 3rd ed. (P.D. Boyer, ed.), Vol. VIII, pp. 555–581 (1973).
218. Kuo, J.F., and Greengard, P., *J. Biol. Chem.* 245, 2493–2498 (1970).
219. Schulman, H., and Greengard, P., *Nature* 271, 478–479 (1978).
220. Kishimoto, A., Takai, Y., Mori, T., Kikkawa, W., and Nishizuka, Y., *J. Biol. Chem.* 255, 2273–2276 (1980).
221. Nemerson, Y., and Furie, B., *CRC Rev. Biochem.* 9, 45–85 (1980).
222. Kornfeld, R., and Kornfeld, S., *Annu. Rev. Biochem.* 51, 631–664 (1985).
223. Kasuga, M., Zick, Y., Blith, D.L., Karlsson, F. A., Haring, H.U., and Kahn, C.R., *J. Biol. Chem.* 257, 9891–9894 (1982).
224. Ushiro, H., and Cohen, S., *J. Biol. Chem.* 255, 8363–8365 (1980).
225. Kaplan, D.R., Hempstead, B.L., Martin-Zanaca, D., Chao, M.V., and Parada, L.F., *Science* 252, 554–557 (1991).
226. Klein, R., Jing, S., Nanduri, V., O'Rourke, E., and Barbaeid, M., *Cell* 65, 189–197 (1991).
227. Benjamin, W.B., and Singer, I., *Biochemistry* 14, 3301–3309 (1975).
228. Auruck, J., Leone, G.R., and Martin, D.B., *J. Biol. Chem.* 251, 1511–1515 (1956).
229. Alexander, M.C., Palmer, J.C., Pointer, R.H., Koumjian, L., and Auruch, J., *J. Biol. Chem.* 257, 2049–2055 (1982).
230. Brownsey, R.W., Hughes, W.A., and Denton, R.M., *Biochem. J.* 168, 441–445 (1977).
231. Gomez, N., Tonks, N.K., Morrison, C., Harmar, T., and Cohen, P., *F.E.B.S. Lett.* 271, 119–122 (1990).
232. Smith, C.J., Rubin, C.S., and Rosen, O.M., *Proc. Natl. Acad. Sci., USA* 77, 2641–2645 (1980).
233. Novak-Hofer, I., and Thomas, G., *J. Biol. Chem.* 259, 5995–6000 (1984).
234. Erikson, R.L., *J. Biol. Chem.* 266, 6007–6010 (1991).

235. Ballow, L.M., Siegmann, M., and Thomas, G., *Proc. Natl. Acad. Sci., USA* 85, 7154–7158 (1988).
236. Sturgill, T.W., Ray, L.B., Erickson, E., and Maller, J., *Nature* 334, 715–718 (1988).
237. Erickson, E., and Maller, J.L., *J. Biol. Chem.* 264, 13711–13717 (1989).
238. Ballow, L.M., Jeno, P., and Thomas, G., *J. Biol. Chem.* 263, 1118–1122 (1988).
239. Ahn, N.G., and Krebs, E.G., *J. Biol. Chem.* 265, 11495–11501 (1990).
240. Haystead, T., Weifel, J.E., Litchfield, E.W., Tsukitani, Y., Fischer, E.H., and Krebs, E.G., *J. Biol. Chem.* 265, 16571–16580 (1990).
241. Anderson, N.G., Maller, J.L., Tonks, N.K., and Sturgill, I.W., *Nature* 34, 651–653 (1990).
242. Ahn, A.G., Seger, R., Bratlien, R.L., Diltz, C.D., Tonks, N.K., and Krebs, E.G., *J. Biol. Chem.* 266, 4220–4227 (1991).
243. Gomez, N., and Cohen, P., *Nature* 353, 170–173 (1991).
244. Seger, R., Ahn, N.G., Boulton, T.G., Yacopoulos, G.D., Panayotutos, N., Radziejewska, E., Erickson, C., Bratlien, R.L., Cobb, M.M., and Krebs, E.G., *Proc. Natl. Acad. Sci., USA* 88, 6142–6146 (1991).
245. Posoda, J., and Cooper, J. A., *Science* 255, 212–215 (1992).
246. Allemain, G. L., Her, J-H., Wu, J., Sturgill, T. W., and Weber, M., *J., Molec. Cell. Biol.* 12, 2222–2229 (1992).
247. Seger, R., Ahn, N. G., Posoda, J., Munar, E. S., Jensen, A. M., Cooper, J.A., Cobb, M. H., and Kuebs, E. G., *J. Biol. Chem.* 267, 14373–14381 (1992).
248. Morrison, D. K., Kaplan, D. R., Escobedo, J. A., Rapp, U. R., Roberts, T. M., and William, L. T., *Cell* 58, 648–659 (1989).
249. Kovacina, K. S., Yonegawa, K., Brautigan, D. L., Tonks, N. K., Rapp, U. R., and Roth, R. A., *J. Biol. Chem.* 265, 12115–12118 (1990).
250. Blackshear, P. J., Haupt, D. M., App, H., and Rapp, U. R., *J. Biol. Chem.* 265, 12131–12134 (1990).
251. App, H., Hazan, R., Zilberstein, A., Ullrich, A., Schlessinger, J., and Rapp, U. R., *Molec. Cell. Biol.* 11, 913–919 (1992).
252. Ohmichi, M., Pang, L., Decker, S. J., and Saltiel, A. R., *J. Biol.*
253. Rapp, U. R., *Oncogenes* 6, 495–500 (1991).
254. Kyriakis, J. M., App, H., Zhang, X-F., Banerjee, P., Brautigan, D. L., Rapp, U. R., and Auruch, J., *Nature* 358, 417–421 (1992).
255. Wood, K. W., Sarnecki, C., Roberts, T. M., and Bienis, J., *Cell* 68, 1041–1050 (1992).
256. Thomas, S. M., DeMarco, M., D'Arcangelo, G., Halegona, S., and Brugge, J. S., *Cell* 68, 1031–1040 (1992).
257. Fischer, E.H., Graves, D.J., Crittenden, E.R.S., and Krebs, E.G., *J. Biol. Chem.* 234, 1698–1704 (1959).
258. Nolan, C., Novoa, W.B., Krebs, E.G., and Fischer, E.H., *Biochemistry* 3, 542–551 (1964).
259. Tessmer, G.W., Skuster, J.R., Tabatabai, L., and Graves, D.J., *J. Biol. Chem.* 252, 5666–5671 (1977).
260. Kemp, B.E., and Pearson, R.B., *Trends Biochem. Sci.* 15, 342–346 (1990).
261. Kennelly, P.J., and Krebs, E.G., *J. Biol. Chem.* 266, 15555–15558 (1991).

262. Shenoy, S., Choi, J-K., Bagrodia, S., Copland, T.D., Maller, J.C., and Shalloway, D., *Cell* 57, 763–774 (1989).
263. Lew, J., Beaudette, K., Litwin, C.M.E., and Wang, J. H., *J. Biol. Chem.* 267, 13383–13390 (1992).
264. Cheng, H-C., Litwin, C.M.E., Hwang, D.M., and Wang, J. H., *J. Biol. Chem.* 266, 17919–17925 (1991).
265. Feizei, T., *Nature* 314, 53–57 (1985).
266. Magnani, J.L., *Methods Enzymol.* 138, 484–491 (1987).
267. Spitalnik, S.L., *Methods Enzymol.* 138, 492–503 (1987).
268. Gould, K., and Nurse, P., *Nature* 342, 39–45 (1989).
269. Draetta, G., Piwnica-Worms, H., Morrison, D., Druker, B., Roberts, T., and Beach, D., *Nature* 336, 738–744 (1988).
270. LePeuch, C.J., Haiech, J., and Demaille, J.G., *Biochemistry* 18, 5150–5157 (1979).
271. Tada, M., and Katz, A.M., *Annu. Rev. Physiol.* 44, 401–423 (1982).
272. Li, C., Wang, J.H., and Colyer, J., *J. Biol. Chem.* 266, 17486–17493 (1991).
273. Wang, J.Y.J., *Molec. Cell. Biol.* 5, 3640–3643 (1985).
274. Hershko, A., Eytan, E., Ciechanover, A., and Haas, A.L., *J. Biol. Chem.* 257, 13964–13970 (1982).
275. Neff, N.T., DeMartino, G.N., and Goldberg, A.L., *J. Cell Physiol.* 101, 439–458 (1979).
276. Kamin, K.E., and Stull, J.T., *Annu. Rev. Pharmacol. Toxicol.* 25, 593–620 (1985).
277. Aksoy, M.O., Mras, S., Kamin, K.E., and Murphy, R.A., *Amer. J. Physiol.* 245, 1255–1270 (1983).
278. Gilman, G., *Proc. Natl. Acad. Sci., USA* 67, 305–309 (1970).
279. Steiner, A.L., Parker, C.W., and Kipnis, O.M., *J. Biol. Chem.* 247, 1106–1113.
280. Brocker, G., Harper, S.F., Teresaki, W.L, and Moylan, R.D., *Adv. Cycl. Nucleot. Research* 10, 1–33 (1979).
281. Konnerly, D.A., Parker, C.W., and Sullivan, T.J., *Anal. Biochem.* 98, 123–130 (1979).
282. Preiss, J.E., Loomis, C.R., Bell, R.M., and Niedel, J.E., *Methods Enzymol.* 141, 294–300 (1987).
283. Berridge, M.J., *Biochem. J.* 212, 849–858 (1983).
284. Rasmussen, H., *Science* 170, 404–412 (1990).
285. Tsien, R.Y., and Rink, T.J., *J. Neurosci. Methods* 4, 13–86 (1981).
286. Blinks, J.R., Mattingly, P.V., Jewell, B.R., van-Leeuven, M., Harrer, G.C., and Allen, O.G., *Methods Enzymol.* 57, 292–328 (1978).
287. Grynkiewicz, G., Poenie, M., and Tsien, R.Y., *J. Biol. Chem.* 260, 3440–3450 (1985).
288. Tsien, R.Y., *Annu. Rev. Neurosci.* 12, 227–253 (1989).
289. Adams, S.R., Harootunian, A.T., Buechler, Y.J., Taylor, S.S., and Tsien, R.Y., *Nature* 349, 694–697 (1991).
290. Castagna, M., Takai, Y., Kaibuchi, K., Sano, K., Kikhawa, V., and Nishizuka, Y., *J. Biol. Chem.* 257, 2847–2851 (1982).
291. Daly, J.W., *Adv. Cycl. Nucleot. Prot. Phosph. Res.* 17, 81–109 (1984).

292. Wells, J.N., Garst, J.E., and Kramer, G.L., *J. Med. Chem.* 24, 954–962 (1981).
293. Tkaez, J., and Lampen, J.D., *Biochem. Biophys. Res. Commun.* 65, 249–287 (1975).
294. Olivecrona, T., Chernick, S.S., Bengtssen-Oliverona, G., Garrison, M., and Scow, R.O., *J. Biol. Chem.* 262, 10740–10759 (1987).
295. Elbein, A., *FASEB J.* 5, 3055–3063 (1991).
296. Semb, H., and Olivecrona, T., *J. Biol. Chem.* 262, 10740–10759 (1987).
297. Hunter, T., and Cooper, J.A., in "The Enzymes, Vol. 17, P.D. Boyer and E.G. Krebs, eds., pp. 191–246, Academic Press (1986).
298. Hank, S.K., Quinn, A.M., and Hunter, T., *Science* 241, 42–52 (1988).
299. Nurse, P., and Bissett, Y., *Nature* 292, 558–560 (1981).
300. Hindley, T., and Phear, G.A., *Gene* 31, 129–134 (1984).
301. Nurse, P., *Nature* 256, 547–551 (1975).
302. Russell, P., and Nurse, P., *Cell* 49, 559–561 (1987).
303. Moreno, S., Nurse, P., and Russell, P., *Nature* 344, 549–552 (1990).
304. Sadhur, K., Reed, S.I., Richardson, H., and Russell, P., *Proc. Natl. Acad. Sci., USA* 87, 5139–5143 (1990).
305. Strausfeld, W., Labbe, J.C., Fesquet, D., Cavadore, J.C., Picard, A., Sudhu, K., Russell, P., and Doree, M., *Nature* 351, 242–245 (1991).
306. Moreno, S., and Nurse, P., *Nature* 351, 194 (1991).
307. Featherstone, C., and Russell, P., *Nature* 349, 808–811 (1991).
308. Huang, K.P., Nakabayashi, H., and Huang, F.L., *Proc. Natl. Acad. Sci., USA* 83, 8535–8539 (1986).
309. Nishizuka, Y., *Nature* 334, 661–665 (1988).
310. Ono, Y., Fujii, T., Ogita, K., Kikkawa, W., Igarashi, K., and Nishizuka, Y., *J. Biol. Chem.* 263, 6927–6932 (1988).
311. Bell, R.M., and Barus, D.J., *J. Biol. Chem.* 266, 4661–4664 (1991).
312. Tonks, N.K., Diltz, C.D., and Fischer, E.H., *J. Biol. Chem.* 263, 6722–6728 (1988).
313. Charbonneau, H., Tonks, N.K., Walsh, K.A., and Fischer, G.H., *Proc. Natl. Acad. Sci., USA* 85, 8695–8701 (1988).
314. Krueger, N.X., Streuli, M., and Saito, H., *EMBO J.* 9, 3241–3252 (1990).
315. Shaw, A., and Thomas, M.L., *Curr. Opinion Cell Biol.* 3, 862–868 (1991).
316. Fischer, E.H., Charbonneau, H., and Tonks, N.K., *Science* 253, 401–406 (1991).
317. Drickamer, K., *J. Biol. Chem.* 263, 9557–9560 (1988).
318. Springer, T.A., *Nature* 346, 425–433 (1990).
319. Stoolman, L.A., *Cell* 56, 907–910 (1989).
320. Chen, W.S., Lazar, C.S., Poonie, M., Tsien, R.Y., Gill, G.N., and Rosenfeld, M.G., *Nature* 328, 820–823 (1987).
321. Kamps, M.P., Buss, J.E., and Sefton, B.M., *Proc. Natl. Acad. Sci., USA* 82, 4625–4630 (1985).
322. Takeya, T., and Hanafusa, H., *Cell* 32, 881–890 (1983).
323. Kmiecik, T.E., and Shalloway, D., *Cell* 44, 65–73 (1983).
324. Piwnica-Worms, H., Saunders, K.B., Roberts, T.M., Smith, A.E., and Cheng, S.H., *Cell* 49, 75–82 (1987).

325. Cartwright, C.A., Eckhart, W., Simon, S., and Kaplan, P., *Cell* 49, 83–91 (1987).
326. McKay, D.B., Weber, I.T., and Steitz, T.A., *J. Biol. Chem.* 257, 9518–9524 (1981).
327. Weber, I.T., Steitz, T.A., Bubis, J., and Taylor, S.S., *Biochemistry* 26, 345–351 (1987).
328. Bubis, J., Neitzel, J.J., Saraswat, L.D., and Taylor, S.S., *J. Biol. Chem.* 263, 9668–9673 (1988).
329. Woodford, T.A., Corell, L.A., McKnight, G.S., and Corbin, J.D., *J. Biol. Chem.* 264, 13321–13328 (1989).
330. Weber, I.T., Shabb, J.B., and Corbin, J.D., *Biochemistry* 28, 6122–6127 (1989).
331. Shabb, J.B., Ng, L., and Corbin, J.D., *J. Biol. Chem.* 265, 16031–16034 (1990).
332. Wendland, M., Waheed, A., Schmidt, B., Hille, A., Nagel, G., vanFigura, K., and Pohlmann, R., *J. Biol. Chem.* 266, 4598–4604 (1991).
333. Shuman, M.A., *Ann. NY Acad. Sci.* 485, 349–368 (1986).
334. Vu, T-K.H., Hung, D.T., Wheaton, V.I., and Coughlin, S.R., *Cell* 64, 1057–1068 (1991).
335. Vu, T-K.H., Wheaton, V.I., Hung, D.T., Charo, I., and Coughlin, S.R., *Nature* 353, 674–677 (1991).
336. Liu, L.W., Vu, T-K.H., Esmon, C.T., and Coughlin, S.R., *J. Biol. Chem.* 266, 16977–16980 (1991).
337. Fehlhammer, H., Bode, W., and Huber, R., *J. Mol. Biol.* 111, 425–438 (1977).
338. Varmus, H., in "Oncogenes and the Molecular Origins of Cancer", R.A. Weinberg, ed., pp. 3–44, Cold Spring Harbor Laboratory Press (1989).
339. Edelman, G.M., Murray, B.A., Mege, R-M., Cunningham, B.A., and Gallin, W.J., *Proc. Natl. Acad. Sci., USA* 84, 8502–8506 (1987).
340. Nagafuchi, A., Shirayoshi, Y., Okazuki, K., Yasuda, K., and Takeiki, M., *Nature* 329, 341–343 (1987).
341. Nose, A., Nagafuchi, A., and Takeiki, M., *Cell* 54, 993–1001 (1988).
342. Sheare, G.T., and Robbins, P.W., *Proc. Natl. Acad. Sci., USA* 83, 1993–1997 (1986).
343. Gill, G.N., in "Oncogenes and the Molecular Origins of Cancer," R.A. Weinberg, ed., pp. 67–96, Cold Spring Harbor Laboratory Press (1989).
344. McCormick, F., in "Oncogenes and the Molecular Origins of Cancer," R.A. Weinberg, ed., pp. 125–145, Cold Spring Harbor Laboratory Press (1989).
345. Fujiyama, A., and Tamanoi, F., *Proc. Natl. Acad. Sci., USA* 83, 1266–1270 (1986).
346. Hancock, J.F., Magee, A.I., Childs, J.E., and Marshall, C.J., *Cell* 57, 1167–1177 (1989).
347. Pellman, D., Garber, E.A., Cross, F.R., and Hanafusa, H., *Nature* 314, 374–377 (1985).

348. Jaffe, S.H., Friedlander, D.R., Matsuzaki, F., Crossin, K.L., Cunningham, B.A., and Edelman, G.M., *Proc. Natl. Acad. Sci., USA* 87, 3589–3593 (1990).
349. Riedel, H., Dull, T.J., Schlessinger, J., and Ullrich, A., *Nature* 324, 68–70 (1986).
350. Ellis, L., Morgan, D.O., Jong, S-M., Wang, L-H., Roth, R.A., and Rutter, W.J., *Proc. Natl. Acad. Sci., USA* 84, 5101–5105 (1987).
351. Maegawa, H., McClain, D.A., Freidenberg, G., Olefsky, J.M., Napier, M., Lipari, T., Dull, T.J., Lee, J., and Ullrich, A., *J. Biol. Chem.* 263, 8912–8917 (1988).
352. Fields, S., and Song, O.-k., *Nature* 340, 245-246 (1989).
353. Chien, C.-T., Bartel, P.L., Sternglanz, R., and Fields, S., *Proc. Natl. Acad. Sci., USA* 88, 9578–9582 (1991).
354. Yang, X., Hubbard, E.J.A., and Carlson, M., *Science* 257, 680–682 (1992).

CONCLUDING REMARKS

The understanding of the processes and roles of co- and post-translational modification reactions is only possible through multi-faceted experimental work. Throughout the text, we chose to show how different approaches can be used to answer specific questions. In some cases, the approach requires a specific probe (chemical or physical), highlighting the need for understanding these reactions at the basic level of enzyme-substrate interactions. Other situations require a probe stable under biological conditions to examine the consequence of protein modification. The examples chosen illustrate these points, but we realize many other cases could have been considered. Lastly, we hope this text will help stimulate others to learn more about this important subject. For some of the modifications discussed in the text, our understanding is at the earliest phase. Many of the conceptual approaches discussed may provide new insights when applied to a different modification reaction.

Index